Rangeland Degradation and Recovery in China's Pastoral Lands

Rangeland Degradation and Recovery in China's Pastoral Lands

Edited by

Victor R. Squires

University of Adelaide
Adelaide
Australia

Lu Xinshi

Beijing Forestry University
Beijing
China

Lu Qi

Chinese Academy of Forestry
Beijing
China

Wang Tao

Director General, Chinese Academy of Sciences
Cold Arid Region Environmental &
Engineering Research Institute
China

and

Yang Youlin

Regional Cooperation Unit of the UNCCD (Convention to
Combat Desertification and Drought)
Bangkok
Thailand

www.cabi.org

CABI is a trading name of CAB International

CABI Head Office
Nosworthy Way
Wallingford
Oxfordshire OX10 8DE
UK

Tel: +44 (0)1491 832111
Fax: +44 (0)1491 833508
E-mail: cabi@cabi.org
Website: www.cabi.org

CABI North American Office
875 Massachusetts Avenue
7th Floor
Cambridge, MA 02139
USA

Tel: +1 617 395 4056
Fax: +1 617 354 6875
E-mail: cabi-nao@cabi.org

A catalogue record for this book is available from the British Library, London, UK.

Library of Congress Cataloging-in-Publication Data

Rangeland degradation and recovery in China's pastoral lands / edited by Victor Squires ... [et al.].
p. cm.
Includes bibliographical references and index.
ISBN 978-1-84593-496-5 (alk. paper)
1. Land degradation--China. 2. Rangelands--Revegetation--Case studies. 3. Range management--Case studies. I. Squires, V.R. (Victor R.), 1937- II. Title.

SF85.4.C6R36 2009
636.08'450951--dc22

2008047548

ISBN: 978 1 84593 496 5

Typeset by SPi, Pondicherry, India.
Printed and bound in the UK by MPG Books Group.

The paper used for the text pages in this book is FSC certified. The FSC (Forest Stewardship Council) is an international network to promote responsible management of the world's forests.

Contents

About the Editors vii

Contributors ix

Preface xi

Acknowledgements xiii

PART I: INTRODUCTION 1

1. **The Context for the Study of Rangeland Degradation and Recovery in China's Pastoral Lands** 3
   Victor R. Squires and Zhang Kebin
2. **Historical Degradation Episodes in China: Socio-economic Forces and Their Interaction with Rangeland Grazing Systems Since the 1950s** 15
   Victor R. Squires and Yang Youlin

PART II: MECHANISMS OF RANGELAND DEGRADATION AND RECOVERY 31

3. **An Analysis of the Effects of Climate Variability in Northern China over the Past Five Decades on People, Livestock and Plants in the Focus Areas** 33
   Lu Qi, Wang Xuequan and Wu Bo
4. **Mechanisms of Degradation in Grazed Rangelands** 45
   Li Xianglin
5. **The Mechanisms of Soil Erosion Processes by Wind and Water in Chinese Rangelands** 61
   Zhi-yu Zhou and Bin Ma
6. **Processes in Rangeland Degradation, Rehabilitation and Recovery** 76
   Victor R. Squires

Part III: Case Studies of Degradation and Recovery 89

7. **Case Study 1: Hulunbeier Grassland, Inner Mongolia** 91
Lu Xinshi, Ai Lin and Lv Shihai
8. **Case Study 2: Horqin Sandy Land, Inner Mongolia** 103
Jiang De-ming, Kou Zhen-wu, Li Xue-hua and Li Ming
9. **Case Study 3: Xilingol Grassland, Inner Mongolia** 120
Jianhui Huang, Yongfei Bai and Ye Jiang
10. **Case Study 4: Ordos Plateau, Inner Mongolia** 136
Yuanrun Zheng and Qiushuang Li
11. **Case Study 5: Hexi Corridor, Gansu** 151
Yuhong Li and Victor R. Squires
12. **Case Study 6: Alashan Plateau, Inner Mongolia** 171
Li Qingfeng
13. **Case Study 7: Qinghai–Tibetan Plateau Rangelands** 184
Ruijun Long, Zhanhuan Shang, Xusheng Guo and Luming Ding
14. **Case Study 8: Northern Xinjiang** 197
Jin Gui-li and Zhu Jin-zhong

Part IV: The Future – How to Prevent the Next Major Degradation Episode 217

15. **Land Tenure Arrangements, Property Rights and Institutional Arrangements in the Cycle of Rangeland Degradation and Recovery** 219
Adrian Williams, Meiping Wang and MunkhDalai A. Zhang
16. **Monitoring and Evaluation as Tools for Rangeland Management** 235
Aijun Liu
17. **How Can the Next Degradation Episode be Prevented?** 247
Victor R. Squires and Yang Youlin

Index 259

About the Editors

Victor R. Squires is an Australian, a former Foundation Dean of the Faculty of Natural Resources, University of Adelaide. His PhD in Range Science is from Utah State University in the USA. He is currently an Adjunct Professor in the University of Arizona, Tucson, USA. He has been an international consultant in dryland management in Iran, North Africa, East Africa and China for more than 20 years and has worked on projects supported by UNDP, FAO, UNEP, IFAD, the World Bank and the Asian Development Bank in many countries. He is author/editor of several books (*Livestock Management in the Arid Zone* and *Drylands: Sustainable Use into the Twenty-first Century*) and numerous research papers and invited book chapters. Dr Squires has recently been named as a recipient of a Science and Technology Cooperation Friendship Award from the Chinese Central Government in recognition of his contribution to China's economic and scientific development. He is one of only 50 recipients of this prestigious award since its inauguration in 1991.

Lu Xinshi is Professor of Range Management at Beijing Forestry University. His research interest is in the management and rehabilitation of grasslands, especially in the Hulunbeier region of Inner Mongolia. He is author of numerous research papers and a book, *Grasslands of China.*

Lu Qi is a Research Professor in the National R&D Centre on Combating Desertification at the Chinese Academy of Forestry, Beijing. His research interest is in desertification processes and drivers of change. His PhD is in Ecology and he has published many papers on land degradation and rehabilitation and is co-editor of the book *Global Alarm: Dust and Sandstorms from the World's Drylands* (UN, 2002).

Wang Tao is Director General of the Chinese Academy of Sciences Cold and Arid Regions Environmental and Engineering Research Institute. Dr Wang is author of many research papers and a major book, *Desert and Desertification in China* (Science Press/Longman Books, 2006).

Yang Youlin is in the Regional Cooperation Unit of the UNCCD (Convention to Combat Desertification and Drought), Bangkok, Thailand. He has a long involvement in desertification issues, first at the Chinese Academy of Sciences Desert Research Institute, Lanzhou and, later, in the State Forest Administration's desertification control unit, Beijing. He is co-editor of the book *Global Alarm: Dust and Sandstorms from the World's Drylands* (UN, 2002).

Contributors

Ai Lin, *Beijing Forestry University, Beijing, China*

Bai Yongfei, *Professor, State Key Laboratory of Vegetation and Environmental Change, Institute of Botany, Chinese Academy of Sciences, Beijing 100093, China*

Ding Luming, *International Centre for Tibetan Plateau Ecosystem Management, Lanzhou University, Lanzhou, Gansu, China*

Guo Xusheng, *International Centre for Tibetan Plateau Ecosystem Management, Lanzhou University, Lanzhou, Gansu, China*

Huang Jianhui, *Professor, State Key Laboratory of Vegetation and Environmental Change, Institute of Botany, Chinese Academy of Sciences, Beijing 100093, China*

Jiang De-ming, *Professor, Institute of Applied Ecology, Shenyang, Liaoning, China*

Jiang Ye, *Professor, State Key Laboratory of Vegetation and Environmental Change, Institute of Botany, Chinese Academy of Sciences, Beijing 100093, China*

Jin Gui-li, *Department of Grassland Science, Xinjiang Agricultural University, Urumqi, Xinjiang, China*

Kou Zhen-wu, *Professor, Institute of Applied Ecology, Shenyang, Liaoning, China*

Li Ming, *Professor, Institute of Applied Ecology, Shenyang, Liaoning, China*

Li Qingfeng, *Professor, Inner Mongolia Agricultural University, Huhhot, Inner Mongolia, China*

Li Qiushuang, *Professor, State Key Laboratory of Vegetation and Environmental Change, Institute of Botany, Chinese Academy of Sciences, Beijing 100093, China*

Li Xianglin, *Professor, Chinese Academy of Agricultural Sciences, Beijing, China*

Li Xue-hua, *Professor, Institute of Applied Ecology, Shenyang, Liaoning, China*

Li Yuhong, *Gansu Water Resources Bureau, Lanzhou, Gansu, China*

Liu Aijun, *Research Scientist, Institute for Rangeland Survey and Design, Inner Mongolia Academy of Animal and Agricultural Sciences, Huhhot, Inner Mongolia, China*

Long Ruijun, *Professor of Range Management, College of Pastoral Agriculture Science and Technology, Lanzhou University, Lanzhou 730020, China*

Lu Qi, *Research Professor, Chinese Academy of Forestry, Beijing, China*

Lu Xinshi, *Professor of Range Management, Beijing Forestry University, Beijing, China*

Lv Shihai, *Chinese Environmental Science Academy, Beijing, China*

Ma Bin, *Zhejiang University, Hangzhou 310029, China*

Shang Zhanhuan, *International Centre for Tibetan Plateau Ecosystem Management, Lanzhou University, Lanzhou, Gansu, China*

Squires, Victor R., *former Dean of Faculty of Natural Resources, University of Adelaide, Australia*

Wang Meiping, *Gansu Agricultural University, Lanzhou, Gansu, China*
Wang Xuequan, *Chinese Academy of Forestry, Beijing, China*
Williams, Adrian, *Formerly Centre for the Management of Arid Environments, Kalgoorlie, Australia*
Wu Bo, *Chinese Academy of Forestry, Beijing, China*
Yang Youlin, *UNCCD, Regional Coordination Unit, Bangkok, Thailand*
Zhang Kebin, *Professor, Beijing Forestry University, Beijing, China*
Zhang MunkhDalai A., *Research Centre for Eco-environmental Sciences, Chinese Academy of Science, Beijing and Bureau of Land and Resources, Hulunbeier City, Hulunbeier 021008, China*
Zheng Yuanrun, *Professor, State Key Laboratory of Vegetation and Environmental Change, Institute of Botany, Chinese Academy of Sciences, Beijing 100093, China*
Zhou Zhi-yu, *Professor, Lanzhou University, Lanzhou, Gansu, China*
Zhu Jin-zhong, *Department of Grassland Science, Xinjiang Agricultural University, Urumqi, Xinjiang, China*

Preface

The purpose of this book is to provide reference material for those responsible for grazing land management in China and its long-term consequences (environmental, social and economic). It responds to the urgent need to collate and review some of the major degradation experienced in China's vast pastoral lands.

In this volume, an outline is presented of the major biological processes and socio-economic influences that operate in selected pastoral rangelands in China. Consideration is given to how these processes and influences can be manipulated to make best use of these important land resources.

Drought/degradation episodes in the rangelands affect not only all components of the resource (domestic livestock, native flora and fauna, soil and biodiversity) but also all those people living and deriving a livelihood from the resource (herder families, rural communities and the government).

In this book, we have confined our analysis to the impact on the resource from a rangeland user's perspective, but we recognize the much wider impacts and urge fellow researchers to take up the challenge of addressing the environmental and social impacts of these major land degradation episodes.

The historical case studies described in the book represent a failure to manage for the extreme climate variability that characterizes north and west China's vast arid rangelands. Thus, they represent an historical 'test bed' for our current scientific understanding of rangelands and government and land user responses.

The early signs of degradation of the forage resource and extreme drought (extensive areas of bare ground, dust and sandstorms, delayed recovery of perennial pastures, death of trees and shrubs) are apparent, but our science is not yet good enough to address the multiplicity of controversial issues such as:

1. The quantification of the resource damage due to livestock in contrast to the effects of extreme climate variability (separating the signal from the noise).
2. The quantification of global warming/greenhouse effects on China's climate in comparison to the natural background variability of the climate system which occurs on interannual, decadal and longer timescales.

The causes of degradation and recovery are fairly well understood. The combination of drought and heavy and prolonged grazing leads to the accelerated death of perennial vegetation, loss of surface soil protection and delayed recovery from drought. Evidence from many counties within China's drylands indicates that several management options and interventions can be successful in arresting and reversing pasture and land degradation.

Scope and Purpose of the Book

This book has been motivated by the proposition that China's arid rangelands will be managed better by a thorough understanding of the mistakes and successes of the past. But this report is not intended as a history of soil and vegetation degradation in China's arid rangelands. Instead, we seek to derive the insight necessary to plan for the future from those with the expertise and first-hand experience in the key regions that are the focus of the book.

Degradation case studies have been recorded in terms of dust storms, floods, animal losses, financial hardship and human suffering. In this context, the drought/degradation events clearly have endangered sustainable land use by placing the future productivity of the resource and viability of herding (and the rural communities) at risk.

At the time of writing, there is strong scientific evidence that the observed increase in atmospheric concentrations of greenhouse gases is driving global warming. However, the future impacts on China's rainfall patterns are uncertain. We believe that the best scientific tools available are the global climate models, but their science is still in its relative infancy and many of the potential drivers of the climate system are yet to be represented adequately or 'parameterized'. Thus, we are racing into an uncertain climatic future, perhaps on a collision course with future climatic extremes. Whether the grazing-based rangelands and the wider community adapt successfully to future climate extremes/variability will depend on how well we use the knowledge of the past degradation (and recovery) as described in this book to avoid repeating the mistakes of the past.

Victor Squires
Adelaide

Acknowledgements

No work of this magnitude is possible without the cooperation and assistance of a large group of people. Individual authors and their employing agencies and their technical assistants and graduate students played their parts. Research grants from the National Science Foundation of China, the Chinese Academy of Sciences and other state-funded research agencies also helped to support the work of the various editors and contributors.

The editors are grateful for the effort provided by Mr Xu Changjiang in interpretation and some translation, and to Dr Zhang Fengchun and his erstwhile colleagues from the Desert Research Institute in Wu Wei, Gansu Province for providing the unpublished data from their field experiments.

The editorial team from CABI helped in many ways to improve and simplify the manuscript to tap this rich vein of knowledge and experience from China and bring it to the wider English-speaking readership.

The various contributors have drawn extensively from the work of others in an effort to combine local expert knowledge held by the lead writers with the published work of specialists into a volume that is both synthetic and integrative. The extent to which this hope is achieved will be due in no small part to the myriad of authors (both Chinese and foreign) whose work has been cited here.

The editors accept responsibility for errors and omissions that may have crept in while endeavouring to present such a wide-ranging work, which covers history, politics and the technical aspects of livestock husbandry and rangeland ecology, as well as the socio-economic aspects that impact on people in China's pastoral rangelands.

Part I

Introduction

The two chapters in this part outline the context for the study of rangeland degradation and recovery in China's pastoral lands, define the terminology and discuss the geographical distribution and site characteristics for the eight degradation and recovery case studies that are the focus of the book. Here we provide an historical overview of the major events and policy changes that have affected the pastoral lands from the beginning of the 20th century to the present day. Major emphasis is on analysing the policy environment since the 1950s and the longer-term implications are outlined.

1 The Context for the Study of Rangeland Degradation and Recovery in China's Pastoral Lands

Victor R. Squires[1] and Zhang Kebin[2]
[1]*University of Adelaide, Australia;* [2]*Beijing Forestry University, Beijing, China*

Synopsis
This book brings better understanding of the causes of major episodes of rangeland degradation and recovery in China's pastoral lands. This chapter sets the context for the analysis, explains its scope and purpose and describes briefly the geographical, social and economic environment in which pastoral land use is conducted in China.

Keywords: rangelands defined; wildlife; watershed; biodiversity; herders; nomadic lifestyle; north China; rotational grazing; carrying capacity; stocking pressure; drought; land degradation; policy; land tenure; user rights

1.1 China's Pastoral Lands

Rangeland is an internationally recognized term that refers to a type of land rather than a type of land use. Rangelands include *grassland*, *steppe* and *desert steppe*. Rangelands have a key role as grazing lands for pastoral use, as a wildlife habitat, as a watershed for China's irrigated and urban/industrial areas and as biosphere reserves. The focus of this book is on rangelands that are used for pastoral purposes.

For the purposes of this study, we define pastoral lands as 'uncultivated land that will provide the necessities of life for grazing and browsing animals and the herder families that depend on them. Therefore, it includes deserts, steppes, forests and natural grasslands and shrublands.' The area of the drier pastoral lands (sometimes referred to as the 'grasslands') in China is about 186 million ha; the exact amount depends on whether the classification is based on climate, soils, drainage effectiveness or vegetation. For practical purposes, most of the 'Three Norths' region (north-west, north and north-east China) is pastoral rangeland, of which 105 million ha is classified as degraded, to a greater or lesser extent.

The pastoral lands of China are characterized by high year-to-year variability in precipitation. This, in turn, results in: (i) variability in plant growth; (ii) uneven provision of nutrition for cattle, sheep, goats, camels, horses and other herbivores (e.g. wildlife); and (iii) limited potential to carry out necessary plant management options such as rest and rotational grazing.

In Parts I and II of this book, we present eight major degradation case studies from across China's pastoral rangelands (Box 1.1). This report is not intended as a history, but uses previous histories and documentation to interpret the causes of degradation and recovery. The main feature of degradation in the documented case studies was the change of land use and the

Box 1.1. Regional degradation case studies in China's pastoral lands (see detail in Part II).

1. **Hulunbeier Grassland, Inner Mongolia** – destruction of grasslands, mobilization of sands, reduced rangeland productivity, relocation of herders, widespread hardship.
2. **Horqin Sandy Land, Inner Mongolia** – mobilization of sand dunes, loss of rangeland productivity, reduced carrying capacity, drying up of lakes.
3. **Xilingol Grassland, Inner Mongolia** – destruction of rangelands, mobilization of sands, reduced rangeland productivity, relocation of herders, widespread hardship.
4. **Ordos Plateau, Inner Mongolia** – widespread land conversion, mobilization of sand dunes, loss of forest cover and perennial vegetation, loss of biodiversity.
5. **Alashan, Inner Mongolia** – the lower reaches of the Hei He valley suffered massive damage, falling water tables, widespread death of forest, loss of rangelands, drying up of lakes, ecological refugees and extreme hardship.
6. **Land reclamation (conversion) in Hexi Corridor, Gansu** – drastic alteration to hydrogeology in Hei He and Shiyang He valleys, leading to rapidly falling water tables, widespread death of riverine forest vegetation (*Populus euphratica*) and protective forestry in shelter belts around the oasis perimeter, ecological refugees and great hardship.
7. **Qinghai–Tibet Plateau rangeland** – destruction of rangeland, reduced carrying capacity, death of livestock, drying up of lakes, loss of biodiversity and widespread hardship.
8. **Land reclamation (conversion) in northern Xinjiang** – damming of rivers, widespread changes in water table, destruction of forests, soil salinity, desert encroachment, land abandonment and widespread hardship.

resultant attempt to carry too many animals, for too long, on areas less suited to prolonged heavy and continuous grazing. This report considers factors that led to excessive grazing pressures, which resulted in degradation. For each of the eight degradation case studies, various factors are considered. The biophysical and the socio-economic contexts within which degradation occurred are reviewed. Economic and political forces and the policy environment had major impacts and were significant factors in managing livestock numbers.

1.1.1 Patterns and distribution of degraded pastoral lands

The arid, semi-arid and dry subhumid areas in China are distributed widely over parts of 471 counties of the Xinjiang Uygur Autonomous Region, Inner Mongolia Autonomous Region, Ningxia Hui Autonomous Region and Tibet Autonomous Region, the provinces of Qinghai, Gansu, Hebei, Shaanxi and Shanxi. Figure 1.1 shows the location of the eight case studies that are the focus of this book.

The distribution of degraded rangeland areas in China extends over several thousand kilometres from east to west. The patterns of degradation in China are complex and diverse because of altitude and substrate and the patterns of land use imposed over past decades. Almost 80% of the arid, semi-arid and dry subhumid areas are affected by degradation to a greater or lesser extent. Rangeland degradation is as high as 56% overall, but in some areas it is worse (Wang, 2006). For example, a remote sensing survey of Inner Mongolia in 1983 showed the area of the degraded rangeland was 21.34 million ha, accounting for 35.6% of the total area of rangeland. By 1995, however, the area of the degraded rangeland had increased to 38.69 million ha – a net increase of 1.74 million ha in 12 years. The annual increase of degraded rangeland is approximately 2% but, as Table 1.1 shows, the rate accelerated in the period up until the 1990s, but in recent times the rate of expansion has been less than the rate of mitigation.

The landscapes vary from magnificent mountains such as the Qilian Shan, Tianshan and the mountains of the Qinghai–Tibet Plateau, above narrow or wide mountain valleys of grass, and shrub vegetation, sand dunes as high as 100 m and shrub deserts occur at lower elevations. At the lower elevation, rangelands are warmer, but the lack of rainfall makes vegetation production very low. Few naturally occurring

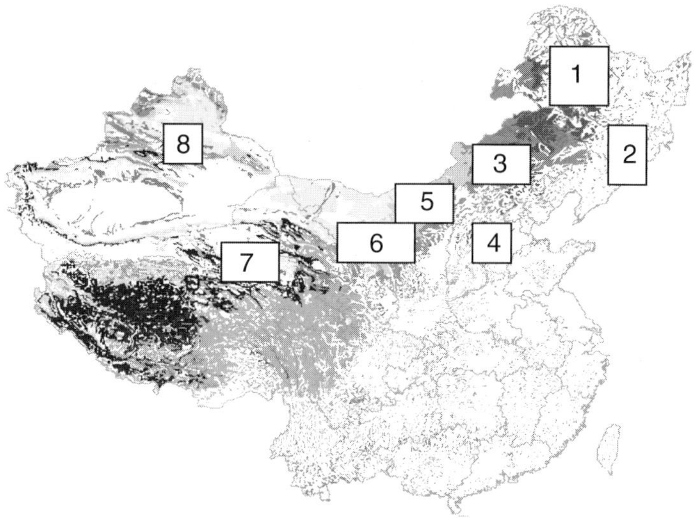

Fig. 1.1. Outline map of China with numbered regions that represent the eight case studies in the pastoral regions where the degradation and recovery case studies documented in this book have occurred. 1, Hulunbeier Grassland; 2, Horqin Sandy Land; 3, Xilingol Grassland; 4, Ordos Plateau; 5, Alashan; 6, Hexi Corridor; 7, Qinghai–Tibetan Plateau; 8, northern Xinjiang.

Table 1.1. Rate of expansion of sandy desertification* in north and north-west China, 1950–2005.

Decades	Rate of expansion (km^2/year)	Remarks
1950s	1560	
1970s	2100	
1980s	2460	
1990s	3436	
2004*	−1234	Expansion is less than mitigation

*Water erosion and other forms represent less than 30%.

trees are present and rarely do shrubs exceed 60 cm in height.

The northern and north-western drylands are within inner Eurasia and under the control of a continental climate all year round. Precipitation decreases gradually from east to west from 400 mm to less than 100 mm. True grasslands (prairies) are represented in the north-east, while steppe and desert rangelands and shrublands dominate the landscape in the north-west. Altitudinal variation of climate in the Qinghai–Tibet alpine area is very significant, which is characterized by low temperature, strong solar radiation, wind and uneven rainfall. Precipitation declines from south-east to north-west on the plain of the plateau and the natural landscape varies accordingly from forest, alpine shrub and alpine steppe to alpine desert.

The purpose of describing these case studies is to derive an understanding of what causes land degradation and what actions and information sources are needed to prevent further degradation episodes. This book is not intended as a history, but uses previous histories and documentation to interpret the causes of degradation and recovery.

Because recovery sometimes occurred decades after the degradation episodes, it has not been possible to *quantify* the extent to which initial productivity and resource condition have been restored. In the case study areas where there has been a considerable loss of soil, irreversible change may well have occurred and the return to initial productivity is unlikely to take place.

The term 'degradation' can be overused, but in the above cases the first-hand observers appear to be in no doubt as to the severity of the damage occurring to the landscape and the vegetation.

The subsequent recovery, as has taken place in some cases, is discussed later in this book. Beginning in Chapter 7 of this book, each case study is assessed in terms of the phases of both *degradation* and *recovery* (see also Chapter 6).

Other examples of degradation and partial recovery that we have not been able to document to the same extent as the eight case studies above include the following.

Mu Us Sandy Land

The Mu Us (also called Mao Wusu) Sandy Land is located south of the loess plateau and was covered by aeolian sands in the Quaternary Period. Soil had formed on the fixed aeolian sands but now surface erosion is obvious and soil humus loss is 30–50%. Sand dunes cover 5–10% in the form of fixed sand dunes, longitudinal sand ridges and vegetated sand mounds and shifting sands. Under the action of wind, the surface is denuded and the gravel percentage is now more than 10%.

Vegetation cover varies from 3 to 50% composed of xerophytes and mesophytes, mainly as steppe or desertified steppe landscapes. Natural steppes have almost disappeared. Sand and gravel-covered lands support bush–grass vegetation (10–30% cover), with many mounds forming around the scattered bushes. Cropping lands are covered by sand accumulation as small dunes or sand sheets. Desert encroachment on settlements, land abandonment and widespread hardship, leading, in some places, to resettlement, are features of the area.

Tarim Basin, Xinjiang

The Tarim Basin is located to the south of the Tianshan Mountains and is an enclosed inland basin and is surrounded by mountains in three directions. The Tarim River is China's largest inland distribution system. It has an arid climate, with great temperature fluctuations. It is one of the most arid regions in China. Human activities in the past 70 years or so and dry climate have accelerated desertification. The damage can be summarized as follows.

CHANGES IN RIVER WATER FLOW, RIVER COURSE REDUCTION; MARSHES DRY UP AND LAKES DISAPPEAR. In the 1950s, Xinjiang had 52 lakes with an area of 5 km^2 each, the total lake area was 9700 km^2, but by the late 1970s the lake area had declined sharply to 4748 km^2. Water flows of the Tarim River reduced year by year and the Kaxgar River ceased its flow into the Tarim River in 1990. In the 1950s, the Yarkant River still had a water volume of 1.0–1.5 billion m^3 to flow into the Tarim River but, as a result of dam construction in its upper reaches, no water flowed into the Tarim River after 1979.

The mean annual water volume of the Hotan River flow into the Tarim River was 1.1–1.2 billion m^3 in the 1950s, but since the 1980s it has reduced to an average 0.8 billion m^3. Later, it fell to 0.4 billion m^3. The longest inland river, the Tarim River, had been dry for 10 years in the lower reaches of Daxihaizi and vegetation had declined; desertification of the 'green corridor' became serious day by day, 160 km of the river course was buried by sand.

Before the 1920s, the 'green corridor' had been prosperous; at that time, the water volume of the Tarim River was greater and there were abundant aquatic plants along both banks. The Luntai Dam was constructed in 1952. As a result of river diversion, the corridor habitat deteriorated. By the 1970s, because the Daxihaizi Reservoir cut off the flow of the Tarim River, the underground water level dropped rapidly and *Populus euphratica* forest vegetation deteriorated over a large area. Rangeland became seriously degraded and desertification developed rapidly in the corridor. In the Alagan region, there was almost no desert 100 years ago and, 40 years ago, there were good natural pastures, but in 1972, following the drying up of the Tarim River, irrigation ceased and the area of abandoned farmland increased rapidly because people cut the shrub forest outside the irrigated field area and the originally fixed and semi-fixed dunes were reactivated by wind action. There were more than 8600 ha of abandoned cropland, of which nearly 2000 ha were buried by the shifting sand.

MOST FOREST AND GRASS VEGETATION IS DEGRADED AND DESERTIFICATION IS EXPANDING. The most concentrated area of *P. euphratica* forest was in the Tarim Basin. In the 19th century, it was described as 'dense flourishing forest'. In 1958, the Tarim River watershed had 400,000 ha of *P. euphratica* forest; in the 1990s, investigation showed that

358,000 ha were lost and only 42,000 ha remained. In addition, 110,000 ha of riparian and riverine vegetation in the middle and lower reaches of the Tarim River had been destroyed and the land had become desertified. The area suffered from a shortage of energy, so firewood was found by cutting *P. euphratica* forest and desert forest. In just three counties, Hetian, Moyu and Luopu, firewood cutting amounted to 300,000 t every year, which caused large areas of barren land.

DESERTIFICATION OF ARTIFICIAL OASES. In the process of developing an artificial oasis, people destroy the forest and open up 'wasteland', destroying the rangeland to make arable land. Because the oasis area is relatively small and fragmented, the surrounding desert vegetation has been destroyed almost completely and there is now no buffer zone between the artificial oasis and the mobile desert. Invasion of shifting sand into the oasis causes direct damage to agricultural and husbandry production at the edge of the oasis and also affects the lives of the people adversely.

Because of water scarcity, some arable land had to be abandoned soon after ploughing and, because of vegetation destruction, the surface soils became loose and vulnerable to wind erosion. Most abandoned cropland became desertified land; for example, in the Tieganlike region in the lower reaches of the Tarim River, people opened up 166,000 ha of forest for farmland in the 1960s and, within 3 years, 3300 ha had already been abandoned. In 1969, abandoned farmland totalled 13,000 ha and it became desertified land covered by shifting sand to a depth of 15–25 cm. Since 1949, about 28,000 ha of farmland within the oasis has become degraded. In mid-May 1986, a strong dust and sandstorm driven by strong wind caused loss of more than 90% of the cotton fields and many crop areas were affected in the oasis hinterland, with a 50% crop loss.

DESERTIFICATION ENDANGERS INDUSTRIAL AND AGRICULTURAL PRODUCTION AND CAUSES HUGE LOSSES. Sandstorms are the biggest recurring natural disaster in Xinjiang. For example, on 9–11 April 1979, the cold front brought sand and dust, which swept across more than 60 cities. The wind force reached 8–9 on the Beaufort scale and, in some areas, reached force 12 (>20 m/s). This caused huge economic losses. On 17–20 May 1986, more than 30 cities were hit by strong winds, the result of which was that 153,000 ha of farmland suffered severely. Damage included the deaths of 16 people and 94,000 head of domestic animals were lost; 800,000 trees, 3000 telephone poles and nearly 2000 houses were blown down. Direct economic losses were enormous.

Pengqu Valley, Tibet

This valley is located in the rain shadow zone of the Himalaya Mountains and belongs to a dry, subhumid climate type. This is an area of 6174.33 km^2, of which 4012.30 km^2, or 64.5%, is affected by dune movement and sand accumulation, 517 km^2, or 8.4%, is affected by water erosion, 81 km^2, or 1.3%, is affected by salinization and 1563 km^2, or 25.3%, is affected by freezing–thawing processes.

The area of land affected by sand disaster is 4012.30 km^2 and it is the most serious type of desertification. Sand dunes are distributed mainly in the middle part and downstream areas of the Pengqu River and its tributaries, like the Jinlongqu River, the Yeruzangbu River and along the valleys of the Duoxiongzangbu River and the Yaluzangbu River. Barchans and barchan chains, transverse dunes, sand mounds, vegetated sand ridges, creeping dunes, semi-fixed and fixed dunes are the main surface landforms. Exposed and semi-exposed gravel and sandy lands are distributed on pediments, alluvial–diluvial plains, river courses, alluvial fans and river terraces.

Water erosion-affected desertified lands are distributed mostly in valleys of the Duoxiongzangbu River, the Yaluzangbu River and at the marginal areas of the lakes and river valleys of Pengqu, which occupies 517.51 km^2. Salinization-affected desertified lands cover 80.90 km^2 and are distributed in low-lying areas and in depressions along the riverbanks and lakeshores in Dingri County, Dingjie County, Sajia County and Lazi County. Desertified lands caused by freezing–thawing processes are the second largest type of desertification and cover an area of 1563 km^2, which is distributed mainly on the slopes of alpine mountain and uplands in the northern foothills of the central Himalaya Mountains and the Lagangguiri Mountains. This type of affected land is manifested in the form of rangeland landslides, mudslides and clay ridges, caused by freezing–thawing processes. In the seriously eroded sections, grass vegetation has blown away and, with it, the thin topsoil. Gravel bedrocks are completely exposed and covered with *Felsenmeer* (sea of rock). Desertified lands in this district are characterized mostly by their patchiness and their distribution at the intersection of the Duoxiongzangbu River and the Yaluzangbu River and along the middle reach and downstream of the Pengqu River valley.

In recent years, the trend of worsening and spreading desertified lands is accelerating and the farmlands and rangelands at the lower reaches of the Pengqu River are being degraded, with rapid sand encroachment and dune movement. For instance, in the Nixia Township of Dingri County, much land is occupied already by shifting sand encroachment and farmlands and grazing lands are rendered useless. As a consequence, villagers have had to migrate to new settlements.

Soil erosion of desertified areas of the Tibet Plateau is serious because higher wind velocity erosion of the plateau has intensified and has become a dominant driving force in the development of sand-affected desertification. Large areas of newly cultivated lands lack effective protection from the wind and the soil structure and texture have deteriorated at an unprecedented rate. As a consequence, sand-affected desertification has occurred and is spreading. The area of arable land has increased from 627.9×10^3 ha in 1952 to 812.0×10^3 ha in 1996.

Biomass is used as the main fuel wood or energy source in Tibet. Following the increase of energy consumption and population growth, shortage of fuel is now more serious. In order to obtain more fuel for daily life, citizens tried to collect all possible available fuels to enable their survival and thus a large area of natural forest or vegetation was destroyed. Approximately 4000–6000 ha of natural bush or forestland were deforested per year due to unwise collection of fuel wood. Along with the destruction of natural vegetation and a decrease in surface vegetative cover, desertification will spread, or become more severe.

Observers and researchers now believe that the first degradation event in each region occurred in the first major drought after the initial intensification of agriculture and the first wave of land conversion (taking rangeland for growing crops and building infrastructure and settlements). While the effects of the above case studies occurred mainly during drought periods (see rainfall anomaly charts), the causes of degradation are to be found in the political and policy environments and social conditions that prevailed in the decades of the 1950–1990s. Policies played a major role, but climatic factors loomed large. Thus, our approach in documenting these case studies is first to analyse the climatic and socio-political forces that have influenced rangelands over the past 60 years.

1.1.2 Components of the rangeland system in China's arid rangelands

China's rangelands span a very wide type of environments from the arid and semi-arid shrublands of western China to the perennial tussock (bunch grass) rangelands in north-east China. Whether the management objective is to have a highly productive rangeland for livestock grazing or to enhance biodiversity, improve ecosystem health, produce high-quality water or sequester carbon, rangeland areas are extremely valuable and worth preserving in good condition. Any use of the pastoral lands needs to be done with the objective of long-term sustainable use in mind.

Herding has been a major form of land use on these extensive arid rangelands for centuries – a tradition that is carried on today, but under a different set of constraints. The traditional ways involved long-distance, seasonal migration and a rotation of pastures, as flocks and herds moved over the vast landscape.

Since the 1950s, government policies were to sedentarize herders in the arid and semi-arid regions and achieve a greater integration of cropping and herding (Brown *et al.*, 2008). There have been successive waves of converting rangeland to cropland, often with dire consequences (Williams, 2002). Over time, cultivation and herding have risen and fallen in response to the seasons and government policies (Halin *et al.*, 2002). The semi-arid environment is brittle, sensitive and risky for agriculture. Agricultural output is not only low and variable, but farming and herding methods in current use accelerate wind erosion and desertification of the land, leading to widespread deterioration in the ecological environment of this region.

Rangeland degradation results in a loss of capacity to produce forage for both livestock and wildlife. It also reduces other rangeland benefits, including: (i) biodiversity values, which have declined in terms of number, variety and range; (ii) watershed protection; (iii) carbon storage; and (iv) air quality. This rangeland degradation is caused by a combination of natural factors (infestation by rodents and insects and changing climatic factors) and human factors, such as inappropriate land-use policies, inadequate rangeland management supervision and overharvesting of rangeland products. The human-induced factors are exacerbated by: (i) overall poor understanding of the functioning and resilience of ecosystems; and (ii) lack of awareness by government officials at various levels of the medium- and long-term environmental impact of interventions being implemented or planned. Since about the 1950s, the rate of degradation of rangelands has accelerated as livestock numbers have risen (see case studies in Part III and Brown *et al.*, 2008).

Data from the State Environment Protection Administration (SEPA) suggest that the total area of degraded rangeland increased by about 95% between 1989 and 1997 (from 645 million ha to 1300 million ha), with a notable acceleration in the middle to late 1990s. The average annual rate for the 1989–1997 period (7.9 million ha/year) is equivalent to about 2% of the total rangeland area per annum. It is very difficult to quantify trends in rangeland degradation due to lack of reliable data, although there are various estimates in the published literature. All experts agree that there is a very high level of rangeland degradation both nationally and regionally and that the situation appears not to be improving, and still may be continuing to deteriorate. There are also anecdotal cases to support this general contention. It should be noted that Xinjiang and Gansu are experiencing rangeland degradation levels well above the average for China.

Further evidence is seen in the changes in preferred livestock in pastoral areas. Cattle were replaced by sheep and later by goats and, in some seriously degraded areas, by camels, as the quantity and nutritive value of the rangeland declined. Commonly, shrubs have been replacing grasses. The perennial forage component provides stability of surface soil cover in terms of reducing soil erosion driven by wind and water. In many situations, the woody component (trees and shrubs) provides fuel. Thus, for each of the major rangelands, it is possible to identify the 'desirable' perennial plants (grasses and shrubs) which form the basis for stability of the rangeland system and resource functioning. Degradation, from the point of view of herders, involves the loss of productivity of this *desirable perennial component* of the plant community (Hodgkinson, 1995). The major causes of the loss of desirable perennial plants in grazed rangelands are:

- severe and/or extended drought;
- a combination of heavy grazing and drought;
- selective grazing and competition from unpalatable grasses and shrubs;
- competition from 'invader' plant species, e.g. poisonous plants; and/or
- lack of recruitment to the plant community.

The productivity of perennial species can also be reduced greatly by soil erosion through the direct loss of soil nutrients (nitrogen and phosphorus) and loss of available moisture by increased runoff and decreased capacity to store moisture. The loss of a small depth of soil through erosion can result in large decreases in nutrient availability and potential productivity. The main cause of accelerated erosion is loss of cover from grazing and grazing-related soil disturbance (Chapter 4).

Sequences of dry or wet years have been a major climatic force in the above degradation case studies. Figure 1.2 is a time series of the 5-year moving average rainfall (expressed as a percentage anomaly from the mean) for the different regions of China where the major degradation episodes have occurred. Drought is an important component of these episodes, revealing the extent of degradation[1] and, at the same time, contributing to further degradation.

Excessive grazing pressure and climatic variability interact to cause the loss of desirable perennial species (grasses and shrubs). Observations over the past 60 years have shown that the combination of heavy use and recurrent persistent drought (see figures above) during what should have been the normal growing season have resulted in the loss of 'desirable' perennial plants. This leads to accelerated soil erosion and further pressure on the grazed resource. Recovery of vegetation generally requires sequences of above-average rainfall years and low (or zero) grazing

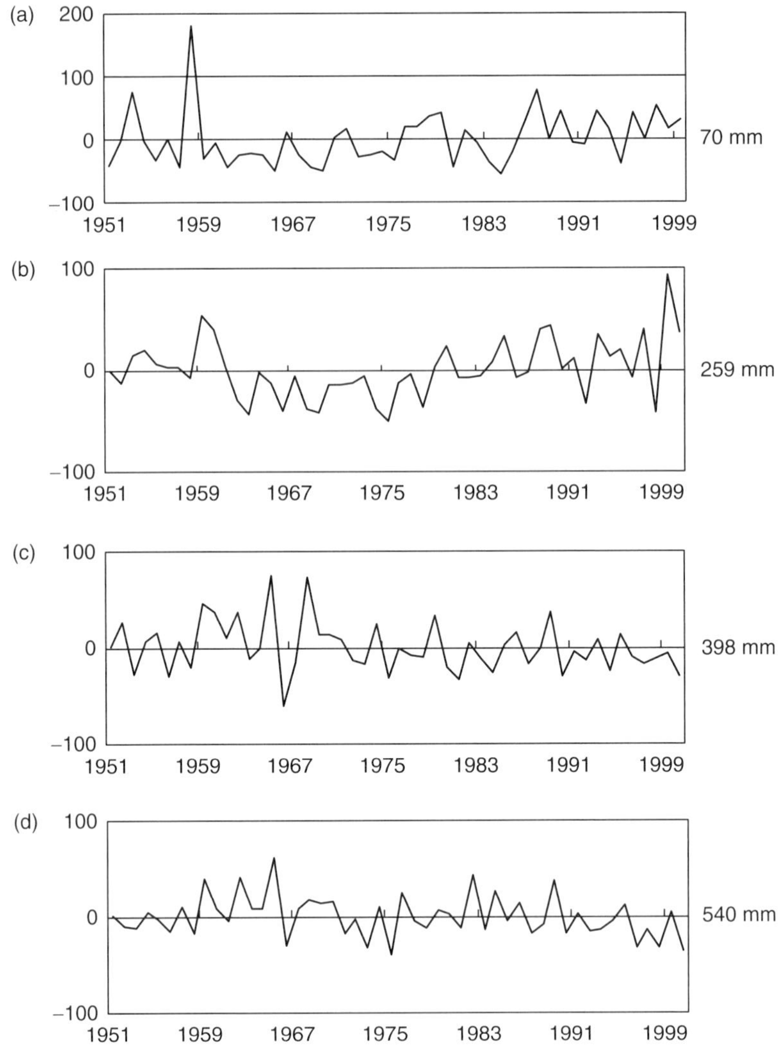

Fig. 1.2. Variation (%) of annual rainfall from its mean value. a, hyper-arid; b, arid; c, semi-arid; and d, dry subhumid areas in the pastoral areas of north and north-west China.

pressure to allow recruitment to plant populations and time for perennial root systems to build up.

The pastoral regions as a whole had major peaks in rainfall during the late 1950s and, in the drier regions, in the late 1990s, when above-average rainfall occurred (see Fig. 1.2). Drought conditions (some quite severe) also occurred, for example, in the far west of China during the decade of the 1960s. A fuller analysis of the time series data is presented in Chapter 3.

A major feature of the climate of several pastoral regions has been the extended periods (3–5 years) of above- and below-average precipitation. These climatic anomalies have resulted in major disturbances to the vegetation, affecting regeneration and mortalities, nutrient pulses, soil surface destruction and reconstruction and, at times, complete biomass utilization by grazing animals.

For the regions discussed in this book, the peaks of rainfall were more pronounced in the 1950s and again in the 1990s and early 21st century. The highest precipitation year was 2003. These periods of above-average rainfall in north-east China (Horqin and Hulunbeier) and in north central China (Ordos Plateau and northern Shaanxi) were most pronounced in the early 21st century. In western China (Hexi Corridor, Alashan and Junggar Basin), there were alternating wetter and drier periods.

There was a drier period in the 1960s/1970s. Major droughts occurred in the 1960s to the early 1970s, during the 1980s, the 1990s and in the late 1990s. Thus, the time series of rainfall indicate that widely separated regions in north-east China and, to a lesser extent, in north central and western China have shared similar periods of high and low rainfall.

The eight degradation case studies chosen for this study cover a wide geographical spread in north and north-west China (Fig. 1.1). They are not the only cases of degradation that have occurred in the past 60 years, but they have been chosen because they are relatively well documented in a number of sources, including published government reports, unpublished files and personal accounts. These sources provide us with the context for the social, political, policy, economic and environmental issues from the time, as well as with data on changes in livestock numbers and assessments of the extent of degradation (see the case studies in Chapters 7–14).

The evidence for rapid and severe degradation is unequivocal. The accounts from the time are graphic in their descriptions of the physical 'horror' of bare landscapes, erosion scalds, gullies and severe sand- and dust storms. Subsequent observations documented the environmental and economic damage caused by loss of palatable plant species and soil loss, and highlighted the human and animal suffering through death and land abandonment.

The emotional and financial plight of herders and their families as a result of severe land degradation and drought leading to land abandonment and forced relocation (as ecological refugees) is less well documented but real none the less (Box 1.2).

We then combine this information with time sequences of climatic forcings, rainfall and simulation of historical biomass potential using present-day methods to build up a composite picture of each degradation case study and the factors that led to it (Chapter 3).

The main feature of degradation in the documented case studies was the change of land use and the resultant attempt to carry too many animals, for too long, on areas less suited to prolonged heavy and continuous grazing. Our analysis considers factors that led to the excessive grazing pressures that resulted in degradation. For each of the eight degradation case studies, various factors are considered. The biophysical and the socio-economic contexts within which degradation occurs are reviewed. Economic and political forces and the policy environment had major impacts and were significant factors in managing livestock numbers.

We also examine current knowledge of the number of phenomena that affect climatic variability in China's rangelands. These influences are complex and current climatological research has shown that there are significant climatic signals at timescales from about biennial to decadal and multidecadal. The future behaviour of the climate system is complicated by the possible presence of changes due to anthropogenic influences (e.g. increasing concentrations of greenhouse gases, ozone depletion, aerosol emissions and land-use change), together with naturally occurring interdecadal variability.

The compilation of long-term weather records can provide a basis for analysing the impacts of climate variability on grazing enterprises. It also provides a context for examining which grazing management options were successful in the face of variability and which, in hindsight, were mistakes.

1.1.3 China's livestock – a major user of rangelands

China has one of the world's largest livestock populations. Given the importance of herding to China's national economy (Tables 1.2 and 1.3), it is not surprising that the factors affecting the survival of livestock during drought have been a cause for concern and the subject of various government-sponsored assistance packages.

Exclusion of grazing in many situations allows natural regeneration (Bao *et al.*, 2002; Shang *et al.*, 2008) and is the preferred approach on extensive low-value rangelands and has been applied globally (Noy-Meir *et al.*, 1989; Pucheta, *et al.*, 1998). Exclosure of livestock for a number of years is a favoured option in many rangeland areas where free grazing has been banned in favour of lot

Box 1.2. Desertification: a human tragedy in the Tibetan Plateau.

The cases of the Shiquanhe Basin and the Great Lake District in Tibet exemplify the degradation processes. In the early 1960s, the Shiquanhe Basin was densely vegetated virgin land with luxuriant natural vegetation. Due to uncontrolled deforestation since the late 1960s, bush vegetation became restricted to the alluvial fan areas at the margins of the basin. Subsequently, large areas inside the basin are now covered by shifting sands and mobile dunes move forward to Shiquanhe Township, and both human settlements and the living environment of the town are impacted seriously by the sand disaster. The lacustrine plain of the Great Lake Basin was vegetated by shrublands, with some wetlands and salinized meadow. However, under the pressures of overgrazing, and other forms of overuse of these rangelands, and due to the permafrost degradation, soil salinization and erosion, meadows, shrublands and steppes were degraded into salinized lands or sand-encroached lands. Under the impacts of the driving forces of desertification in this region, meadows and alpine range are gradually degrading from alpine meadow to desert steppe, semi-exposed sand-gravel lands and exposed sand-gravel lands. Degradation varies in its intensity, characteristics and evolutionary processes and there is a mosaic of sand encroachment and salinized lands that is growing in area year by year.

Desertification is worsening with great speed and the annual spread rate of desertified land caused by sand movement and sand accumulation is 0.3% in the central east part and 0.3% in the western part of north-west Tibet. It is classified as severely affected. Strong winds and dust storms are frequent and, on average, there are 113 wind-dust days per year. The desertified area caused by sand movements and sand drifting is the sand-generating source area of dust and sandstorms. For example, in mid-February 1997, there was a dust and sandstorm in Rikeze District and one person was killed and many people were wounded, 76 houses collapsed, 87 telephone poles and 870 trees were pushed over and 10,000 livestock were injured. During February to March of 1997, 23 houses collapsed and 1089 head of livestock were killed by dust and sandstorms in the Shannan District.

Mudslides are an associated natural disaster during the water erosion process. On 6–9 July 1997, a large landslide and mudslide occurred near the village of Dongga Township, in the Cuona County of Shannan Prefecture, and 8.27 million ha (Mha) of cultivated lands were washed away, 6.87 Mha of forest and 15,000 trees were blown down, 4.8 Mha of rangeland were buried and two reservoirs, three bridges, 3 km of paved road and 7 km of flood dyke were destroyed. In addition to direct economic loss, there were great hardship and suffering. The sand disaster threatened settlements and towns and affected the living environment adversely. Shifting sand accumulated and buried settlements and houses; streets were covered by sands to a depth of 0.20–0.60 m. Much land was covered by shifting sand encroachment and farmlands and grazing lands were rendered useless. As a consequence, villagers had to migrate to new settlements. The Changsuo Development Zone of Dingri County was another example and lands were abandoned due to desertification and sand movement and all local development programmes were cancelled or postponed. Three townships of Nima County on the north Tibet Plateau were moved to their new locations due to the wind–sand disasters. In total, 748 villages have suffered under the impact and threats of sand movement and shifting sand disaster and it is estimated roughly that the annual costs for clearing the accumulation of sand on the streets, relocating villages and constructing new settlements will be several tens of millions of dollars. About 20% of all settlements and houses and 30% of livestock enterprises in the central west part of Naqu Prefecture are under threat from shifting sand and some facilities will even have to be abandoned because of serious sand invasions.

Table 1.2. China's livestock population in the pastoral lands (in thousand head). Source: National Bureau of Statistics (NBS) (2000).

	Cattle	Horses	Donkeys	Mules	Camels	Goats	Sheep
Total for six pastoral provinces and regions[a]	29,608	3,907	3,637	1,502	329	38,032	91,454
Contribution by pastoral provinces to total (%)	23.32	43.83	38.91	32.14	99.70	25.67	69.76

[a]The pastoral provinces and regions include the Inner Mongolia Autonomous Region, Xinjiang Uygur Autonomous Region, Tibet Autonomous Region, Qinghai Province, Sichuan Province and Gansu Province.

Table 1.3. Output of the principal livestock products from China's pastoral lands. Source: National Bureau of Statistics (NBS) (2000).

	Meat (thousand t) (cow, sheep and goat)		Milk (thousand t) (cow and sheep)	Wool				Other fibre (t)	
				Fine	Semi-fine	Carpet	Total wool	Cashmere	Camel and yak
Total for six pastoral provinces and regions[a]	814	959	2,084	71,694	22,629	105,507	172,336	5,971	9,914
Contribution by pastoral regions (%)	16.11	38.16	27.61	62.83	45.63	47.29	60.86	58.65	48.65

[a]The pastoral provinces and regions include the Inner Mongolia Autonomous Region, Xinjiang Uygur Autonomous Region, Tibet Autonomous Region, Qinghai Province, Sichuan Province and Gansu Province.

feeding. Experience from north and west China indicates that exclusion can increase productivity of degraded rangeland (Wang *et al.*, 1996; Bao *et al.*, 2002), but concerns are being expressed about the undesirable effects of long-term grazing bans on ecosystems that have co-evolved with grazing animals (Yi *et al.*, 2004). There is also pressure from pastoralists to reopen the regenerated areas. Before this can be done, it is necessary to generate guidelines for reasonable grazing pressure (based on field experiments) and develop mechanisms to ensure that there is enforcement of rules about stock numbers and entry and exit times.

It is against this background then that we examine, analyse and comment on the pattern of rangeland degradation and recovery in China's pastoral lands. The process of degradation is still occurring in many areas within northern and north-western China. Our analysis should be helpful to policy makers and land managers to help prevent further cases of severe and damaging degradation and also to help plan and execute programmes to extend successful recovery measures to other areas where similar conditions prevail.

Addressing rangeland degradation might start with the consideration of whether or not grazing land allocations are adequate and equitable (Chapter 15) and whether stocking rates are appropriate to the land type and its condition (Chapter 4).

The aim of this book is to understand the causes of major rangeland degradation events and the processes in China's pastoral rangelands, so that there is a greater level of preparedness in the future that should lead to a reduced probability of repeating the same mistakes. Analysis of the recovery, where it is occurring either through natural regeneration or through technical interventions, can help us replicate and scale up successful approaches and restore rangeland productivity.

Note

1 The so-called 'crucible of drought' – a time when weaknesses in the system are revealed.

References

Bao, Y.T., Li, Y.M. and Yang, C. (2002) Comparison of plant community characteristics under different gradients of grazing intensity. *Pratacultural Science* 19(2), 13–15 (in Chinese).

Brown, C.G., Waldron, S.A. and Longworth, J.W. (2008) *Sustainable Development in Western China: Managing People, Livestock and Grasslands in Pastoral Areas.* Edward Elgar, Cheltenham, UK and Northhampton, Massachusetts, 294 p.

Halin, Z., Xueyong, Z., Tonghui, Z. and Ruilian, Z. (2002) Boundary line on agro-pasture zigzag zone in North China and its problems on eco-environment. *Advances in Earth Science* 17, 739–747 (in Chinese, with English abstract).
Hodgkinson, K.C. (1995) A model for perennial grass mortality under grazing. In: West, N.E. (ed.) *Rangelands in a Sustainable Biosphere. Proceedings of the IVth International Rangeland Congress*, Volume 1. Society for Range Management, Denver, Colorado, pp. 240–241.
National Bureau of Statistics (NBS) (2000) www.stats.gov.cn/english/.
Noy-Meir, I., Gutman, M. and Kaplan, Y. (1989) Responses of Mediterranean grassland plants to grazing and protection. *The Journal of Ecology* 77(1), 290–310.
Pucheta, E., Cabido, M., Diaz, S. and Funes, G. (1998) Floristic composition and above ground net plant production in grazed and protected sites in a mountain rangeland in central Argentina. *Acta Oecologica* 19(2), 97–105.
Shang, Z.H., Ma, Y.S., Long, R.J. and Ding, L.M. (2008) Effect of fencing, artificial seeding and abandonment on vegetation composition and dynamics of 'black soil land' in the headwaters of the Yangtze and the Yellow Rivers of the Qinghai–Tibetan Plateau. *Land Degradation and Development* 19(5), 554–563.
Wang, T. (2006) *Desert and Desertification in China.* Science Press/Longman Books Co. Ltd, Beijing.
Wang, W., Liu, Z. and Hao, D. (1996) Research on the restoring succession of the degenerated landscape in Inner Mongolia. 1. Basic characteristics and driving force for restoration of the degenerated pasture. *Acta Physiologica Sinica* 20(5), 449–459 (in Chinese).
Williams, D.M. (2002) *Beyond Great Walls: Environment, Identity, and Development on the Chinese Rangelands of Inner Mongolia.* Stanford University Press, Palo Alto, California, xii, 251 pp.
Yi, R., Lin, Y. and Zhong, C. (2004) Relationship between botanical composition and grazing intensities in Xilingole rangelands, Inner Mongolia. *Ecological Science* 23(1), 12–15 (in Chinese).

2 Historical Degradation Episodes in China: Socio-economic Forces and Their Interaction with Rangeland Grazing Systems Since the 1950s

Victor R. Squires[1] and Yang Youlin[2]
[1]*University of Adelaide, Australia;* [2]*UNCCD, Regional Coordination Unit, Bangkok, Thailand*

Synopsis
This is not intended as a history of China's past over the last 70 years. Instead, in this chapter, we seek to derive from history the insights necessary to plan for the future. The clear message that emerges is that the policy environment in the early years of the creation of New China often had unintended consequences and that the implications of these policies were not fully comprehended. The legacy of these past mistakes is with us today. We were motivated by the proposition that China's pastoral rangelands would be better managed in the future if there were a greater understanding of how they were shaped in the recent past. By better understanding the mistakes made and the successes achieved, we might avoid ignoring the lessons of history.

Keywords: policy; pastoral nomadism; ecosystem management; human impacts; climate variability; population pressure; migration; resettlement; social systems; household responsibility system; land tenure; markets; natural disasters; collectivization; decollectivization; Cultural Revolution; land conversion; land reform

2.1 Relevance of Ecological History to Environmental Management

Managing ecosystems without any knowledge of their history may result in future disaster. Simple description of the environment or observation of the environmental variables (monitoring, observation and experiment) over a few years is essential, but inadequate, for detecting rates, directions and magnitudes of change in highly complex and dynamic systems (both biophysical and socio-economic). This is particularly so in north and north-west China, where climatic variability is high and extreme weather events may be more important than long-term averages (means) in determining what is actually taking place at any one place at a particular time. Any period of observation is likely to be unrepresentative of the longer term and may or may not contain rare events. Shorter-term change or stability must be placed in the context of longer-term trends or variability.

In order to predict the outcomes of management actions, studies of present-day patterns must be complemented by studies of current processes, but both kinds of study need to be extended by investigation of the patterns and processes in the past, leading up to the present day. Predictions are dependent on explanations of how the complex interactions between processes (and the patterns they create) have changed over time. Environmental pattern at any one time is both

 Rangeland Degradation and Recovery in China's Pastoral Lands (eds V.R. Squires *et al.*)

the outcome of preceding interactions between processes and what was already there and the template on which processes act in the immediate future.

To deal with ecological problems, the most compelling reasons for studying environmental history arise from the following considerations:

- Many problems are inherently historical; for example, the impact of rapid land reclamation in the decades immediately after the foundation of 'New China' (1950s onwards) on pre-existing environments in north and north-west China. While written records can provide much information about what was there, reconstructions of the medium past are needed in order to perceive the kind and magnitude of induced changes.
- Instrumental records (meteorological data, etc.) are invariably too short to predict the return periods of less frequent or rare events, particularly those that are of greatest magnitude (and impact). This applies to disturbances such as floods or earthquakes, as well as to stresses applied over a longer period, such as droughts or severe freezing.
- Many phenomena have been recorded rarely or discontinuously in space and time, or not at all. Often, monitoring is initiated only when something is perceived as a problem or becomes a political issue (such as the increased frequency and severity of dust and sandstorms (Yang *et al.*, 2002)). Archival records can often be used to reconstruct the history of such variables or events, both quickly and economically.
- The environment is complex and dynamic, with many interacting processes. The past can provide a record of rare, and often critical, concatenations of variables or events, such as droughts, severe winters, floods, land-use change and episodes of accelerated soil erosion. The historical record can be used to distinguish events, or sequences of events, that may not be of the greatest magnitude but may be of most significance in their effects on species, communities or ecosystems; for example, a severe freeze during a drought may have a far greater effect than a freeze or a drought alone.
- While ecosystems may remain stable for long periods, there may be considerable shorter-term variation within those periods. To manage ecosystems, we need to know the limits of reversible variation and, if possible, the thresholds of irreversible change and the likely agents of change. The time-span required to cover the range of variability depends on the longevity of dominant or critical species.
- Short-term variability may mask long-term trends and only the historical record can provide this information, particularly in an environment such as north and north-west China where short-term climatic variability is so high.
- Predictions from modelling and time series analysis will prove inadequate, or even dangerously misleading, if boundary conditions are changing. We need evidence for rates and directions of change and how these have altered over time in, for example, climate and shifts in species composition in terrestrial ecosystems and changes in biodiversity.
- If, as is predicted, the earth experiences rapid, global environmental change over the next century, any changes will be superimposed on pre-existing dynamics, not a static world. To predict effectors of global change, we need to know what the dynamics are, including longer-term trends, limits of variability and frequency and magnitude of rare events.

2.2 A Region Transformed: Human Impact on the Rangelands of North and North-west China

The loss or transformation of the vegetation cover and the loss or replacement of plant species have been a continuing process that has been accelerated over the past few decades as a result of changes to land-use intensity, human population pressures and climatic oscillations.

The rangelands of north and north-west China have historically supported a viable pastoral economy, as well as wildlife populations, but recent social, political and economic changes are affecting both pastoralists and wildlife (Miller, 1998a, 1999a). In this chapter, we analyse changes in the socio-political and policy environment that have impacted on the rangelands and the people who use them. To assist the reader, we have divided the analysis into a series of phases that begin early in the 20th century, but the main focus is on the period after the 1950s.

2.2.1 The period before 1956

Historical records indicate that animal husbandry developed very early in all of the focus areas (Chapters 7–14) that were first inhabited by nomadic peoples belonging to ethnic minorities. Traditional range management relied heavily on livestock mobility, but by the 1940s transhumant pastoralism had all but disappeared in some areas – due primarily to the influx of new populations. The landless farmers and refugees of war who immigrated were not livestock farmers and were not familiar with transhumant practices. Instead, they led a settled life on farms and set up mixed farming systems incorporating crop farming and animal husbandry. As the rangelands were vast and abundant and the population scarce, the mechanisms for resource use introduced by these migrants were highly appropriate at that time.

An initial land reform was carried out before the 1950s under which some areas of rangeland were redistributed from landlords to individual farmers or herdsmen. A second wave of land reforms, after the Land Revolution (1950–1952), redistributed all remaining agricultural and grazing lands in the same way. On the ground, however, range management practices remained very much the same throughout the 1911–1956 period, with property rights remaining vested with individual users, lineages or the village community as a whole (Chan *et al.*, 1992). Authority for delegating grazing rights for specific plots rested with community chiefs, village elders and heads of clans and individual owners of the rangelands.

In some areas that were newly settled after 1949, there was no detailed regulated system of grazing (such as rotational, seasonal or deferred grazing) in place, while in others the customary practices to regulate livestock and access to water and forage had been developed over centuries. In principle, everybody in the village was free to use the range, while a tradition of overlapping grazing existed between neighbouring villages. Outsiders could use grazing land only with the permission of the village committee (or committee head), which was also responsible for the resolution of conflicts over grazing and water use inside the village or between neighbouring villages.

At the time, these arrangements were deemed appropriate for a relatively abundant resource base whose productivity was highly variable because of erratic rainfall and which benefited more from flexible boundaries than from fixed ones. Questions to elderly farmers/herders in Gansu and Xinjiang about the period before collectives revealed that the rangelands were so abundant and vast that they simply did not need additional regulations such as deferred or rotational grazing. The absence of these forms of grazing management implies that the rangelands in many areas could renew themselves continually without the need for formal management interventions. Furthermore, the needs that the rangelands had to satisfy were not yet determined by market demand, but rather were adjusted constantly and limited within the local social system itself (Ho, 2005). It was only after the Communist Revolution in 1949 and the introduction of the People's Communes in 1958 that certain characteristics of the regulatory framework proved to be destructive.

From the last century of the Qing Dynasty to the establishment of the rural cooperatives in 1956, the rangelands were generally owned by landlords or small communities, but commonly used by livestock farmers. However, after the foundation of the People's Republic of China, grazing areas became more restrictive, as rangelands increasingly became the target of reclamation (conversion) for crop farming. During the period from 1949 to 1956, more than 80,000 km^2 of rangeland were opened up in this way. The way in which rangelands were managed – which had remained almost unchanged until that time – was also about to be radically altered.

After the land reforms of 1950–1952, the Chinese government set itself the task of developing and collectivizing agriculture – a process which passed through several overlapping stages.[1] It was the establishment of the Higher Agricultural Producers' Cooperatives (HACPs) in 1956 in pastoral areas for the collectivization of rangeland use and ownership and the People's Communes in 1958, on a backdrop of increasing livestock numbers and the Cultural Revolution campaigns of the 1960s and 1970s, that would prove calamitous for the grazing areas.

During China's People's Communal period (roughly 1960–1983 in this region), all pastoralists were organized into People's Communes. Livestock numbers increased dramatically (see Chapter 3). In most Chinese People's Communes during this time, no livestock were owned privately and compensation to pastoralists was in 'work points' rather than cash. The communes in many provinces and counties attempted to develop irrigation for growing hay and to cultivate varieties of exotic grasses to improve productivity. Administrative centres were built, houses and corrals for livestock operations were built and wells were dug. Growing hay and irrigating many rangelands failed because of saline soils and the harsh climate. Ultimately, the commune system was abandoned and considered a failure, but there is little doubt that the land-use changes during this period resulted in large areas of rangeland degradation that can still be seen today.

When the People's Communes were disbanded and the 'household responsibility system' was established in 1983, livestock were distributed to the membership and seasonal grazing lands were allocated to all households. Seasonal ranges were allocated to households based partly on their history in the region.[2] A family-based 'Household Contract Responsibility System' (HCRS), which offered farmers more managerial freedom by linking rewards directly to production and efficiency, was implemented.

The HCRS policy practice model was extended subsequently to grazing areas and, in 1985, the State Rangeland Law of the People's Republic of China[3] was promulgated, under which rangeland could be contracted to collectives or individuals. The law prohibits certain 'harmful' activities and empowers local governments 'to stop anyone from farming a rangeland in violation of the provisions of the present law, to order the person to re-vegetate the overgrazed or degraded rangeland and grazing lands, and to pay a fine if serious damage has been done'.

The success of previous agricultural reforms was not, however, mirrored in the livestock sector. Today, the Rangeland Law is unenforceable in some parts of China, while the contract system for rangelands has failed miserably (Williams, 1996a,b; Thwaites *et al.*, 1998). Far from promoting the sustainable use of the rangelands, the new system tended to enhance rangeland and steppe degradation, with economic freedom acting as a stimulus for individuals to increase production, whatever the long-term implications for the rangeland. The situation as it stands raises a number of questions about the implementation and consequences of the HCRS and Rangeland Law in the livestock sector (Zhang, 2006). In particular, it raises doubts about the wisdom of extending to the livestock sector policy measures designed for crop farming, without taking into account the inherent differences between production systems (Ho, 2005).

2.2.2 The period 1956–1966

Between 1956 and 1966, rangeland management remained relatively unaffected by political campaigns, and animal husbandry went through a stable development. After the People's Communes had been established in 1958, the government of most provinces and autonomous regions sought to increase agricultural productivity, with special emphasis on the livestock sector. The ruminant population rose rapidly in this period. For example, between 1958 and 1965, the total population of sheep and goats in Ningxia increased by 91.5% (GTZ, 1990). Human populations in most of the focus areas rose rapidly too. The combined increase in human and animal populations created an unprecedented level of land scarcity and increased the need (both perceived and real) for the regulation of rangeland use.

At the same time, new grazing techniques such as deferred and rotational grazing were introduced and land reclamation and the digging or wild collection of medicinal herbs became more strictly regulated and, in most cases, prohibited. During this period, institutions such as the Animal Husbandry Bureau (AHB) of the Ministry for Agriculture attempted to develop new approaches for sustainable use of rangelands.

Small-scale experiments, including the use of fencing, were carried out in counties where animal husbandry was most important. However, the scale of these experiments was very modest and many areas that were fenced were opened again during the period of the Cultural Revolution from 1966 to 1969. Other measures introduced to improve rangeland use included the sinking of new wells to mitigate the concentration of grazing around existing wells and pathways and the establishment of artificial forage areas to help alleviate fodder shortages in winter and spring.

In many ways, these attempts to improve (and sustain) livestock productivity through capital investments and improved common property management were destined to fail in their objectives. This was true for several reasons.

Unclear management responsibilities

Attempts to effect 'collective maintenance, collective management and collective usage' by the production team conflicted with the existing property rights structure. In principle, it was the commune rather than the individual user (i.e. members of the production team) which owned the rangeland. Therefore, the term 'collective' did not refer to one institutional level, but three: the production team, the production brigade and the commune. Ownership of the rangeland was vested in the commune, the ownership of livestock in the brigade, while the team was charged only with herding the flock. Under these arrangements, individual herdsmen or rangeland users had no interest in using the range in a sustainable way, as it was not perceived as being their own.

Ambiguous territorial boundaries

The tradition of overlapped grazing prevailed throughout the time of the People's Communes. The attempts to introduce enclosure not only contradicted this tradition but also, arguably, were unsuited to the management of Chinese rangelands, which are characterized by highly variable productivity. These rangelands benefit more from flexible arrangements than from rigid ones and it is hardly surprising that fencing experiments failed. If formal agreement did exist between the communes over the various boundaries of the rangeland under their jurisdiction, the boundary rules would still be void, as the communes lacked the authority structures to enforce them.

Open group membership

In essence, everyone was automatically a member of a commune and designated to a certain production team. There was no restriction on the number of sheep, nor were participants excluded from the team if rangeland or steppe were threatened with overgrazing. In some cases, capital investments were made to augment the productivity of the pastures. Yet, as there were no rules for allocation, or for setting the boundaries of the natural resource, the rangelands continued to be squandered, and these capital investments were in vain.

Poor enforcement of rules

The enforcement of rules was left to Grassland Monitoring Stations (GMSs) under the AHB. These institutions were short of technical and financial resources and were seriously understaffed. They were also subject to the political situation, which was reflected in the number of reorganizations, mergers and disbandments that took place (GMSs were abolished completely between 1967 and 1978). Within the brigade or team, there were no formal structures that could have taken on the job of enforcing sanctions on resource use.

Poor communications

External information was scarce in the commune as information had to be filtered through the village cadres – putting them in a more privileged position than the peasants (Croll, 1994). This situation was hardly conducive to the horizontal flow of information between users needed for effective natural resource management in common property management systems. In sum, overlapped grazing actually persisted throughout the period of the communes. The only difference was that the previous system of rangeland allocation within the villages was replaced by the institutional structure of the communes. But the communes could never be effective in the management and protection of rangeland because of the organizational set-up described above. The authority for the enforcement of herding rules should have been vested in the production team, and not in the commune or brigade.

The signs were negative. The available area of rangeland per ruminant had decreased dramatically and the property regime that ensued was one of open access, or nobody's land. Unfortunately, there was also no chance for any improvement in rangeland management; a period of political instability and of great destruction to China's rangelands was about to begin.

2.2.3 The policy reforms: 1966–1978

During the Great Proletarian Cultural Revolution, the so-called 'grain-first' policy, which had its origins in the greatest famine in human history, was formulated (Smil, 1993). After the famine (which was itself the result of the negative policy of the 'Great Leap Forward' of 1958), the Chinese government became preoccupied with attaining self-sufficiency in cereal production. This led to the conversion of vast areas of wasteland, forests, rangelands, steppe and even desert and gobi lands, much of which was unsuitable for agriculture. The rhetoric of 'planting crops in the middle of lakes and on the top of mountains' was accompanied by serious rangeland/steppe degradation and a fall in forage production.[4] Many grazing animals died from starvation. For example, one-third of the combined herd in Yanchi County, Ningxia Hui Autonomous Region died between 1966 and 1976, while Ningxia's ruminant population fell by 28.5% over the same period (GTZ, 1990).

Policies for the livestock sector were geared to increasing ruminant numbers, rather than their productivity or quality. GMSs and Veterinary Stations were disbanded and the piece-rate system and household sideline activities were branded as 'capitalist tails' and abolished. As for the grazing areas, the Chinese government was now looking for new ways of managing them – and was leaning more and more towards privatization, as the limitations of the People's Communes became increasingly apparent. Yet the situation had changed fundamentally from that in 1949. The total area of viable rangeland had declined dramatically since the negative policies of the 1960s – allowing scarcely any room for political manoeuvre. The need for sustainable use and successful management of the rangelands was more urgent than ever.

2.2.4 The period 1978–present

After the initial successes of rural reform in the early 1980s, privatization and decentralization became regarded as magic spells for agriculture. After the abolition of the People's Communes, an attempt was made to apply the HCRS model, used for contracting agricultural land, to grazing lands. Under Article 4 of the 1985 Rangeland Law, 'all rangeland assigned to a collective for long-term use may be leased under a contract to a collective or an individual'. The government thus sought to redistribute responsibility for rangelands to individual herdsmen or livestock farmers.

In principle, too, tenants are bound by the regulations contained in the Rangeland Law and can be held responsible for any damage done to their leased plots. In reality, however, the experience of contracting rangeland to individual households or collectives has been a failure.[5] The failure of the HCRS has seriously undermined the effectiveness of the Rangeland Law, as its premise – the leasing of rangeland to collectives, households or joint households – has proved to be untenable. The roles of the Bureau of Animal Husbandry (BAH) and GMSs therefore amount to the supervision of vast areas of 'nobody's' grazing land – a role they are incapable of fulfilling.

To implement the HCRS for the grazing areas, rangeland areas were classified as 'good', 'average' or 'poor' and then distributed equally to households, taking the productivity of the plot into account. In many rangeland areas, this parcelling out of land proved to be a rather arduous task – partly because there was more rangeland to deal with and partly because of the mobile manner of grazing. Nevertheless, attempts were still made to distribute rangeland to individual households (see Chapter 15).

In most places, the provincial BAH is responsible for the overall supervision of rangeland management. Rangeland management stations (GMSs) are responsible for the management and protection of rangelands. In addition, a special police force – the Rangeland Police – was set up to enforce the Rangeland Law. Being part of the so-called 'economic police force', the Rangeland Police can impose fines, but cannot arrest or detain people, or carry weapons. They are stationed at the county GMS.

Few farmers/herders see the promulgation of the Rangeland Law as significant, and fewer still are acquainted with its contents. More pressing has been the increasing difficulty of finding pasture for their livestock. Transgressions of the Rangeland Law are frequent, although no accurate figures are available. In practice, farmers in breach of the law are rarely fined by the

Rangeland Police, and the nominal penalties imposed tend to be merely symbolic. Despite the presence of formal user rights, liability rights and inalienability rights, as well as institutions to enforce these regulations, rangelands in most of the focus areas were essentially open-access systems.

The State Rangeland Law does not contain specific regulations limiting the number of livestock on a given plot, nor does it provide a solid basis for users to assume responsibility for managing the natural resource. Questions have to be asked, therefore, about the political motives behind its promulgation. Is it merely a symbolic law? Or did the motives of the policy makers who supported it change between the time the law was drafted and its execution? In China, where politics are at best 'opaque', these questions are difficult to answer.

There are several reasons why the HCRS system failed in places in north and north-west China. The most obvious was the way in which a blanket HCRS policy was applied universally, regardless of variations in local conditions. Local opposition to the HCRS reforms, which came mainly from officials whose careers were rooted in the collectivist system, increasingly became quiet as the reforms gathered momentum.

By the time the HCRS had become generally accepted, opposition to reform had become stigmatized as 'leftist obstruction'. As a result, the new agricultural policies were imposed nationally without considering local variations (White, 1993). Most importantly, policies were transferred wholesale from agricultural areas to rangelands, despite the obvious technical and physical differences between the systems. The HCRS went against 'traditional' grazing strategies, which required flexible boundaries (and indeed overlapping boundaries) to function. Past experiences with fences were also ignored: the failure of deferred grazing experiments in the 1950s and 1960s had come about when plots, closed from grazing by one village, were used by flocks from surrounding villages, a problem that still occurs today in areas fenced under the various national programmes such as 'return degraded pasture lands to rangelands'.

The lack of appropriate property rights structures under the HCRS has also contributed to the failure of reform (Chapter 15). The existing system of property rights has its origin in a fundamental disagreement between reformists and conservatives over the role of the market. Reformists pushed for private land ownership and a de facto land market, while conservatives tried to strengthen those state institutions undermined by rural reforms and re-establish certain principles of central planning (see White (1993) for more on this). Property rights under the HCRS therefore turned out to be something of a political compromise. In pastoral areas, where clear property rights are necessary for effective range management, this political compromise is inadequate. The current system has all the trappings of a state property regime, with property rights being vested in the village collective and rangeland management and protection being overseen by external bodies (the BAH and county GMSs). It is difficult to envisage how users can take any interest in making long-term investments to maintain the rangelands when property rights are not vested in organizations which they regard as their own.

It was not true that China's rangelands became degraded because of common property regimes under the People's Communes; rather, it was because the sort of common property regime instituted by government was nothing more than 'eating from the big rice pot'. All too often, common property systems are confused with open-access systems. As Bromley (1991) reminds us, common property resource management implies private property for a group – members of which have rights as well as duties regarding the resource. It is exactly these rights and duties that the People's Communes failed to create or enforce, mainly because of a political fear to vest property rights clearly in either individual users or groups of users.

Under the current system, livestock and its products are owned privately, but the land tenure is owned by the state. Livestock operations are legally restricted to grazing only in designated rangeland at designated times, but there is some latitude in summer range and in emergencies (e.g. when snow is excessive, the county grazing bureau can allow livestock to use other areas). Many seasonal ranges are not fenced; instead, most are based on recognizable geographical features. Spring ranges are grazed from late February to mid-July. All spring rangelands are centred on lambing structures made of adobe and mud. Summer rangelands are grazed from mid-July to late September and herders find summer grazing areas the least limiting.[6] Autumn rangeland is intermediate in elevation and is grazed from late September to mid-November. Winter rangeland is grazed between mid-November and late February and is the second most rigidly defined and defended. Winter encampments are generally stationary through the entire winter season.

Certainly, many of the past grazing problems have been associated with changes in the pastoral system and increased numbers of herders and livestock using these rangelands. However, because of the current situation (relatively large numbers of livestock for the available management, absentee owners, many inexperienced herders and no apparent land ethic of either contract herders or herd owners), we believe rangeland sustainability in most areas of China is threatened by current livestock policy and management patterns.

The socio-economic changes in China during the 20th century have resulted in the growth of both human and livestock populations using the rangelands. This has altered traditional pastoral systems. The consequence of the high number of contract herders, and the fact that even some herding families lack a long history in the local area, is evidence that there is little 'traditional ecological knowledge' in herding practices and that 'ties to the land' have been severed.

Herding has become a job, instead of a lifestyle, and many of the new contract herders have no previous experience of the severe weather patterns (severe droughts and extreme winters) that can occur. These rangelands continue to support significant wildlife in some areas and are an important livestock production system for the whole country (Tables 1.2 and 1.3). However, we believe the present pastoral system consisting of large numbers of inexperienced herders in many regions will require a more active management and monitoring programme by land managers (at all levels of government) to ensure sustainable use of these rangelands for both pastoralists and the wildlife that use these areas.

The reasons for rangeland degradation in China are too often couched in technical and demographical terms – with the institutional environment being ignored. In practice, however, technical considerations about deferred and rotational grazing, carrying capacity and stocking rates have little meaning if they do not adequately incorporate institutional arrangements which provide the incentives for collective action.

Rangeland degradation in China's vast 'Three Norths' region cannot be blamed solely on population growth, overgrazing or reclamation of marginal land. Rather, it has its roots in the failure of successive Chinese governments to create conditions under which collective management could be effective. The initial nationalization of China's rangelands undermined the legitimacy of local customary rights systems over the use of the range. As the central and local government failed to encourage mutual cooperation, the management of rangelands evolved into an open-access system.

Two sets of factors were most important in creating this situation: property rights structures and institutional arrangements. But we can usefully consider other root causes of rangeland degradation.

2.3 Factors That Have Shaped Degradation in China's Arid Rangelands

Biodiversity of China's arid rangelands has suffered in the past 50 years or so. Efforts to restore biodiversity will depend on having a better understanding of how herders behave and on the implementation of effective management schemes. The failure to view rangeland management as, fundamentally, an economic and social issue might well be considered the root of the problems that made biodiversity conservation objectives virtually impossible to achieve in north-west China (and elsewhere).

There are few signs that there will be widespread success in the management of China's arid rangelands. The most fundamental lessons – that all rangelands are vulnerable to overgrazing and that the long-term yield from most arid rangelands is modest – have not found their way into the general dialogue on rangeland management. While there is some understanding that we have passed the point of sustainable yield of most arid rangelands in north-west China, there is little action to get back to that point. Nor will there be while ever large numbers of herder households are so poor and so dependent on grazing systems that have been altered drastically by successive waves of inward migration and loss of prime grazing areas through land conversion (see the individual case studies in Part III).

No amount of effort to restore ecological integrity to the arid rangelands can succeed in the face of rising populations and the lack of a cohesive policy on alleviating rural poverty among the predominantly ethnic minorities.

2.3.1 Attitudes towards herders

In China, there are a number of cultural factors that weigh against nomadic pastoralists and their traditional modes of livestock production. One factor is the suspicion of 'warrior pastoralists' who, before the development of a strong state presence, could raid the surrounding agricultural peoples.[7] The Great Wall of China, for example, was constructed as a defensive bulwark against horse-riding barbarians from the Mongolian steppes (Zhang and Borjigin, 2007). While an actual threat from the 'barbarians' is no longer rife, trepidation towards herders is still prevalent among some people in China and it influences attitudes towards herders.

A second factor is disapproval of the herders' way of life from many government officials, who often view herders as 'backward', 'ignorant' and 'lazy'. The migratory movement of herders' livestock is often viewed as 'wandering' and an unsound type of use of the rangeland, instead of a purposeful and productive means of making efficient use of rangeland forage. The structure of the herders' flocks is believed widely to be 'irrational' and uneconomic. Negative stereotypes about herders are common in China, where they are often seen as not being 'modern' and 'progressive' and are 'in the way' of development (Miller, 1998b,c).[8]

A third factor is ethnicity, for most pastoral people in China are ethnically different from settled farmers and from most people living in urban areas. Herders are minorities in China and speak a different language, dress differently and have different beliefs from the majority Han people. As Salzman (1994) noted, these cultural gaps between herders and non-herders often result in a gulf of distaste and disapproval, which may greatly inhibit a balanced, realistic assessment of the herder population by others, including those holding policy-making power. Much of the problem stems from the inability of traditional Chinese society, which is based on labour-intensive agriculture, to accommodate the flexibility and mobility that make herding possible on the rangelands (Light, 1994).

The concept of herders' irrational rangeland and animal husbandry management resulting in rangeland degradation is grounded in the widespread belief in China that herders keep large livestock herds for reasons of social power and prestige. The 'tragedy of the commons' idea has also been invoked in China to illustrate the irrational and destructive aspects of traditional herding. On the other hand, a number of pastoral specialists working in China disagree with these assumptions. Research has found that many of the traditional herding strategies and practices are rational and ecologically and economically sound, given the constraints under which pastoralists operate (Williams, 1997a; Wu, 1997; Miller, 1999a,b). A fresh, objective assessment of herding systems on the basis of these research findings should be made before disregarding them completely as outmoded.

Since 1978, with economic liberalization, China has adopted a more pragmatic, and less ideological and ethnocentric, approach to the question of herding (Light, 1994). Herders now own their animals and enter into contracts for the use of land. The government in the pastoral areas actively shows its support for livestock production and assistance to herders affected by severe snowstorms. However, considerable effort is being made to settle herders, to allocate fixed parcels of rangeland to them and to fence the rangeland in a 'top-down' type of approach with little participatory involvement of the herders themselves (Williams, 1996a,b).[9]

2.3.2 Social dimensions

Herders have played an important role in the rangeland ecosystems of western and northern China for thousands of years. As such, the social dimension of rangeland ecosystems should be an important aspect of research, management and development in the arid rangelands but, unfortunately, it is not. In China, the integration of social and natural scientific research has been impeded by both organizational divisions between academic disciplines and the intellectual assumption that views humans as being separate from their natural environment. As a result, rangeland researchers have usually neglected such issues as the effects of traditional herding systems on rangeland ecology, the dynamics of herd growth and risk management strategies among herders and the impact of large-scale migration of inland settlers into pastoral areas to convert rangelands to cropland. Despite the extensive area of rangeland used by herdsmen or livestock farmers and the large pastoral population, little information is available in China about nomadic pastoral production systems in the country and misconceptions abound with regard to herders and their way of life.

The inattention given to the pastoral areas and the herding system means that the ethnic herders often remain obscure, marginalized on the fringes of Chinese geography, scholarship and national economic priorities, and dwarfed by the numbers and political centrality of inland majority Han farmers. Whereas income levels have increased significantly in most agricultural areas of China since the 1980s, real incomes in most pastoral areas have not increased much (Longworth and Williamson, 1993). In part, this is because comparative price advantages have moved in favour of crop production relative to animal husbandry. For example, the price of grain tripled in the 1980s relative to prices from the 1950s, but the price of sheep increased only by a factor of 2.6 (Zhou, 1990; Williams, 1997b). Grain prices have continued to be held high under a state subsidy scheme until recent times. The returns to farmers who grow wheat, for example, had fallen dramatically until recently (2008), when prices of both meat and grain have risen dramatically.

The socio-economic changes that affect herder communities profoundly are both effects and causes of the degradation of rangelands (Zhang, 2006). The coherence and homogeneity of herder communities have disintegrated as new employment opportunities and production practices widen income differentials between households and disperse economic interests and labour resources among many sectors besides livestock.

2.3.3 Land tenure, privatization and common property resources

Since 1985, China has moved towards the establishment of clearly defined (individual) private property rights to land in the pastoral areas by parcelling and privatizing rangeland to individual herders on long-term contracts. Privatization of rangeland was intended to be an initial step in a series of adjustments to 'rationalize' the animal husbandry sector (Brown *et al.*, 2008). A premise of the rangeland contract system was that herders were deterred from investment in land improvements because of uncertainty as to whether they could appropriate the benefits (Banks, 1997). Policy makers assumed that private enclosures would force independent households to confront the contradictions between forage supplies and livestock numbers[10] (Williams, 1996a,b; Wu and Richard, 1999). The rangeland contract system was based on the assumption that, through better definition of property rights and the introduction of individual land tenure, land tenure security would be improved. This, in turn, would give herders the incentive to manage their lands in a sustainable way and invest in rangeland improvements (Banks, 1997). Setting the rules on herd size, for example, has particular difficulties. Unlike other controls, which may close rangelands to grazing, or limit grazing to animals of particular kinds or seasons of use, where breaches are obvious, with flock size limits on individual householders it is often not evident to the casual or even the official observer that breaches are occurring. Despite the most rigorous checks, misreporting of herd size is considered to be common in many quota-based systems. Enforcing a version of individual property rights alone (such as under the State Rangeland Law) still has not prevented overgrazing or arrested the spread of desertification (Williams, 1996a,b; Brown *et al.*, 2008).

In Inner Mongolia, it has been found that privatization and fenced enclosures have actually compounded grazing problems for many herders by intensifying stocking rates on highly vulnerable communal rangeland, exacerbating wind and soil erosion (Williams, 1996a,b; CID, 1997; Humphrey and Sneath, 1999). Land-use intensification throughout the rangelands is fragmenting landscapes into simpler, discrete units. The result is a reduction in the scale of landscape–

animal–human interactions. Fragmentation of the landscape reduces biocomplexity and simplifies ecosystems by disconnecting interdependent spatial units into separate entities, thus compartmentalizing important ecosystem functions into isolated pieces. See Chapter 15 for a detailed analysis of the issues related to land-user rights, tenure and policy.

2.4 Root Causes of Rangeland Degradation in China

The root cause of much of the desertification in the Three Norths region of China is due to sedentarization, and the most serious cases are associated with agropastoral systems where the best grazing land has been taken for cropping. This fragmentation can occur regionally, in the form of altered land tenure and/or enterprise size, particularly where there is an effort to intensify production. The result is a reduction in the scale of landscape–animal–human interactions, which has focused initially on capturing the higher-quality resources (water, grazing, cropping lands). There is evidence that this reduction in complexity has an impact on ecosystem function and system resilience, which may lead to dysfunction in ecological communities, enterprise economics and social structures. The history of rangelands in developed countries suggests that this process invariably reverses itself and consolidation of land begins to occur once more. By this stage, the higher-value (usually, most agriculturally valuable) parts of the landscape have been excised and the reconsolidation occurs in the residual rangelands or grazing lands.

Landscapes function as complex integrated systems. The movement of livestock, money and materials among different elements of the landscape results in many emergent properties. Landscape elements become connected by virtue of movement-mediated interactions. This connectivity is important for ecosystem viability, but modern land tenure systems fragment the landscape into small parcels (e.g. under the State Rangeland Law). Fragmentation reduces biocomplexity and simplifies ecosystems by disconnecting interdependent spatial units into separate entities, thus compartmentalizing important ecosystem functions into isolated pieces.

The problem of tenure and property regimes in the pastoral areas of China is a significant one (NRC, 1992; Ho, 1996; Williams, 1996a,b; Banks, 1997). Addressing the problem will require both a more realistic assessment of the rangeland resources and for the herders' own traditional land tenure arrangements and objectives to be considered in the design and implementation of land tenure reform. Thus, there will have to be a devolution of some authority over the assignment, monitoring and enforcement of rangeland use rights to village-based institutions or herder groups (Banks, 1997).

China is facing a dilemma regarding the effective establishment of privatization of land tenure in the context of its extensive pastoral areas. There are high transaction costs associated with the privatization of rangeland, including the high private costs of monitoring and enforcing boundaries relative to the benefits (given the low productivity of the rangelands). They also include high public costs associated with the delineation of boundaries, the adjudication of disputes and the monitoring and enforcement of contractual provisions relating to rangeland management. More importantly, the privatization of rangeland tenure leads to herders' loss of flexibility of herd movements and, consequently, a means to manage environmental risk in a spatially variable climate (Banks, 1997).

From the action of 'opening up wasteland in Mongolia' in the Qing Dynasty, to 'migrate farmers to cultivate the arable land areas at the boundary' during the period of the Republic of China, through several phases of large-scale cultivation of rangelands and construction of state farms in the first 50 years of the People's Republic of China, an unprecedented destruction of ecological environments has resulted along the farming–nomadic interface zone and now large-scale, cultivation-related desertification looms (Sun and Liu, 2001; Enkhee, 2003).

There are two serious consequences once a farming production style is introduced into a farming–nomadic interface zone on a large scale and the herdsman turn into farmers. First of all, accelerated desertification occurred over large areas and developed at a rapid speed once the thin loam soil was broken and the thick loose Quaternary sand was exposed to erosion (Zhang *et al.*, 2007). Secondly, rapid population growth, characteristic of the traditional farming economy, causes an accelerated consumption of timbers for construction and of fuel plants (so forests were denuded). This, in turn, placed great stress on the environment, which led to further degradation of natural ecological systems and further acceleration of desertification. One massive wave after another of land conversion and cultivation along the marginal zone has led to the retrogression of sustainable use of rangeland (Bao, 2003). It has developed into the largest, most prolonged and most significant ecological disaster in China's history and needs to be examined profoundly (Zhang and Borjigin, 2007). The expansion of agriculture based on pigs, irrigation (artificial oases) and other ideas from southern and eastern China meant that traditional herders were forced on to smaller and smaller areas of good land and that the landscape became fragmented as the more favoured areas were excised for cropping.

Over 90% of the 331 million ha (Mha) of usable rangeland in the arid, semi-arid and dry subhumid zones suffer from moderate to severe degradation at a time when livestock numbers are rising to meet the demand for meat, and other livestock products are also rising. Urbanization and rising standards of living throughout the PRC are the drivers for this. These pressures to produce more are likely to put the rangelands under greater stress and contribute to more frequent and severe dust storms. Efforts to relieve the grazing pressure on rangelands include an expansion of water-demanding artificial pastures and the building of indoor feeding sheds. The long-term implications of this system of meat production have not been considered. In a water-scarce region such as the arid rangelands, it would seem to be more relevant to use the scarce water for domestic consumption or for the production of high-value crops. A reform of the water-pricing policy to reflect the real cost (and value) of water would soon force the rational use of water (see Chapter 11).

There is a reluctance to acknowledge that much of the current land degradation is due to poor land-use decisions and flawed development strategies over a long period of time, rather than climate changes or other natural factors. In the PRC, the arable land per capita is 0.11 ha. The shrinking arable land area and increasing demand for agricultural produce puts pressures on farmers to extract higher yields from their land. This leads inevitably to increased soil erosion. There is a clear connection between land degradation and poverty. Almost 90% of rural people living in poverty are located in areas suffering from soil erosion. In the arid, semi-arid and dry subhumid areas, rapidly increasing livestock populations exacerbate the spread of deserts and contribute directly to the increasing frequency and severity of dust/sandstorms.

Notes

[1] For more details on rural institutions during the period 1949–1956, see Chen and Buckwell (1991).

[2] Many areas have experienced waves of inward migration, some long ago and others in more recent times.

[3] Commonly referred to as the Rangeland Law.

[4] There was a 170,000 ha (40% of the total area of rangeland) increase in the desertified area of Yanchi County between 1962 and 1976 (NRC, 1992). A marked decline in productivity in rangeland areas followed. In Inner Mongolia, a decrease of 40–60% was reported and, in Xinjiang, there was a 50% decrease in the period 1965–1975 (Hu *et al.*, 1992, p. 76).

[5] Perhaps the feeling is summed up as 'the household contract responsibility system is just something on paper, the actual delimitation of land has failed. Therefore, the situation in the rangelands is now one of "eating from the big rice pot". Nobody feels responsible for the rangelands any more.'

[6] Although this is changing. Many herders now report that summer rangelands are under threat, as livestock numbers increase and pastures degrade through incursions by poisonous or unpalatable plants, for example, in the Nalati rangelands in Xinyuan County, Xinjiang.

[7] For a more detailed discussion on these cultural factors affecting pastoralists generally, see Salzman (1994).

[8] There is a long-standing derogatory perspective that views the national minorities as ignorant and backward. The Marx–Lenin–Mao model of hierarchical social evolution holds that different types of economic activity correspond to different levels of social advancement. Hunting and gathering is the most primitive

form, followed by mobile pastoralism, then by sedentary agriculture and, lastly, by the industrial society (Williams, 1997a,b).

[9] See also 'Grapes of Wrath in Inner Mongolia', a May 2001 report from the US Embassy in Beijing that sees parallels between the 'dust bowl' in the USA in the 1930s and the massive flood of ecological refugees from China's rangelands when they collapse.

[10] The State Rangeland Law formulated in the mid-1980s was based on the assumption that desert rangelands in north-west China were deteriorating due to lack of stewardship. Its implementation was aimed at reducing livestock numbers and constraining herd mobility. Pastures were allocated to individual households and large areas were demarcated (using fences). The purpose was to convert a traditional herder way of life into an 'efficient' livestock enterprise.

References and Further Reading

Banks, T. (1997) Land tenure and sustainable agriculture in marginal environments: the case of Western China. Paper presented at the 41st Annual Conference of the Australian Agricultural and Resource Economics Society, 20–25 January 1997, Gold Coast, Australia.

Bao, Q. and Dong, H. (2003) Ecophilosophy dimension: population ecogeneration and sustainable development. *China Population, Resources and Environment* 13(4), 9–12 (in Chinese).

Bao, Y. (2003) *The History and Future of Rangeland Animal Husbandry of Inner Mongolia.* Inner Mongolia Education Press, Huhhot, China, 195 pp. (in Chinese).

Barfield, T. (1989) *The Perilous Frontier: Nomadic Empires and China.* Basil Blackwell, Oxford, UK.

Borjigin, J. (2002) *Comments on the History of Nomadic Culture.* People's Press of Inner Mongolia, Huhhot, China, 300 pp. (in Chinese).

Bromley, D.W. (1991) *Environment and Economy: Property Rights and Public Policy.* Blackwell, Oxford, UK.

Brown, C.G., Waldron, S.A. and Longworth, J.W. (2008) *Sustainable Development in Western China: Managing People, Livestock and Grasslands in Pastoral Areas.* Edward Elgar, Cheltenham, UK and Northhampton, Massachusetts, 294 pp.

Chan, A., Madsen, R. and Unger, J. (1992) *Chen Village under Mao and Deng.* University of California Press, Berkeley, California.

Chen, L.Y. and Buckwell, A. (1991) *Chinese Grain Economy and Policy.* CAB International, Wallingford, UK.

CID (1997) Improvement of Northern China Rangelands Ecosystems, ADB TA No. 2156-PRC, Main Report. Unpublished Report, Asian Development Bank, Manila.

Clarke, G. (1998) Socio-economic change and the environment in a pastoral area of Lhasa Municipality, Tibet. In: Clarke, G. (ed.) *Development, Society and Environment in Tibet.* Papers presented at a Panel of the 7th International Association of Tibetan Studies, Graz, 1995. Verlag de Osterreichischen, Vienna, pp. 1–46.

Croll, E. (1994) *From Heaven to Earth.* Routledge, London.

Enkhee, J. (2003) A historical retrospect on rangeland desertification: cultural dimension of development. *Journal of Inner Mongolia University (Humanities and Social Sciences)* 35(2), 3–9 (in Chinese).

Gai, S. and Gai, Z. (2002) *Disappeared Cultures: Enlightenment for Modern People.* Inner Mongolia Press, Huhhot, China, 596 pp. (in Chinese).

Gegenguva, O. (2002) Mongolian ecological culture in the context of ecological ethics. *Journal of Inner Mongolia University (Humanities and Social Sciences)* 34(4), 3–9 (in Chinese).

Goldstein, M. (1992) Nomadic pastoralists and the traditional political economy – a rejoinder to Cox. *Himalayan Research Bulletin* 12(1–2), 54–62.

Goldstein, M. and Beall, C. (1989) The impact of China's reform policy on the nomads of Western Tibet. *Asian Survey* 24(6), 619–641.

Goldstein, M. and Beall, C. (1990) *Nomads of Western Tibet: The Survival of a Way of Life.* University of California Press, Berkeley, California.

Goldstein, M., Beall, C. and Cincotta, R. (1990) Traditional nomadic pastoralism and ecological conservation on Tibet's northern plateau. *National Geographic Research* 6(2), 139–156.

GTZ (Guojia Tongjiju Zonghesi) (1990) *Quanguo ge Sheng, Zizhiqu, Zhixiashi Lishi Tongji Ziliao Huibian* (Compilation of National Historical Statistics of all Provinces, Autonomous Regions and Cities Directly Under the Central Government). Zhongguo Tongji Chubanshe, Beijing.

Ho, P. (1996) *Ownership and Control in Chinese Rangeland Management since Mao: the Case of Free-riding in Ningxia.* Pastoral Development Network 39c. Overseas Development Institute, London.

Ho, P. (2005) *Institutions in Transition: Land Ownership, Property Rights and Social Conflict in China.* Oxford University Press, Oxford, UK.

Hu, S.T.P., Hannaway, D.B. and Youngberg, H.W. (1992) *Forage Resources of China.* Pudoc, Wageningen, The Netherlands.

Humphrey, C. and Sneath, D. (eds) (1996) *Culture and Environment in Inner Asia, Volume I: The Pastoral Economy and The Environment.* White Horse Press, Cambridge, UK.

Humphrey, C. and Sneath, D. (1999) *The End of Nomadism? Society, State and the Environment in Inner Asia.* Duke University Press, Durham, North Carolina.

Jian, L. and Lu, Q. (eds) (1998) *Rangeland Management and Livestock Production in China.* Reports of the Sustainable Agriculture Working Group, China Council for International Cooperation on Environment and Development. China Environment Science Press, Beijing.

Levine, N. (1998) From nomads to ranchers: managing pasture among ethnic Tibetans in Sichuan. In: Clarke, G. (ed.) *Development, Society and Environment in Tibet.* Papers presented at a Panel of the 7th International Association of Tibetan Studies, Graz, 1995. Verlag de Osterreichischen, Vienna, pp. 69–76.

Li, J. (1998) China's agricultural resource and its sustainable development. In: Jian, L. and Lu, Q. (eds) *Rangeland Management and Livestock Production in China.* Reports of the Sustainable Agriculture Working Group, China Council for International Cooperation on Environment and Development. China Environment Science Press, Beijing, pp. 40–50.

Li, O., Ma, R. and Simpson, J. (1993) Changes in the nomadic pattern and its impact on the Inner Mongolian steppe rangelands ecosystem. *Nomadic Peoples* 33, 63–72.

Light, N. (1994) *Qazaqs in the People's Republic of China: The Local Processes of History.* Indiana Center on Global Change and World Peace, Occasional Paper No. 22, Bloomington, Indiana.

Longworth, J.W. and Williamson, J.G. (1993) *China's Pastoral Region: Sheep and Wool, Minority Nationalities and Rangeland Degradation. Rangeland Degradation and Sustainable Development.* CAB International, Wallingford, UK.

Ma, R. (1993) Migrant and ethnic integration in the process of socio-economic change in Inner Mongolia, China: a village study. *Nomadic Peoples* 33, 173–192.

Meng, C. (1999) *Rangeland Culture and Human History.* International Cultural Press, Beijing, 996 pp. (in Chinese).

Miller, D. (1998a) Rangeland privatization and future challenges in the Tibetan Plateau of Western China. In: Jian, L. and Lu, Q. (eds) *Proceedings of the International Workshop on Rangeland Management and Livestock Production in China*, Reports of the Sustainable Agricultural Working Group, China Council on International Cooperation on Environment and Development (CCICED), 28–29 March, 1998, Beijing, China. China Environmental Science Press, Beijing, pp. 106–122.

Miller, D. (1998b) Tibetan pastoralism: hard times on the plateau. *Chinabrief* 1(2), 17–22.

Miller, D. (1998c) Conserving biological diversity in Himalayan and Tibetan Plateau rangelands. In: *Ecoregional Co-operation for Biodiversity Conservation in the Himalaya*, Report on the International Meeting on Himalaya Ecoregional Cooperation, 16–18 February 1998, Kathmandu, Nepal. UNDP and WWF, New York, pp. 291–320.

Miller, D. (1999a) Nomads of the Tibetan Plateau rangelands in Western China, Part Two: Pastoral production. *Rangelands* 21(1), 16–19.

Miller, D. (1999b) Nomads of the Tibetan Plateau rangelands in Western China, Part Three: Pastoral development and future challenges. *Rangelands* 21(2), 17–20.

Neupert, R.F. (1999) Population, nomadic pastoralism and the environment in the Mongolian plateau. *Population and Environment* 20(5), 413–441.

NRC (1992) *Grasslands and Grassland Sciences in Northern China.* Committee on Scholarly Communication with the People's Republic of China, National Research Council. The National Academies Press, Washington, DC, 230 pp.

Nyberg, A. and Rozelle, S. (1999) *Accelerating China's Rural Transformation.* The World Bank, Washington, DC.

Salzman, P. (1994) Afterword: reflections on the pastoral land crisis. *Nomadic Peoples* 34/35, 159–163.

Sheehy, D. (1992) A perspective on desertification in North China. *Ambio* 21, 303–307.

Simpson, J.R. and Li, O. (1996) Feasibility analysis for development of northern China's beef industry and grazing lands. *Journal of Range Management* 49(6), 560–564.

Smil, V. (1993) *China's Environmental Crisis: An Inquiry into the Limits of National Development.* M.E. Sharpe Inc., New York.

Sun, J. and Liu, T. (2001) Desertification in northeastern China. *Chinese Quaternary Sciences* 21(1), 72–78 (in Chinese).

Thwaites, R., de Lacy, T., Li, Y.H. and Liu, X.H. (1998) Property rights, social change, and rangeland degradation in Xilingol Biosphere Reserve, Inner Mongolia, China. *Society and Natural Resources* 11, 319–338.

Walker, B. (1993) Rangeland ecology: understanding and managing change. *Ambio* 22(2–3), 80–87.

White, G. (1993) *Riding the Tiger: The Politics of Economic Reform in China.* Macmillan, London, 286 pp.

Williams, D.M. (1996a) The barbed walls of China: a contemporary rangeland drama. *The Journal of Asian Studies* 55(3), 665–691.

Williams, D.M. (1996b) Rangeland enclosures: catalyst of land degradation in Inner Mongolia. *Human Organization* 55(3), 307–313.

Williams, D.M. (1997a) Patchwork, pastoralists, and perception: dune sand as a valued resource among herders of Inner Mongolia. *Human Ecology* 25(2), 297–318.

Williams, D.M. (1997b) The desert discourse of modern China. *Modern China* 23(3), 328–355.

Wu, N. (1997) *Ecological Situation of High-Frigid Rangeland and Its Sustainability: A Case Study of the Constraints and Approaches in Pastoral Western Sichuan, China.* Dietrich Reimer Verlag, Berlin.

Wu, N. and Richard, C. (1999) The privatization process of rangeland and its impacts on the pastoral dynamics in the Hindu-Kush Himalaya: the case of western Sichuan. In: Eldridge, D. and Freudenberger, D. (eds) *People and Rangelands: Building the Future*, Proceedings of the VIth International Rangeland Congress, 19–23 July 1999, Townsville, Australia. VI International Rangeland Congress, Inc, Aitkenvale, Australia, pp. 14–21.

Xue, Y.K. (1996) The impact of desertification in the Mongolian and the Inner Mongolian rangeland on the regional climate. *Journal of Climate* 9(9), 2173–2189.

Yang, Y.Z. (1993) *Minority Areas in Gansu Province.* Gansu Nationalities Press, Lanzhou, Gansu, PRC, 644 pp. (in Chinese).

Yang, Y., Squires, V.R. and Lu, Q. (eds) (2002) *Global Alarm: Dust and Sandstorms from the World's Drylands.* UN, Beijing, 265 pp.

Zhang, M.D. and Borjigin, E. (2007) Mongolian nomadic culture and construction of ecological culture. *Ecological Economics* 62, 19–26.

Zhang, M.D., Wang, X.K., Sun, H.W. and Feng, Z.W. (2007) HulunBuir sandy grassland blowouts: influence of human activities. *Journal of Desert Research* 27(2), 214–219.

Zhang, Q. (2006) May they live with herds: transformation of pastoralism in Inner Mongolia, China. MSc thesis, University of Tromso, Norway.

Zhou, L. (1990) Economic development in China's pastoral regions: problems and solutions. In: Longworth, J. (ed.) *The Wool Industry in China.* Inkata Press, Melbourne, Victoria, Australia, pp. 43–56.

Part II

Mechanisms of Rangeland Degradation and Recovery

The four chapters in this part examine the way in which degradation begins and proceeds under the conditions prevailing in northern and north-western China. Climatic conditions have changed and evidence has been collated from 50 years of records from 175 monitoring sites. Grazing impacts are explained and the vital role of managing stocking pressure and its timing and intensity of grazing is outlined. Soil erosion processes are analysed and, finally, the mechanisms of degradation and recovery are discussed.

The distinction is made between restoration and rehabilitation at the landscape level and the implications for whole rangeland ecosystems are considered. Root causes of rangeland degradation are discussed against the background of the pressure–state–response model, which avoids the tendency to 'blame the victim'. Rather, it explains the actions of the land users as a response to pressures exerted by the physical environment, the policy and the regulatory environment. The recovery phase is considered and the keys to successful recovery are analysed, including technical and policy interventions (including the regulatory framework) and the importance of community participation. The lessons learnt and the development of a set of guiding principles are discussed.

3 An Analysis of the Effects of Climate Variability in Northern China over the Past Five Decades on People, Livestock and Plants in the Focus Areas

Lu Qi, Wang Xuequan and Wu Bo
Chinese Academy of Forestry/China National R&D Centre on Combating Desertification, Beijing, China

Synopsis
Statistical data on annual temperature and precipitation from 173 meteorological stations of the main pastoral regions of China from 1951 to 2004 were collected for analysis of long-term trends of air temperature, precipitation, aridity, the interdecadal features and spatial variation of these regions. The analysis concludes that, over the past years, there is a distinct warming trend, with a definite shift to warmer temperatures in the late 1980s. The coldest period was in the 1960s, the warmest period in the middle/late 1990s and the warmest year was 1998. Annual precipitation varies significantly, with a general rising trend but with a relative shortage period of rain/snow in the 1960s/1970s and relatively ample rain/snow in the 1990s and early 21st century. The highest precipitation year was 2003. The impact of these climatic changes on people, livestock and plants is examined.

Keywords: warming; precipitation; cycles; aridity; deficit; interdecadal; dust and sandstorms; evapotranspiration; soil erosion

3.1 Background

The geographical position of arid and semi-arid areas of north China determines the dry mid-latitude climate, which is dominated by continental polar air masses for much of the year. The climatic continentality of north China is prominent because of the large distance from the open sea and the weak marine influence due to mountain ranges in the east and south-west. Most rainfall occurs in summer as a result of sporadic invasions of maritime air masses and therefore annual precipitation in these areas varies widely. The wettest years receive several times more rainfall than the driest years. Another pronounced feature of the climate of north China is the frequency of high wind during the long and dry seasons of winter and spring, when vegetation is withered and dormant. This situation greatly facilitates wind erosion of sandy surfaces. There are 30–210 days a year in north China when the wind velocity is higher than 5 m/s, the threshold velocity required to transport particles of sand and silt, and 20–80 days a year with wind velocities of 20 m/s, which can cause severe deflation. The wind regime shapes the landscape of north China into progressive transition from gobi, through sandy desert land to loess from north-west to south-east.

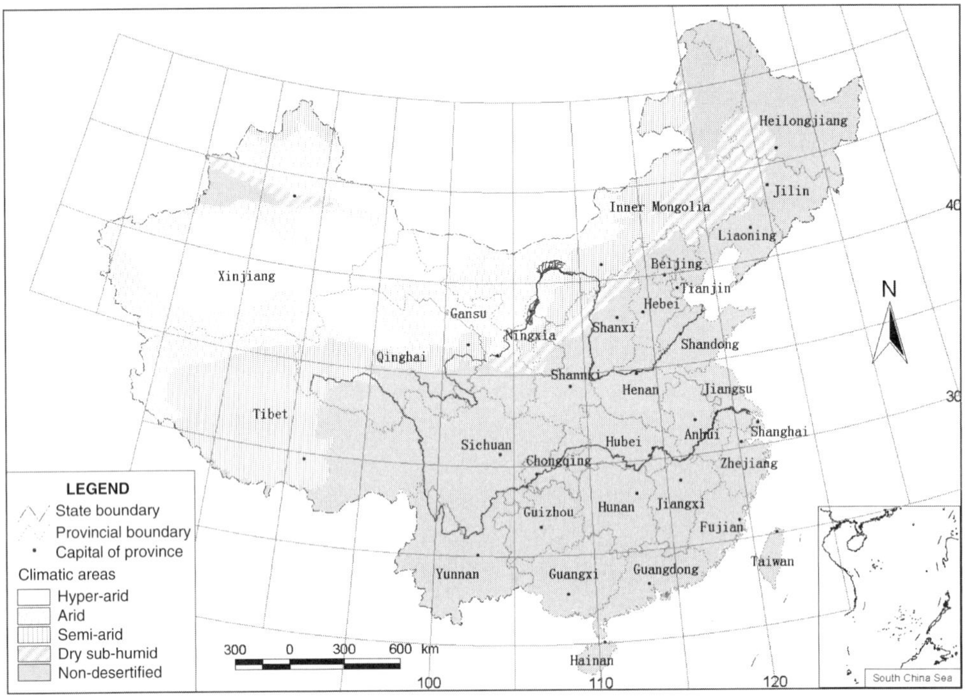

Fig. 3.1. Areas of potential desertification in China. The desert areas (hyper-arid) are excluded from the UN classification of desertification. The major areas at risk are the semi-arid and dry humid areas (Wu *et al.*, 2007).

3.2 Climate Variability

The climatic fluctuations in China vary between different areas (Fig. 3.1). There was a general trend towards increasing climatic aridity in north China from the 1960s through the 1990s. A pervasive and significant temperature rise in north China over the past decades has been observed. During this period, however, rainfall in north China remained average or declined in many areas. For example, mean annual precipitation in Inner Mongolia was constantly under the normal range over the period of 1963–2000 (Fig. 3.2). According to Le Houérou's estimate (Le Houérou *et al.*, 1988; Le Houérou, 1992), a one-degree increase in temperature would cause an annual potential evapotranspiration (PET) rise of approximately 5.25%. As a result of declining rainfall, the P:PET ratio decreases and indicates a rise in land aridity, definitely leading to a reduction in effective soil moisture, a condition favouring accelerated desertification. Nevertheless, this kind of climatic fluctuation has a trivial effect on land degradation, as compared with anthropogenic impacts.

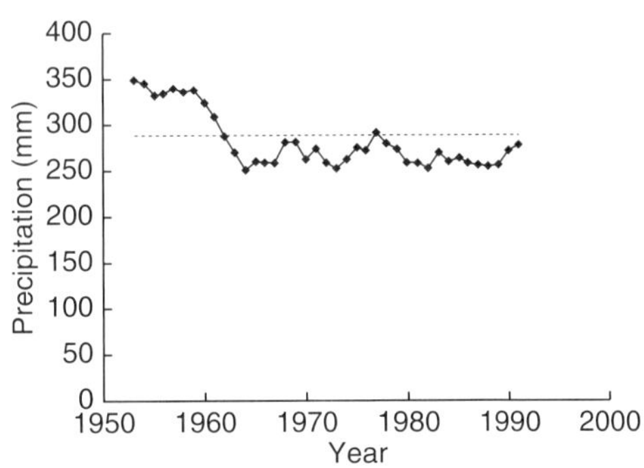

Fig. 3.2. Annual precipitation in Inner Mongolia from 1951 to 1996. The solid line represents the 5-year running mean and the dashed line the long-term annual mean (after Li, 2001).

To explore the relationship between land degradation and climate change in China, statistical data regarding annual temperature and precipitation from 173 meteorological stations in the main arid regions of China from 1951 to 2004 were collected for analysis of long-term trends of air temperature, precipitation, aridity, the interdecade features and spatial variation of these regions (Fig. 3.3). The analysis concludes that over the past years, the main arid regions of China have become obviously warmer, with a definite shift to warmer temperatures in the late 1980s. The coldest period was in the 1960s, the warmest period in the middle/late 1990s and the warmest year in 1998. Annual precipitation varies significantly, with a general rising trend and with a relative period of shortage of rain/snow in the 1960s/1970s and relatively ample rain/snow in the 1990s and early 21st century. The highest precipitation year was 2003. The annual moisture level rises and falls, with a generally slightly wetter trend. The number of spring sandstorm/high wind days fluctuates and is cyclic, with a downward trend. The frequency and severity of sand- and dust storms peaked in the period from the end of the 1950s to the first

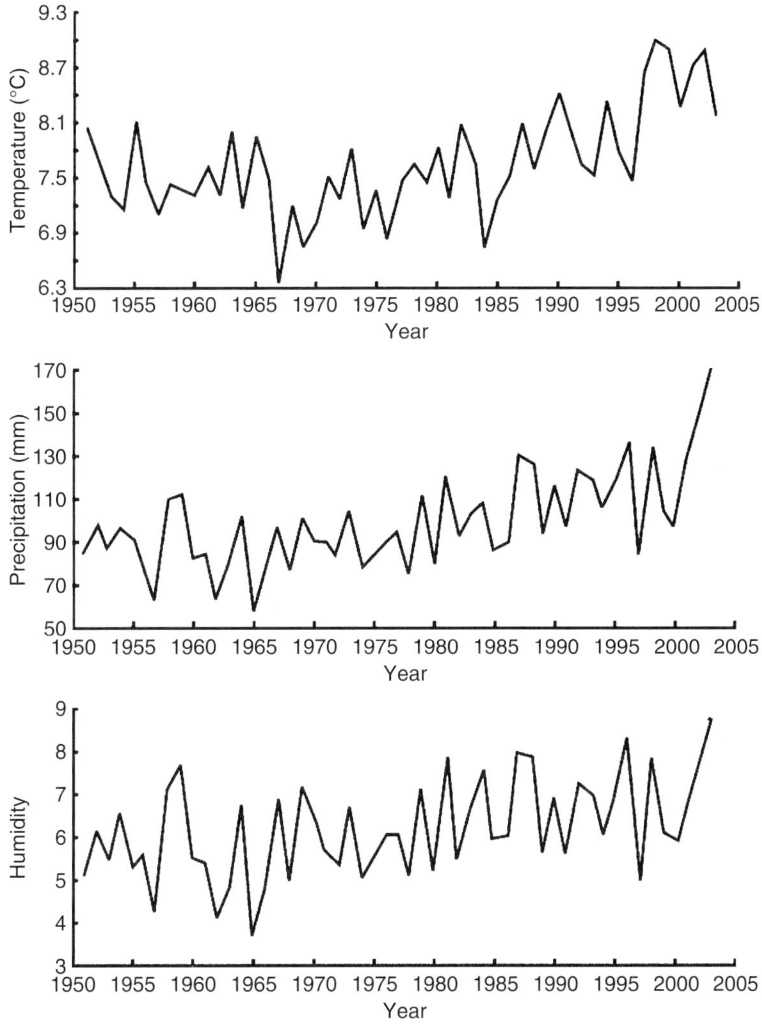

Fig. 3.3. Air temperature, precipitation and humidity indices of arid land over the past 50 years (Zhu, 2005).

half of the 1980s. There has been a trough period since the middle 1980s, except in 2001, which showed a modest rise.

The air temperature of northern Xinjiang, the Qinghai–Tibet Plateau and the Qaidam Basin over the past 50 years is obviously on a fluctuating rise, while the precipitation and humidity indices are on a similar rising trend, especially during the early 21st century, with the exception of the Three Rivers[1] Headwater Area, with periods of declined rainfall since the 1990s (Qian and Zhu, 2001). The air temperature of eastern and southern Xinjiang, Gansu, to the west of the Yellow River, the Alxa Plateau in western Inner Mongolia and Hami-Badain Jaran in Xinjiang fluctuates, but shows a rising trend over the past 50 years. The precipitation and humidity indices are also on a rising trend, with the exception of the end of the 1990s and in 2004 with less rain. The air temperature in the Mu Us Desert, the Hetao Plain and the Ulanqub Plateau of Inner Mongolia, Bashang of the northern Heibei Province and the Hunshandake Sandy Land in the far northern part of China has basically risen over the past 50 years, while the precipitation and humidity indices vary with a low rising trend, except for the period around the end of the 1990s, which had below-average annual precipitation.

The air temperature of the northern loess plateau and the agriculture/animal husbandry transition zone over the past 50 years fluctuated but showed a rising trend, with varying precipitation and humidity indices for a relatively dry period of below-average annual precipitation level from the late 1990s to the early 21st century (Fig. 3.4). There was a relatively wet period in 2003 and a relatively dry period in the western part and in the eastern agriculture/animal husbandry transition zone, but a relatively wet period in the central loess plateau in 2004. The air temperature of the Horqin Sandy Land has risen for the past 50 years, with a slight decreasing trend of both the precipitation and humidity indices, especially since the 1990s (Fig. 3.5). The air temperature of the Hulunbeier Sandy Land over the past 50 years has risen, while the precipitation and humidity indices have varied. It was relatively dry in the 1990s/early 21st century, wet in 2002 and 2003 and dry again in 2004 (Fig. 3.6).

Water deficiency is a primary problem in arid and semi-arid areas. Inadequate rainfall results in a paucity of surface and groundwater in north China. For example, the natural average annual discharge of the Yellow River, the second longest river in China, which stretches across northern China, is less than one-sixteenth of that of the Yangtze River, the longest river in China. River runoffs in north China have decreased (Fig. 3.7) because of climatic changes and increasing human demands (Ren *et al.*, 2002).

Zero flow occurred in many major rivers of north China in the 1970s and 1980s. Some naturally perennial rivers, such as the lower Yellow River, the lower Yongding River (one of the source rivers of the Hai River) and the lower Hutuo River (one of the source rivers of the Hai River) in the North China Plain, have turned into seasonal rivers. The Yellow River had zero flow for 20 out of 26 years from 1972 to 1997 (Fig. 3.7) and both the length of reach and period of zero flow grew longer during that time. The consequences of zero flow include land aridity, vegetation diminution and further desertification. Also, a continuous decline in regional groundwater levels occurred. Many inland lakes either diminished or vanished completely.

In numerous regions of north China, such as the Yellow River drainage area, the Tarim River drainage area in Xinjiang, the Shiyang River drainage area in Gansu, the Ruoshui River drainage area in Inner Mongolia and the Hai River drainage area in Hebei, groundwater levels dropped considerably, causing vegetation deterioration (see Chapter 11 for more details).

3.3 Major Regional Development Initiatives

Obviously, climate variability alone cannot account for the severe land degradation observed in recent decades in north China. The primary causes of current land degradation in some areas were identified as intensified and irrational human activities (Wang *et al.*, 1991), such as in the Mu Us Sandy Land, the Bashang area and the Horqin Sandy Land. More intensive exploitation of natural resources can be expected with the rapidly increasing population living on the land in north China. Increasing population pressures, plus the need for economic development, have resulted in intensified human activities such as overgrazing, overcultivation, excessive firewood

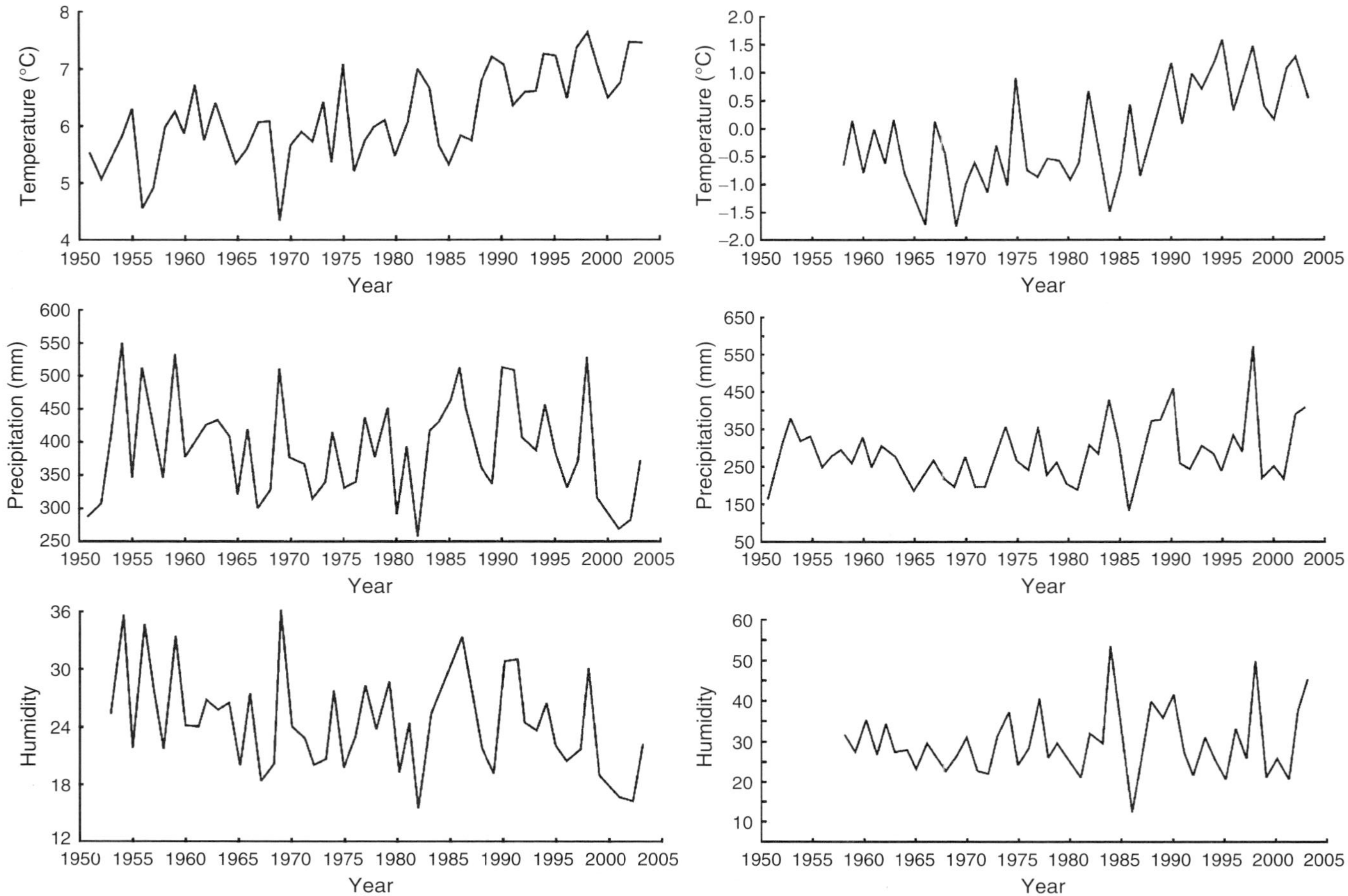

Fig. 3.4. Air temperature, precipitation and humidity indices of the Junggar Basin (Xinjiang, left) and Hulunbeier (Inner Mongolia, right) over the past 50 years (Zhu, 2005).

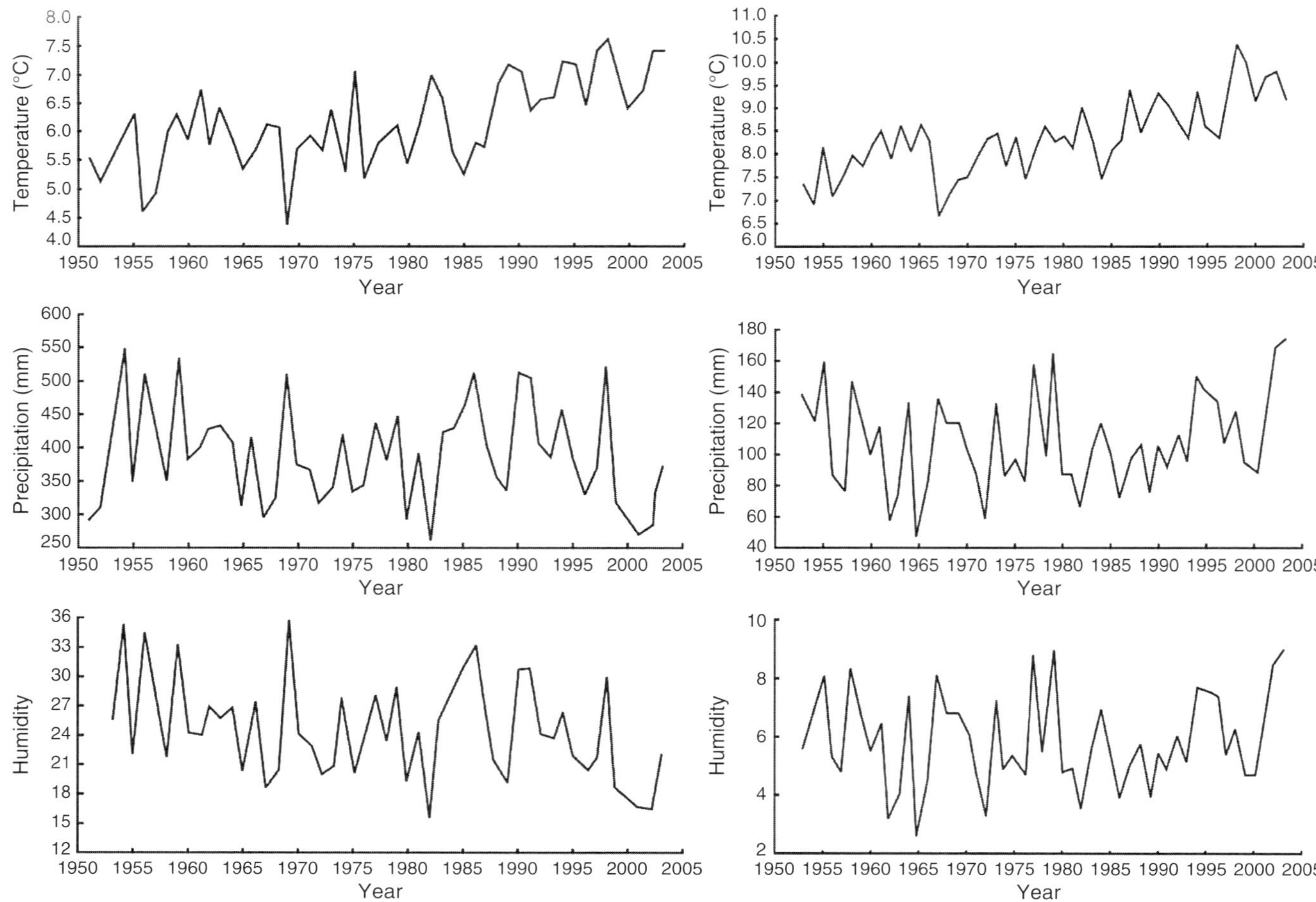

Fig. 3.5. Air temperature, precipitation and humidity indices of the Horqin (north-east Inner Mongolia, left) and Alashan (western Inner Mongolia, right) over the past 50 years (Zhu, 2005).

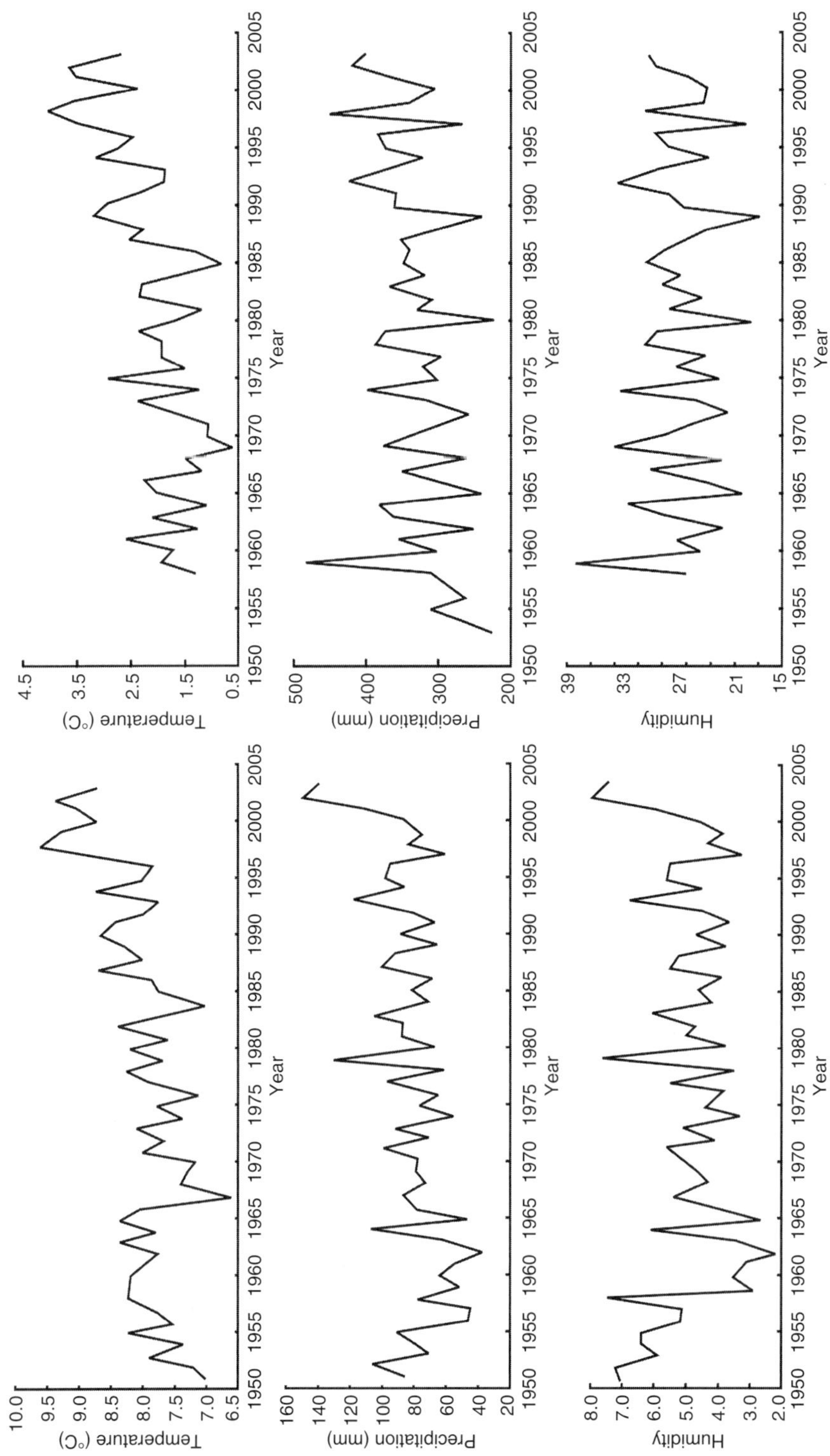

Fig. 3.6. Air temperature, precipitation and humidity indices of the Hexi Corridor (Gansu, left) and Xilingol Hungsandake (Inner Mongolia, right) over the past 50 years (Zhu, 2005).

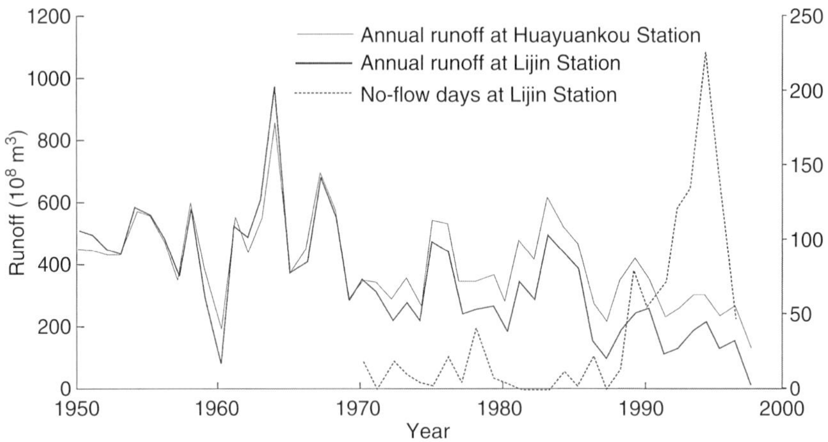

Fig. 3.7. Annual runoff and no-flow days through the lower reaches of the Yellow River from 1950 to 1997.

gathering, irrational use of water resources and environmental neglect. Industry and mining were given priority. All of these factors contributed to land degradation in northern China. For example, in Inner Mongolia the population is now nearly four times that of 50 years ago and the livestock population has increased from 10 million, 50 years ago, to 60 million today. The region has experienced two periods with a rise in livestock population, 1950–1965 and 1985–2000 (Fig. 3.8).

The livestock population densities of grazing animals were far beyond the carrying capacity on many of the rangelands of north China, resulting in poor range condition in these areas (Chen and Tang, 2005). Animal populations on rangelands in arid and semi-arid areas have exceeded carrying capacity by 30–70% (Liu *et al.*, 2002). Overgrazing may lead to a progressive reduction in the vegetation cover and increased wind erosion and runoff, which are conducive to desertification. Another consequence of overgrazing is the destruction of native forage plants, which are then replaced either by annuals having little forage value or by unpalatable and toxic species. For example, *Cynanchum komarovii*, a toxic annual, was associated with the rangeland subjected to heavy grazing in the Mu Us Sandy Land, north central China. The expansion of cultivated areas experienced two climaxes during 1950–1960 and 1990–2000 (Fig. 3.9).

Cropping practices in arid and semi-arid areas usually expose bare soils to wind erosion much of the year and soils lose productivity and are left barren after several years' cropping (Fig. 3.10). Cultivation can accelerate the wind erosion rate of fixed aeolian sandy soil by several hundreds of times that of undisturbed soil. Firewood gathering is severe in locales short of energy. Natural vegetation provides more than 50% of everyday fuel demand in the Mu Us Sandy Land. In some regions, local people have not yet run out of the firewood that was collected with such disregard over 20 years ago! Besides the increase in human and animal populations, other causes of human-induced desertification are irrational land-use practices and exploitation of natural resources. 'Putting grains in command', an unsound land development policy, led to large-scale land clearing for cultivation before 1980 and resulted in the rapid expansion of desert (Chapter 2). After 1980, when Chinese economic reform was initiated, short-term economic returns were prioritized, while disregarding environmental conservation and sustainability. Human economic activities aggravated the situation of water shortage in north China. The development of irrigated farmland and industrial mining operations increased consumption of surface water and exploitation of groundwater (Chapter 11). Within Inner Mongolia, for example, irrigated farmland has increased from 0.35 million ha (Mha) 50 years ago to 2.07 Mha today, which has aggravated the water shortage situation. Coal mined in Inner Mongolia climbed from 10.8 billion kg in 1980 to 51.6 billion kg in 1997 and, although it

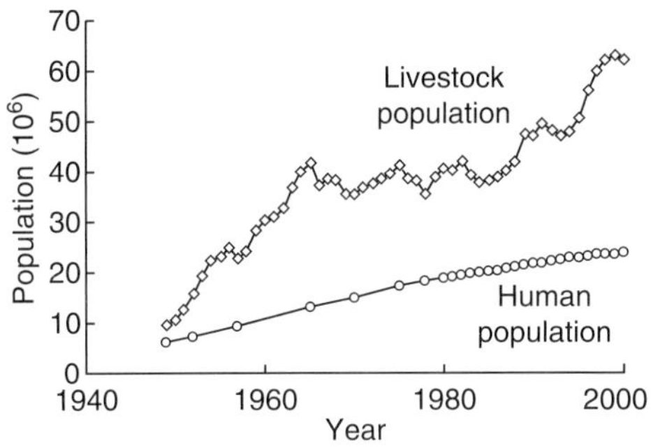

Fig. 3.8. Evolution of the human and livestock population in Inner Mongolia from 1949 to 2000.

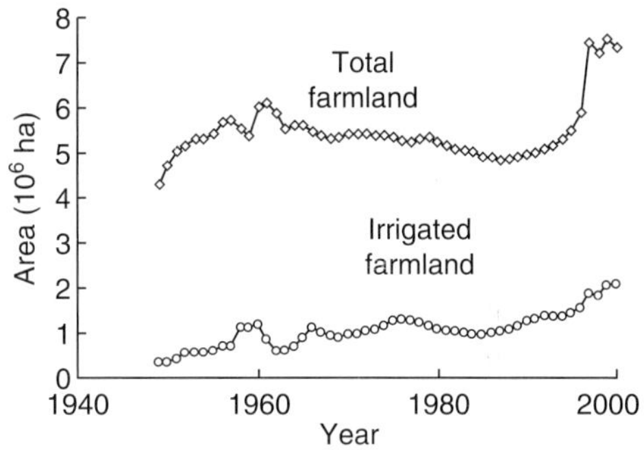

Fig. 3.9. Variation of the area of total farmland and irrigated farmland in Inner Mongolia from 1949 to 2000.

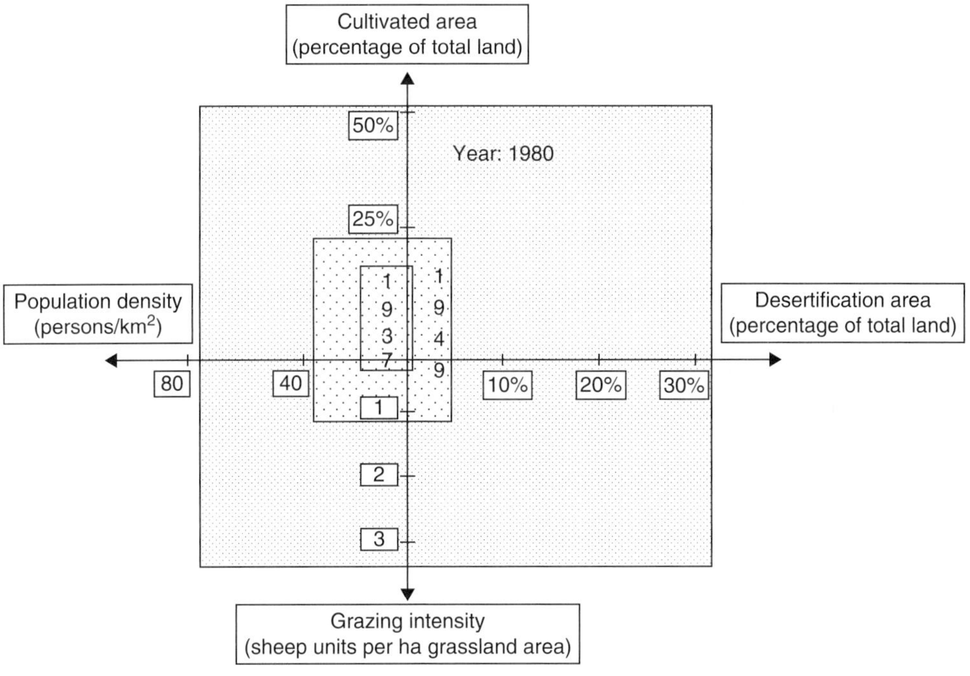

Fig. 3.10. Dynamics between people, livestock, cropland and desertification.

declined a little, for a time it remained as high as 45.1 billion kg (in 2000) and, in recent times, energy shortages have accelerated coal production (Fig. 3.11). This has put great demand on groundwater. The recent price rise of coal and other energy sources has stimulated further mining.

Human disturbance to dry sandy soils can increase supplies of surface materials for transport. Vegetation is efficient at trapping dust. The erroneous methods of land use, such as conversion of land from lakes and sloping land, deforestation and destruction of vegetation cover, causes land to lose dust traps and seed banks. There are negative consequences from land degradation. The deterioration of vegetation will make the soil more susceptible to wind and water erosion, which takes the scarce soil organic matter

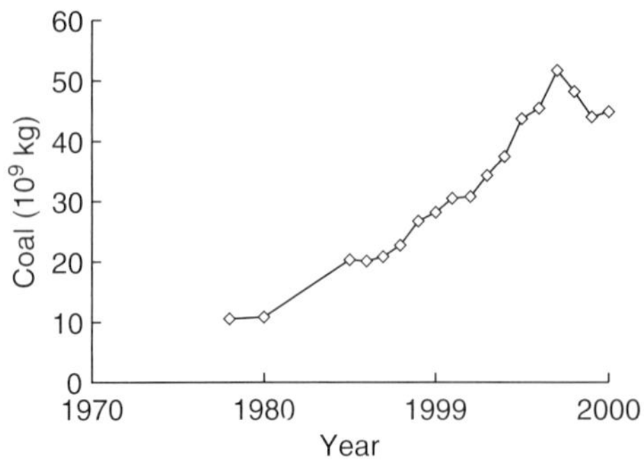

Fig. 3.11. Coal production in Inner Mongolia from 1978 to 2000.

and mineral nutrients away and, in turn, prevents vegetation recovery. As a result, land degradation will proceed in steps, which are increasingly difficult and costly to reverse.

Although it is virtually impossible to separate anthropogenic impact from that of climatic change, as the two processes often work together, many research findings have suggested the dominant contributions of human activities to land degradation in north China (Lu *et al.*, 2004; Wang *et al.*, 2004). About 94.5% of desertification on sandy steppe or reactivated vegetated dunes is associated with human disturbance. Evidence of human-induced desert expansion over historical time comes from examining the aeolian grain size and sedimentation patterns. Remarkable instances of human-induced desertification of semi-arid steppe grassland have been documented dating back to the Han Dynasty about 2000 years ago. Many research findings concluded that climate-dominated desert formed through the geological ages, while human activities have controlled accelerated land degradation processes in recent decades. Much degradation actually occurred as a 'blistering' process on rangeland, which was climatically steppe and which is now covered with masses of mobile sand dunes. Among the human causes of land degradation, conversion of grassland is believed to be the most important. The tide of land conversion from the late 1950s to the 1960s caused about 100,000 km^2 of severe land degradation.

The natural vegetation is well adapted to its climate and it is human interventions that make the land system more vulnerable to climatic variability. This situation is very similar to the American dust bowl of the 1930s, which was due to agricultural devastation.

3.4 Glacial Retreat

Several important mountain ranges occupy territory in northern Xinjiang and in the Hexi Corridor of Gansu. Many glaciers occur there. The Qilian Mountains of the Hexi Corridor are located on the northern fringe of the Tibetan Plateau and have 2815 glaciers, occupying about 1930 km^2. Glacier meltwater and more abundant precipitation in the mountains supply a fair amount of water to oases scattered across the huge desert area, where the scant precipitation cannot otherwise support human inhabitants. Water from glaciers in both Xinjiang and Gansu is significant as water reservoirs, especially in the arid inland regions such as the Taklimakan and Gobi Deserts.

Along with global warming, the glaciers are melting and shrinking in Xinjiang and Gansu (Liu *et al.*, 2003). In Xinjiang from 1963 to 2000, the area of glaciers in the Tianshan Mountains was reduced on average by 12.5% (Chapter 14). The changes of matter counterbalance of No. 1 Glacier during 1959–1985 averaged −94.5 mm/year, whereas during 1986–2000 it increased to −358.4 mm/year – a 2.8-fold increase (Fig. 3.12). Accordingly, melting runoff from the No. 1 Glacier also increased greatly; during 1958–1985 it averaged 508.4 mm/year, whereas during 1985–2001 it was 936.6 mm/year by the same calculation method. Thus, it can be seen that the temperature has increased rapidly since the 1980s and, consequently, has accelerated the rate of glacier melting.

Similar results have been reported for glaciers in the Qilian Mountains of Gansu (Sakai *et al.*, 2006) and in Tibet (Pang *et al.*, 2007). For example, the total area of ice glaciers in the Hexi Corridor region of Gansu is 1657 km^2, with a storage capacity of 80.13 billion m^3, accounting for 84% of the total glacier area of the Qilian Mountains. The glaciers are considered as an important 'long-term regulatory reservoir' for the Hexi Corridor. Runoff from melting glaciers accounts for more than 10% of the annual replenishment of rivers in the Hexi Corridor (Li *et al.*, 2002). However, in recent years, snowlines

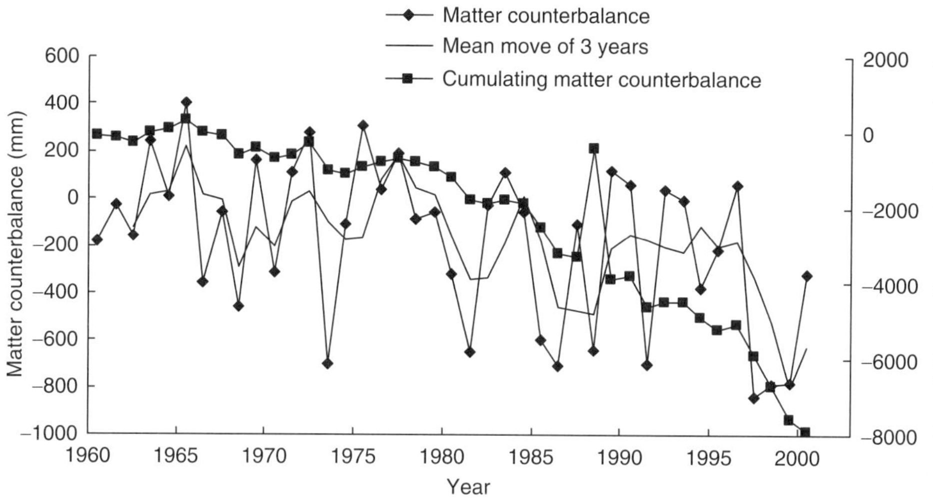

Fig. 3.12. Changes of glacier matter counterbalance and cumulating matter counterbalance in the headwater region of the Urumqi River in the Tianshan Mountains since 1959.

have moved up and glaciers are retreating; wetlands and lakes in mountainous regions are disappearing. It is estimated that, in the case of a temperature rise of 3°C in the Qilian Mountains, snowlines will retreat a further 500 m and most of the glaciers will melt in just a few years.

The July 1st Glacier is located in the western region of the Qilian Mountains in Gansu Province. Area shrinkage and surface lowering have accelerated in the past 15 years. In the 19 years between 1956 and 1975, the glacier has retreated 68.3 m (at an average annual rate of 3.16 m/year), but recent data suggest that this rate has accelerated to 5.9 m/year. Meltwater discharge from the glacier in the past 17 years has increased due to glacier shrinkage, about 50% of that occurring between 1975 and 1985. The recent accelerated glacier shrinkage has been attributed to increasing temperature (Shi, 2006).

Note

[1] The Yangtze, Yellow and Mekong Rivers all rise here.

References

Chen, Y. and Tang, H. (2005) Desertification in north China: background, anthropogenic impacts and failures in combating it. *Land Degradation and Development* 16(4), 367–376.

Le Houérou, H.N. (1992) Climate change and desertization. *Impact of Science on Society* 166, 183–201.

Le Houérou, H.N., Bingham, R.L. and Skerbek, W. (1988) Relationship between the variability of primary production and variability of annual precipitation in world arid lands. *Journal of Arid Environments* 15(1), 1–8.

Li, J. (2001) Temporal variation of droughts in northern parts of China. *Agricultural Research in the Arid Areas* 19, 42–51.

Li, S., Cheng, G. and Li, Y. (2002) *Reasonable Utilization of Water Resource and Eco-environmental Protection in Hexi Corridor.* Yellow River Water Conservancy Press, Zhengzhou, China.

Liu, L., Zhang, F. and Zhao, Y. (2002) A predicting analysis of comprehensive productivity of China's rangeland resources from 2000 to 2050. *Acta Prataculturae Sinica* 11, 76–83.

Liu, S., Sun, W., Shen, Y. and Li, G. (2003) Glacier changes since the Little Ice Age maximum in the western Qilian Shan, northwest China, and consequences of glacier discharge for water supply. *Journal of Glaciology* 49(164), 117–124.

Lu, Q., Yang, Y., Wang, S. and Liu, T. (2004) *The Revelation: Combating Desertification in China*. Science Press, Beijing.

Pang, H.X., He, Y. and Zhang, N. (2007) Correspondence: Accelerating glacier retreat on Yulong Mountain, Tibetan Plateau, since the late 1990s. *Journal of Glaciology* 53(181), 317–319.

Qian, W.H. and Zhu, Y.F. (2001) Climate change in China from 1880 to 1998 and its impact on the environmental condition. *Climatic Change* 50(4), 419–444 (in Chinese).

Ren, L., Wang, M., Li, C. and Zhang, W. (2002) Impacts of human activity on river runoff in the northern area of China. *Journal of Hydrology* 261, 204–217.

Sakai, A., Fujita, K., Duan, K., Pu, J., Nakawo, S. and Yao, T. (2006) Five decades of shrinkage of July 1st glacier, Qilian Shan, China. *Journal of Glaciology* 52(176), 11–16.

Shi, Y.F. (2006) Recent and future climate change in north-west China. *Climatic Change* 80(3–4), 379–393 (in Chinese).

Wang, T., Li, X.Z. and Li, Q.S. (1991) A preliminary study of present land desertification in Bashang Plateau, Hebei Province. *Journal of Desert Research* 11, 39–45.

Wang, T., Wu, W., Xue, X., Sun, Q.W., Zhang, W.M. and Han, Z.W. (2004) Spatial–temporal changes of sandy desertified land during last 5 decades in northern China. *Acta Geographica Sinica* 59, 203–212.

Wu, B., Su, Z. and Chen, Z. (2007) A revised potential extent of desertification in China. *Journal of Desert Research* 27(6), 911–917.

Zhu, L. (2005) *Dynamic of Desertification and Sandification in China*. Chinese Agriculture Press, Beijing.

4 Mechanisms of Degradation in Grazed Rangelands

Li Xianglin
Chinese Academy of Agricultural Sciences, Beijing, China

Synopsis

The mechanisms of degradation in grazed rangelands (loss of perennial grasses and shrubs), invasion by toxic and unpalatable plants, rodent outbreaks and the impact of periodic prolonged drought on plant growth and survival under intense grazing pressure are discussed. The components of the grazing systems in China's rangelands are outlined. The question of whether better grazing management is a dream or an economic and ecological imperative is also considered.

Keywords: geographical distribution; environmental services; transhumance; agropastoral integration; feed balance; drought impacts; afforestation impacts; noxious plants; rodents

4.1 Components of the Grazing Systems in China's Rangelands

In China, rangeland is defined as the land covered mainly by natural herbaceous vegetation, or with sparse shrubs or trees concurrently present in the community, which provide food for livestock and habitats for wildlife, as well as environments, organic products and other functions for humans (Hu, 1997). With a total area of nearly 4 million km^2, rangelands are the largest single component of China's 9.6 million km^2 of total land. Rangelands are found mainly in north and north-west China, covering vast areas of temperate and cold semi-arid to arid zones through the Tibetan Plateau and northern China to the Asian steppe (Hu and Zhang, 2001). Generally, rangelands are too arid and/or too cold to support croplands or dense forests and so contribute mainly livestock to the country's human carrying capacity. Rangelands have many uses other than as a source of feed for livestock and are of great environmental importance. They are usually important hydrological catchment areas, are important as wildlife habitats for the *in situ* conservation of plant and other genetic resources and are used frequently for sport and tourism. The rangelands also provide an important buffer against the impact from desert. The great mass of mountains formed by the Tibetan–Qinghai Plateau and the Tianshan ranges is the source of most of China's rivers.

4.1.1 The pastoral regions

The rangeland systems are generally characterized by sparse vegetation, limited social infrastructure, poor communications and a harsh climate. Extensive grazing is the most important approach to utilizing the rangeland resources. Herding communities are generally 'minority nationalities', including Mongolians, Kazakhs, Kyrgyzes and Tibetans. Grazing systems are important, both environmentally and as a source of livelihood for the herders. Both transhumant and agropastoral systems are common and involve both full-time nomads and settled farmers, who take their stock to summer pastures. Six provinces/autonomous regions, namely Tibet, Inner Mongolia, Xinjiang, Qinghai, Sichuan and Gansu, are known as major pastoral areas for their large rangeland area and number of livestock. They account for 75% of the national total rangeland area (Table 4.1) and 70% of the national total of grazing animals.

Sheep and cattle are the major grazing animals, with sheep kept mostly in temperate regions, typically in north and north-west China, and cattle being important in all systems. Yaks can be found only on the Tibetan Plateau, covering most of the Tibet Autonomous Region and Qinghai Province and parts of Sichuan, Yunnan and Gansu Provinces, with an elevation between 3000 and 5000 m above sea level (Chapter 13). Goats are the most widely distributed livestock species, since they can adapt to a wide range of climatic and feeding conditions across the country. The data in Table 4.2 present a brief profile of grazing systems, as compared with mixed crop–livestock farming systems, in temperate and humidal subhumid, tropical–subtropical agroecological zones in China. This description gives a general picture of the livestock systems, although it is based on data available in the 1990s (Conner *et al.*, 1996; Li *et al.*, 2008).

4.1.2 Grazing systems in north and north-west China

Extensive grazing occupies a vast area of land and is the most important approach to utilizing the rangeland resources in the pastoral

Table 4.1. Rangeland area in China. Source: Liao and Jia (1996).

Province	Rangeland area (Mha)	% of national total	Theoretical carrying capacity (SSU)[a]	% of national total
Inner Mongolia	78.8	20.1	44.2	9.9
Tibet (Xizang)	82.1	20.9	27.1	6
Gansu	17.9	4.6	11	2.5
Sichuan	21	5.3	46.4	10.3
Qinghai	36.4	9.3	29	6.5
Xinjiang	57.3	14.6	32.2	7.2
Other provinces	99.3	25.3	259	57.6
China	392.8	100	448.9	100

[a] Sheep stock unit. One SSU is defined as a ewe of 40 kg live weight with its lamb(s), equivalent to a daily feed consumption of 5–7.5 kg fresh forage (about 2.5 kg DW).

Table 4.2. Human population, pasture, arable land and livestock population in China allocated to livestock systems/agroecological zones. Source: Conner *et al.* (1996); Li *et al.* (2008).

		Livestock systems/agroecological zones[a]					
	Total	LGA	LGT	LGH	MR/IA	MR/IT	MR/IH
Human population (million)	1114.3	3.0	87.3	180.3	158.4	323.7	361.6
Pastureland (Mha)	400.0	13.0	172.0	78.0	5.9	60.8	70.3
Cropland (Mha)	96.6	0.5	15.0	9.0	34.4	26.2	11.5
Cattle (million head)	77	2.0	6.7	19.4	7.3	9.9	31.7
Sheep (million head)	113.5	2.6	24.8	11.5	31.5	17.6	25.5
Goats (million head)	97.8	6.3	40.7	16.3	6.5	23.4	4.6

[a] LGA, livestock only, grassland-based, arid–semi-arid; LGT, livestock only, grassland-based, temperate; LGH, livestock only, grassland-based, humid–subhumid tropics and subtropics; MR/IA, mixed farming, rainfed/irrigated, arid–semi-arid; MR/IT, mixed farming, rainfed/irrigated, temperate; MR/IH, mixed farming, rainfed/irrigated, humid–subhumid tropics and subtropics.

areas. Uncontrolled grazing on communal pastures prevailed until the Long-term Grassland Use Contract System was put into practice in the 1980s (Li *et al.*, 2007). Communal grazing still exists, though to a lesser extent, in remote summer pastures or open pastures, especially in Gansu's Hexi Corridor (Chapter 11), Tibet (Chapter 13) and parts of Xinjiang (Chapter 14).

In the north-west desert–mountain areas (Xinjiang and western Gansu), where there is a great variation in topography and climate, herders generally employ seasonal grazing systems by which animals graze in the basins in winter, move in transhumance to the mountains in spring, to the high mountains in summer and return to the basins in late autumn. Two-season (warm and cold) or three-season (winter, spring/autumn and summer) grazing systems are employed, depending on the environmental conditions. Accordingly, the rangelands are divided into winter pasture, spring/autumn pasture and summer pasture based on topography and climate. They are strict seasonal grazing systems and animals spend 1–2 months travelling from winter to summer pasture.

On the Qinghai–Tibetan Plateau, although animals graze where the elevation is above 3000 m, transhumance is also common and the grazing land generally is divided into two types of seasonal grazing belts: the lower pasture for cold-season grazing and the higher pasture for warm-season grazing. The summer–autumn grazing is roughly from June to late October or early November. The herd then moves to the winter–spring pastures for about 200–230 days from November (sometimes sooner) to May. In general, the two-season grazing system is adopted mainly in open plateau areas. During the warm season, a yak herd may be moved every 30–40 days, depending on pasture availability and herd size. With more complex topography, three-season grazing systems are also employed (Long, 2003).

The grazing systems in the north differ from those in the north-west and the Tibetan Plateau. In Inner Mongolia and the north-east, where grasslands are relatively flat and the environment is relatively simple, pastures are grazed in any season, whenever water is available, without seasonal restrictions on grazing and animals are moved rotationally following a predetermined range and schedule.

4.1.3 Challenges associated with extensive grazing

Feed deficiency and livestock loss

Feed deficiency during the winter is the key problem of the systems that depend on natural grazing year-round. The increase in livestock number, coupled with the invasion of crop cultivation into pastures (Chapter 3 and case studies), results in reduced availability of winter grazing and causes considerable livestock losses, which are often worsened by catastrophic snows. Before vegetation turns green in spring, both the quantity and quality of the forage remaining on these pastures are at their lowest for the year. Animal intake of feed during the winter–spring period is much lower than in summer–autumn grazing. As a result, as much as a third of the live weight the animal achieved in the previous autumn is lost during the late winter to early spring period (Long, 2003). This, in turn, leads to a large number of livestock in the herd becoming sick or weak. Many of the sickest or weakest animals may die, especially if heavy snowstorms occur in the absence of sufficient feed supplementation and at the very time when most new calves and lambs are born. The severe seasonal imbalance with natural grazing has long been recognized and a management strategy of 'seasonal livestock farming' has been developed, involving reserving feed (often hay) and marketing animals in autumn so as to keep only a minimum number of animals (mainly breeding animals) for wintering (Ren *et al.*, 1978).

Rangeland management

Intensification, involving enclosure of family grazing lands, upgrading of stock breeds, improvement of pastures, integration of forage crops and supplementary feeding, has been taking place with apparent success in agropastoral systems. The introduction of new winter feed supply, through irrigated hay production, has improved the overwintering condition and survival of grazing animals. The government has launched a series of policies to encourage the settlement of pastoral nomads. However, as a traditional culture of the pastoral communities, transhumance is still maintained in many mountainous areas with excellent alpine pastures for summer grazing. The survival of pastoral nomads indicates that many strategies of animal husbandry and grassland management developed centuries ago are well adapted to the spectrum of environment conditions (Miller and Craig, 1997). Nevertheless, because of rangeland degradation and other problems caused by overgrazing on common grazing lands, there is an obvious need for a 'co-management' strategy that encourages participation of the local community in the management of common natural resources in a consensus-based manner with decision-making power being shared among the various actors (Yan, 2007).

Market opportunities

The major interaction of grazing with other production systems is through the land-use pattern and the market. The importance of grazing systems in terms of sustaining food production has declined as interaction with crop cultivation has turned favoured lands into mixed systems which supply similar animal products. Market forces, biophysical constraints and environmental concerns are putting a ceiling on the potential for intensification of grazing systems. Therefore, the market share of livestock products from grazing systems is declining as compared with other livestock production systems.

Rangeland degradation

Perhaps the most serious challenge facing the grazing systems is rangeland degradation. It has been clear that rangeland degradation across northern China is not only threatening the livelihoods of herders but also causing severe environmental problems, notably desertification and dust storms. The airborne desert dumps itself on many cities and moves on to neighbouring countries. It has also been evident that the vegetation of the upper catchments of the main rivers is under severe pressure and is often seriously degraded. This decreases infiltration and speeds up runoff, thereby increasing flooding and soil erosion. It also increases the silt load, with consequent damage and cost to agriculture and structures far downstream. Rangeland degradation also leads to the loss of important plant and other genetic resources. The ecological service value of rangeland systems has recently attracted great attention in China (Xie *et al.*, 2001; Guo *et al.*, 2004) and it is argued that rangelands should be classified into, and managed as, different systems according to their key functions in providing ecological service value or livestock production (Guo *et al.*, 2006).

4.2 Mechanisms of Degradation in Grazed Rangelands

4.2.1 Plant species response to grazing

Many rangelands have historically supported native grasslands, originally dominated by perennial grasses. These grasses are useful for soil stabilization through their cover of the ground year-round and extensive root systems, are productive and also are very palatable to livestock. As the intensity and the duration of grazing increases, the perennial grasses are consumed, trampled and decline (Jiao *et al.*, 2006). Their roots suffer from loss of aboveground parts and soil compaction. In addition, when they are heavily grazed (particularly at flowering and seed set times), they produce fewer seeds, decreasing recruitment of new individuals into the population. The decline of the perennial grasses is followed by increased abundance of less palatable annual (or short-lived perennial) grasses and herbs (Jiang *et al.*, 2003). Space has been freed by the decline of the native perennials and the annuals spread themselves quickly through re-seeding. These annuals (and short-lived perennials) often produce abundant seed and have effective dispersal mechanisms, having the life history trait characteristics of such 'weedy annuals'.

Wang and Li (1999) provided a description of the response of typical steppe plant species to grazing pressure, based on a 6-year experimental study on degraded grassland in Inner Mongolia dominated by *Artemisia frigida* and short grasses (Table 4.3). It was found that grasses with different life types and vegetative reproduction properties differed in their response to pressure, which formed the basis of community succession induced by grazing. Species with strong stolons, such as *A. frigida* and *Potentilla acaulis*, were well adapted to heavy grazing; *Leymus chinensis* and *Agropyron cristatum*, vegetatively regenerating mainly through rhizomes, and *Stipa krylovii*, a bunch grass, were adapted to light grazing; *Carex duriuscula*, with

Table 4.3. General model of the response to grazing of perennial plant species in typical steppe of Inner Mongolia.

Species	Main form of regeneration	Grazing pressure
Artemisia frigida	Stolon	Heavy
Potentilla acaulis	Stolon	Heavy
Leymus chinensis	Rhizome	Light
Agropyron cristatum	Rhizome	Light
Stipa krylovii	Tiller	Light
Carex duriuscula	Tiller	Moderate
Cleistogenes squarrosa	Tiller	Moderate
Melissitus ruthenica	Branching	Light
Kochia prostrata	Branching	Light

rhizomes and vigour tillering, and *Cleistogenes squarrosa* (bunch grass) were adapted to moderate grazing; *Melissitus ruthenica* and *Kochia prostrata*, with branching growth, were adapted to light grazing. It was also observed that the frequency of short grasses in the community decreased gradually with increasing grazing. The degraded grassland, dominated by *A. frigida* and short grasses, converged ultimately to a community dominated by *P. acaulis*.

Miniaturization of individual plants

Overgrazing also leads to miniaturization of individual plants, particularly palatable perennial plants (Wang, 2000a,b). Miniaturization is generally characterized by a group of morphological traits: a stunted height, shortened and reduced leaf size, shortened internodes, hard stems and leaves and shallower distribution of the root system. As a result of overgrazing, miniaturization of individual plants is the basic cause of decreasing community productivity and regenerating ability. This gives opportunities for the population of invading or unpalatable species to increase and eventually fill the gaps left by the suppressed plants. It is a negative feedback mechanism and an important indication of degradation succession of grassland communities.

Reduced soil seed bank

The soil seed bank is central to the restoration of plant communities. Seeds are able to remain viable in the soil for different periods of time, depending not only on species but also on soil conditions. The longevity of a seed is decisive for how long a species is present in the seed bank after the supply of fresh seeds has been interrupted, for example, after grassland species have been replaced by shrubs and herbs. A study in Inner Mongolia (Zhan *et al.*, 2005) on the soil seed bank of an *S. krylovii* steppe community at two sites with different management regimes, one enclosed pasture without grazing since 2001 and another grazed continuously for more than 20 years, showed that continuous grazing reduced the seed bank of the rangeland significantly and the scarcity of seeds of some important steppe species, combined with the unbalanced distribution of seeds among species, might inhibit the restoration process of the *S. krylovii* steppe. This suggests that artificial re-seeding and other management strategies should be employed to assure good seed supply for the restoration of degraded steppe communities.

Loss of perennial species

The consequences of permanent loss of species or species groups from plant communities are poorly understood, although there is increasing evidence that individual species effects are important in modifying ecosystem properties (Wardle *et al.*, 1999). Many perennial species are adapted to survive under difficult conditions and they are able to persist for long periods. Persistence is essential as it will ensure the long-term viability of grassland and pasture resources in the landscape.

In the Horqin sandy rangeland of Inner Mongolia, rapid vegetation degradation has been observed (Zhao *et al.*, 2003). The degradation process started with a rapid decrease in vegetation coverage, canopy height and aboveground standing biomass when most parts of stems and leaves were grazed. The vegetation coverage, height and aboveground standing biomass in heavily grazed lands were reduced by 82.1%, 94% and 97.9%, respectively, as compared with those without grazing for 5 years. With increased grazing pressure, perennial as well as certain palatable annual species were replaced gradually by the annuals of poor palatability and the percentage of unpalatable annuals eventually reached 86%. Deterioration of vegetation was accelerated by increased wind erosion and bared land area. This effect is obviously different from the degradation process of meadow grassland, where wind erosion is much less significant (Chapters 5 and 9).

The loss of palatable perennial species is common in overgrazed pastures of various types, though in different rates dependent on the environment and the degree of overgrazing. Loss of perennial species was also observed in typical steppe (Wang and Li, 1999; Liu and Li, 2006), alpine meadow (Jiang *et al.*, 2003; Dong *et al.*, 2005), lowland meadow steppe in north-eastern (Yang *et al.*, 2001) and arid rangeland in north-west China (Jiao *et al.*, 2006), along with changes in species diversity.

Rangeland degradation caused by inappropriate management practices, coupled with climate change, has a profound environmental impact. Extensive grazing still occupies a vast area of land, though there is an increasing trend towards intensification and industrialization. Meanwhile, the human population pressures on land and government subsidies are causing higher-quality (higher-rainfall) grazing lands to be converted to low-quality croplands, which is a primary reason for the increasing wind erosion on croplands. During the past decades, the degree of pasture degradation rose alarmingly and desert was gaining rapidly on pasture (Yang *et al.*, 2005). It is estimated that nearly 90% of the rangeland area in China is degraded to a certain extent and roughly half of the area is classified as moderately to severely degraded. As a result, pasture productivity is estimated to have reduced by 30–50%, with a direct economic loss of over US$8 billion each year. Many factors contribute to grassland degradation but the major cause is overexploitation of rangeland resources as a function of the increase in the human population and subsequent increase in livestock numbers. The situation is worsened by frequent, prolonged drought and increasing pest damage (particularly by rodents).

4.2.2 Drought and plant growth

The climate zones

China has three climatic zones: east monsoon zone (EMZ), north-west arid and semi-arid zone (NASZ) and the Qinghai–Tibetan Plateau zone (QTPZ) (Sun and Shi, 1994). China's rangelands are located mostly in the NASZ and QTPZ. The NASZ is in inner Eurasia and is controlled by a continental climate all year round. Precipitation decreases gradually from east to west from 400 mm to less than 100 mm. Steppe rangeland and desert dominate the landscape. The QTPZ generally is characterized by a low temperature, strong solar radiation and winds, uneven rainfall and a significant vertical variation of climate and landscape. The precipitation declines from south-east to north-west on the plateau and the natural landscape varies accordingly from forest, alpine shrub and alpine steppe to alpine desert. The climate features of representative cities in northern China pastoral areas are shown in Table 4.4.

Climate change

There has been accumulated scientific evidence that the globe has been undergoing a profound climate change, which has a significant effect on rangeland ecosystems. In north-east China, the aridification trend has become more serious since the 1970s; the drought index in north China also reached a high value during the 1990s (Qian and Zhu, 2001). Ding *et al.* (2007) have summarized the main results and findings of Chinese scientists

Table 4.4. Climate features of pastoral areas in north and north-west China.

Location	Mean temperature (°C)			Frost-free days	Annual precipitation (mm)	Max. snow thickness (cm)
	January	July	Annual			
Huhhot	−12.5	22.2	6.2	117	401.6	30
Lhasa	−2.1	15.3	7.5	135	424.1	12
Lanzhou	−6.1	22.1	9.3	159	316.0	10
Xining	−7.7	17.2	5.9	128	367.5	14
Yinchuan	−8.4	23.3	8.7	152	139.8	11
Urumchi	−12.7	23.7	6.6	161	276.1	44

in the past years. It has been clear that China's average annual mean surface air temperature has increased by 1.1°C over the past 50 years and by 0.5–0.8°C over the past 100 years, slightly higher than the global temperature increase for the same periods. Northern China and the winter season have experienced the greatest increases in surface air temperature (Chapter 3). It is evident that north-west China has undergone an increased interdecadal variability of precipitation and that north China has suffered a severe drought. Some analyses show that frequency and magnitude of extreme weather and climate events have also undergone significant changes in the past 50 years or so. Climate change has affected China's ecosystem severely, particularly in the arid and semi-arid regions with vulnerable ecological backgrounds (Chapter 3). Open shrub and desert steppe may be the two most affected systems (Wu *et al.*, 2007).

Future climate change may cause drastic increase in production costs and investment need, increased potential in aggravation of desertification, shrinking grassland area and reduced productivity that result from increased frequency and duration of drought occurrence due to climate warming. Climate change has impacted on the natural ecosystems in China. For example, massive glacier areas in north-west China and the Qinghai–Tibetan Plateau have melted at an alarming rate over the past decades (Jin *et al.*, 2005; Shi *et al.*, 2006; Pang *et al.*, 2007), with global warming believed to be the culprit (Chapter 3). China's remote Xinjiang region is home to nearly half of the nation's glaciers, which supply the rest of the country and other parts of Asia with water. However, they have shrunk by 20% and snowlines there have receded by about 60 m since 1964. There is evidence of an increase in the frequency of extreme hydrological events, such as drought in the north and flood in the south. The arid continental river basins are particularly vulnerable to climate change (Chapter 11).

Rangeland productivity response

Many reports on global change have predicted major change in the temporal and spatial pattern of temperature and precipitation, which may have significant effects on temperate grasslands in arid and semi-arid regions. The responses of rangelands to environmental changes, especially the amount and timing of precipitation, can be very different. Some studies indicate that drought may result in degradation of ecosystem function in net ecosystem exchange (NEE), even changing the ecosystem from a carbon sink to a carbon source.

Huang *et al.* (2006) measured the carbon dioxide flux during the 2005 growing season in Xilin River Basin of Inner Mongolia using the eddy covariance technique to quantify net ecosystem carbon exchange in *L. chinensis* steppe and its response to precipitation. Only 126 mm precipitation fell during the growing season, far less than average, indicating the steppe was in an extreme drought condition. It was found that the maximum half-hour average CO_2 uptake was $-0.38\,mg/m^2/s$ in 2005, which was half that in typical growing seasons. Moreover, the ecosystem was a CO_2 source most of the growing season, releasing about 0.05 mg $CO_2/m^2/s$ at night-time. It was concluded that the seasonal pattern of CO_2 uptake followed that of above-ground biomass closely and was affected strongly by soil temperature and soil water content. The ecosystem emitted 372.56 g CO_2/m^2 during the growing season in 2005.

Wang *et al.* (2005) studied the response of the vegetation system to climate change in an area on the Tibetan Plateau located in the transition zone between alpine meadow and alpine steppe. The floral assemblages in the area were investigated to identify the migration of the boundary between alpine steppe and alpine meadow caused by climate warming and drying. The results of the research indicate the arid zone has been expanding at a rate of 14.2 km^2 every decade, accompanied by a degeneration of vegetation in the broader central plateau area. The spread of the alpine steppe has resulted not only in a decrease in the overall vegetation coverage, but also in a decrease in the total biomass of the surface vegetation in the area. The total biomass of the alpine steppe community in the study area is 77% of that of the alpine meadow community. Based on the observed responses of floral carbon isotopic compositions to varying climate conditions in this study and the findings of the previous studies in the same area, the researchers conclude that the change to a warmer and drier climatic environment is the main cause of the progressive degeneration of the vegetation in the studied area. The simulations on future changing vegetation indicate that the vegetation degeneration will be accelerated if the warming and drying trend continues. This means that the warming and drying is causing a set of processes that reduce plant biodiversity and vegetation productivity on the Tibetan Plateau.

The impact of afforestation

Since 1978, China has pursued one of the most ambitious conservation programmes in the world: the Shelter Forest System Project in north China aiming at desertification control through large-scale afforestation in arid and semi-arid areas (Li, 2004; Cao, 2008). From 1999 to 2005, 2.6 million ha (Mha) of rangelands were planted with trees (SFA, 2006). However, it is argued that these costly efforts have yielded little success in the arid and semi-arid regions of China, but have had significant impacts on soil moisture, hydrology and vegetation coverage (Cao, 2008; Liu, 2008). Despite the fact that the area of afforestation is increasing rapidly as a result of the afforestation projects, the area of degraded land has continued to expand and the severity of desertification has continued to intensify (Yang *et al.*, 2005; Cao, 2008). There is a plenty of evidence of the failure of large-scale afforestation projects carried out by the government in the arid and semi-arid north and north-west of China (Liu, 2008).

Drought is a major constraint to the re-vegetation of arid and semi-arid regions in northern China. Soil moisture is generally deficient in planted forests because of low annual precipitation, and this has led to large-scale mortality of plantations during drought years (Wang *et al.*, 2003; Xu *et al.*, 2006). Previous research in these regions (Cao *et al.*, 2007) has revealed that, in contrast with natural grassland and forest, for which water use was historically in equilibrium with the water supply, soil moisture content to a depth of 6 m in afforestation areas has decreased by 32–37%.

An inverse relationship often exists between the soil water balance and afforestation of grassland and farmland (Vitousek *et al.*, 1997) because of the large amounts of soil moisture consumed by fast-growing trees. This moisture cannot be replenished during the rainy season; thus, reserves of soil water are depleted, the woody vegetation eventually dies because of water stress and desertification ensues. The decrease in soil moisture in afforestation plots, combined with reduced sunlight under the tree canopies (which affects the growth of understorey vegetation adversely), has led to decreased vegetation cover in the afforestation plots. Because dense steppe vegetation can absorb more of the wind's momentum than less dense plant communities or bare soil, this vegetation can effectively control wind erosion (Liu and Li, 2006).

4.2.3 Perennial grasslands

Overview of vegetation types

The vegetation types of natural grasslands in China can be grouped generally into two broad categories: temperate steppes and meadows.

Three major types of temperate steppes are recognized: meadow steppe, needlegrass steppe and needlegrass steppe–dwarf shrubs, or subshrubs. Meadow steppes are distributed in the subhumid climate zone, with dominant species being *Aneurolepidium chinese* (or *L. chinensis*), *S. baicalensis* and *Filifolium* spp. in temperate and *Bothriochla ischaemum* and *Themeda triandra* var. *japonica* in warm-temperate steppe. The needlegrass steppes are found mainly in the semi-arid areas of the Inner Mongolia Plateau and the loess plateau, dominated by xerophytic tussock grasses such as *S. grandis*, *S. krylovii*, *S. brebiflora*, *S. bungeana* and *C. squarrosa*. The steppe of needlegrass–dwarf shrubs, or subshrubs, is a transition between desert and steppe. This type of vegetation consists mainly of xerophytic grasses such as *S. gobica*, *S. glareosa*, *S. klemenzii* and *C. mutica*, mixed with subshrubs such as *A. xerophtica*, *Ajania trifolia* and dwarf shrubs such as *Caragana stenophylla*, etc. In addition, there are also alpine steppes, which are composed mainly of cold-xerophytic dense tussock grasses such as *S. purpurea*, *S. subsessiliflora* var. *basiplumosa* and *C. moorcroftii*, etc.

Plants of temperate meadows are mostly mesophytes. *Phragmites communis*, *Calamagrostis epigeios* and many forbs are common on neutral or calcareous soils; *Sanguisorba* spp. and *Vicia* spp. grow on acid soils. Saline meadows contain many species of halophytic grasses and herbs such as *Aeluropus littoralis*, *A. dasystachys*, *Achnatherum splendens* and *Scorzonera mongolica* var. *butjae*. Alpine meadows are distributed mainly on the Qinghai–Tibetan Plateau. They are composed mainly of numerous species of *Kobresia*.

Recovery of perennial grassland

Over the past two centuries, perennial grass cover has declined and shrub density has increased in many arid grasslands (Asner *et al.*, 2004). It has been realized generally that shifts from perennial to annual grasses are more obvious in heavy grazing regimes (Cingolani *et al.*, 2003; Friedel *et al.*, 2003). As a characteristic of desertification, these changes in vegetation are thought to have often occurred following prolonged periods of intense grazing by domestic livestock. Based on this understanding, the removal of livestock grazing by fencing has been the most important approach to the restoration of degraded rangelands.

In many historically grazed arid grasslands, the subsequent removal of livestock grazing for 20 years has not resulted in increased grass cover (Asner *et al.*, 2004). The apparent stability of vegetation following the cessation of livestock grazing has led to the hypothesis that arid grasslands exist in one of two alternate stable states: grassland or desertified shrubland. While the conversion to shrubland can occur rather rapidly following intense overgrazing, the recovery of perennial grasses is often presumed to be difficult or impossible, even with livestock removal. Recent research (Asner *et al.*, 2004; Valone and Sauter, 2005), however, has suggested there may be time lags of decades in the response of perennial grasses to removal of livestock in historical grassland ecosystems dominated by shrubs. This means that recovery of perennial grasses at severely desertified sites is possible, but require a long enough timescale.

4.2.4 Perennial shrublands

The shrubby communities

Scrubs in China can be divided into two groups: the primary one that is found in alpine and subalpine areas, saline lands, strands and deserts, and the secondary one that is formed after deforestation. In northern and north-western China, temperate deciduous scrubs such as *Caragana* spp., *Salix* spp. and *Artemisia* spp. are commonly found on semi-arid sand dunes, while *Tamarix* spp. are found on arid saline soil.

There are diverse shrubby communities distributed on various desert ecosystems in north-west China. Naturally occurring shrubs, such as those in Alashan (western Inner Mongolia), Ganzhou and Subei (Hexi Corridor, Gansu), have an important role to play. Many areas have useful (palatable) shrubs that react to grazing in various ways (Fig. 4.1.).

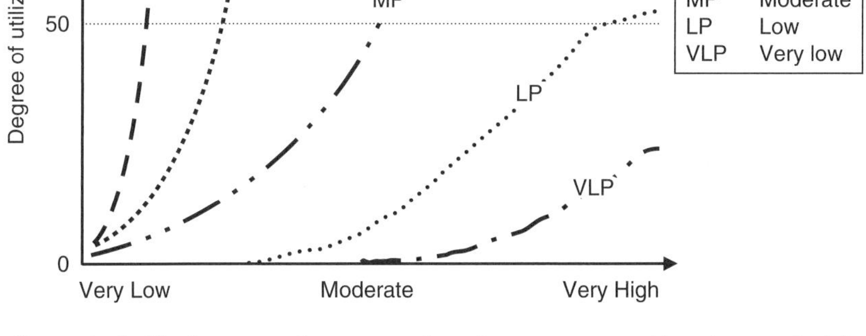

Fig. 4.1. Conceptual utilization curves for a range of grazing pressures and five classes of relative palatability of perennial shrubs.

Dwarf semi-shrubs are found on hills and gobi plains with an extremely arid climate. Only a few low-growing xerohalophytic plants such as *Sympegma* spp. and *Iljinia* spp. occur sparsely in the cracks of rocks on hills, which are usually devoid of vegetation. Dwarf hairy subshrubs mixed with ephemeral forbs are seen on loess-like and salt-free soils, occurring on the lower slopes of mountains in the Junggar Basin, Xinjiang Province. Succulent dwarf subshrubs, dominated by halophytes such as *Kalidium* spp., *Halostachys belangeriana*, *Halocnemum strobilaceum* and *Suaeda* spp., are distributed extensively on saline soils in arid regions. Shrubs and subshrubs like *Artemisia* spp. and *Caragana* spp. are widespread on the sand dunes and *Ephedra przewalskii*, *Zygophyllum* spp., *Nitraria spharerocarpa* and *Calligonum* spp. are very common on the gobi deserts. *Haloxylon ammodendron* and *H. persicum* are representative plants of the semi-arboreal sandy deserts. The former, growing on the bottom of sand dunes, is a halophyte. The latter, a non-saline plant, is limited to the slopes or ridges of stabilized or semi-stabilized sand dunes in the Junggar Basin.

The alpine dwarf-shrubby tundra is seen locally on the arid soil of the summits of high mountains, in the north-east of the temperate zone, and it consists of dense evergreen shrubs, belonging mainly to the Arctic species of *Ericaceae*. Alpine broadleaf semi-sclerophyllous thickets, consisting largely of *Rhododendron* spp. mixed with *Sinarundinaria* spp., and alpine broadleaf deciduous thickets composed mainly of *Salix* spp., *C. jubata* and *P. fruticosa* are distributed on the high mountains of the eastern Qinghai–Tibetan Plateau. The alpine deciduous creeping subshrubs mixed with herbaceous plants occur principally on the summits of mountains in the north-western part of the plateau. The alpine creeping dwarf semi-shrubs dominated by *Ceratoides compacta* are found on sand-gravelly deserts in the north-western part of the Qinghai–Tibetan Plateau.

Invasion of shrubs into arid and semi-arid grasslands

In arid and semi-arid regions, species interactions are important factors structuring the diversity of plant communities (Aguiar and Sala, 1994; Holzapfel and Mahall, 1999; Tielborger and Kadmon, 2000; Hochstrasser and Peters, 2004). Therefore, in arid land ecosystems, invasion or extinction of facilitating or competing dominant plant species may have a particularly strong effect on various measures of the temporal stability of subdominant plant species. Community instability or compositional community instability can be defined as the gain or loss of species or changes in species abundances that result in large directional changes in community composition and diversity (Collins, 2000). Therefore, numerous hypotheses and conceptual models dealing with rangeland desertification or degradation processes recognize that the invasion of shrubs in grasslands is the most striking feature of the variation of vegetation patterns in grassland degradation or desertification processes in arid and semi-arid regions. The invasion of shrubs in grasslands may increase the heterogeneity of the temporal and spatial distribution of primary vegetation and soil resources. As a result, the biological processes of the soil–vegetation system are concentrated increasingly in the 'fertile islands' under shrub canopies and the soil between shrubs turns gradually into a bare area or moving sand under the influences of prolonged wind and water erosion (Ludwig *et al.*, 2005).

Research on the Qinghai–Tibetan Plateau (Li *et al.*, 2006) indicated that, with the development of desertification, there was a significant degradation in the physical properties of the soil: a remarkable decrease in the content of silt and silt clay, yet an increase in the content of sands in the soil. This resulted in a decrease in soil water-holding capacity. Soil organic matter is also reduced with desert development, leading to destruction of the stability of the physical structure of the soil and the loss of nutrients from surface and subsoil layers. In response to changes in soil properties, the species composition, diversity and abundance, community structure and plant life forms were also changed. Consequently, with desert development, herbaceous species, especially grasses, were lost from the community composition and replaced by xerophytic shrubs or semi-shrubs. Finally, psammophytic annual plants dominated the vegetation composition, while shrubs were maintained at a low coverage.

The role of shrubs in combating desertification

Although the replacement of grasses by shrubs is generally considered as a process of rangeland degradation, the shrub vegetation invading patches of bare ground can serve as starting points for the restoration of degraded or desertified rangelands in arid areas and can be used for desertification control and ecosystem restoration. This has been clearly demonstrated by the success in combating desertification at Shapotou, north-western China (Li, 2005). Nearly 50 years of succession of artificial sand-binding vegetation at the site resulted in a reversal of desertification. In this process, the establishment of the artificial vegetation begins with installing sand barriers and planting xerophytic shrubs under a condition of less than 200 mm annual precipitation (no irrigation). Then a spatial heterogeneity develops.

Redistribution of precipitation by the canopy of xerophytic shrubs led to litter accumulation and cryptogamic crust development and accelerated soil-forming processes under the shrub canopies. This created a favourable condition for the invasion and colonization of annual and perennial plant species. With the depletion of soil moisture in the deep layer in the sand stabilization area, the cover of shrubs in the sand-stabilizing vegetation declined and the dominance of shrubs in the communities decreased and disappeared gradually from the vegetation composition. This, in turn, reduced the spatial heterogeneity of soil nutrient distribution. The propagation of numerous cryptogams on fixed sand surface and the colonization of annual and perennial grasses (increasers) further promoted the succession and restoration of the vegetation towards grass-dominated vegetation, which was similar to the primary vegetation types of the adjacent steppe-desert and desert-steppe.

4.2.5 Invasion by toxic plants

Rangelands in China have had domestic stock grazing for thousands of years and many of them have been overgrazed in the past decades. As a result, the plant composition has changed greatly from the original ecosystems. Many rangelands have been degraded severely to states that are susceptible to invasion by harmful weeds and toxic plants. The invasive weedy or toxic plants impact livestock production by lowering the yield and quality of forage, interfering with grazing, poisoning animals, increasing the costs of managing and producing livestock and reducing land value. They also impact wildlife habitat and forage, deplete soil and water resources and reduce plant and animal diversity.

Overgrazing and degradation of alpine rangelands on the Qinghai–Tibetan Plateau is always associated with an invasion by toxic plants: the more severe the overgrazing, the greater such an invasion (Long, 2003). The toxic plants commonly found include: *Stellera chamaejasme*, *A. inebrians*, *Aconitum szechenyianum*, *A. rotundifolium* and some seasonally toxic herbage such as *Ranunculus* spp., *Pxytropis* spp., *Gentiana* spp., *Pedicularis* spp. and *Senecio* spp. The seasonally toxic plants are avoided by animals during the growing season but are grazed when dry. Species of *Senecio* were found to be the predominant toxic plants causing the death of large numbers of yak on overgrazed pastures in Bhutan (Winter *et al.*, 1992, 1994).

Locoweed is one of the most common poisonous plants in the world and also is the most important toxic plant in China's grassland. The distribution range of the locoweed is extending continuously and it has become, or is becoming, a dominant species in some areas, reducing grassland productivity and causing poisoning or death of grazing animals on rangelands (Li, 1993, 2003). Locoweeds seriously threaten the grassland and livestock farming.

There has been a long history of introduction of non-native species, especially those with perceived beneficial impacts. With rapid economic development, including an explosive growth in international trade and transportation, there has been increased potential for new introductions of alien plants. Currently, alien species are widespread in the country; occur in many ecosystems; represent most major taxonomic groups; and are introduced unintentionally, as well as intentionally, for cultivation (Xie *et al.*, 2000). According to an initial preliminary survey and calculation conducted in 1995 on species of plants in the habitats on farmlands, pasturelands and waters, at least more than 58 exotic plant species brought about damage in agriculture and forestry in China (Ding and Xie, 1996; Ding and Wang, 1998). The report of a survey by Qiang and Chao (2000) indicated that there were 108 species of weed identified as exotic weeds and 15 of them were believed to be countrywide or regional exotic weeds.

4.2.6 Interactions between grazing and rangeland pests

An indirect associated cause of rangeland degradation is the high population of pest species in some areas of the northern and western pastoral rangeland regions. At present, the rangeland area infested annually with pest species is reported to be 40 Mha. Rodents and rabbits infest 30 Mha and insect pest species infest 10 Mha. These small herbivores not only consume substantial forage in competition with livestock herbivores but, as ground-burrowing animals, induce loss of soil structure and facilitate soil erosion in areas with high-density populations.

Rodents

There is often a rapid increase in the number of rodents following rangeland degradation, which is, in turn, accelerated by increased damage by rodents through consuming both aerial biomass and the roots of plants. Rodents also dig up much soil, which covers the surface of nearby swards. Pika (*Ochotona curzoniae*) and Chinese zokor (*Myospalax fontanierii*) have been recognized as the most damaging rodents that invade and destroy degraded meadows, but the alpine steppes and deserts are rarely attacked by these small animals. Pikas are active during the day and *M. baileyi* at night. The density of the pikas tends to increase on the alpine meadow in line with increasing degrees of sward degradation. However, the largest numbers of pikas (up to 150/ha) are also found on medium-degraded meadow (Long, 2003).

The case study by Wei *et al.* (2007) gave an example of the interaction between rodents and degradation on a pasture in the Naqu County of Tibet. The mounds and pits, as a result of erosion induced by burrowing rodents, covered up to 7% of the total area. *Lancea tibetica*, *Lamiophlomis rotata* and *P. bifurca* were the dominant species in eroded pits and *Kobresia pygmaea*, the dominant species at the undisturbed sites, became a companion species at the eroded sites. In the process of erosion, the original vegetation was covered by soil ejected by the pika, then the mounds were gradually eroded by wind and rain and, finally, erosion pits formed. The proportion of the plants with feeding value was more than 94% in the rodent-infected area, much higher than that in the uninfected area.

Despite the fact that plateau pika populations may reach high densities, and correspondingly reduce forage for domestic livestock (yak, sheep, horses), and they may be responsible for habitat degradation, some researchers (Smith and Foggin, 1999) argue that the plateau pika is a keystone species because it: (i) makes burrows that are the primary homes to a wide variety of small birds and lizards; (ii) creates microhabitat disturbance that results in an increase in plant species richness; (iii) serves as the principal prey for nearly all of the plateau's predator species; and (iv) contributes positively to ecosystem-level dynamics. The plateau pika should be managed in concert with other uses of the land to ensure the preservation of China's native biodiversity, as well as long-term sustainable use of the pastureland by domestic livestock.

Locusts

Locusts are potentially the most destructive pest insects in the world. The desert locust (*Schistocerca gregaria* Forskal) and the migratory locust (*Locusta migratoria* L.) are the two major species. According to a study by Tanaka and Zhu (2005), the outbreaks of the migratory locust in the Jiminay County of Xinjiang in north-west China caused a reduction in grassland production by 42%. In 2004, huge numbers of hatchlings appeared after thawing and grew in the grazing land and pastures. Little rain fell in the spring. On 22 May, hopper density reached 1500 individuals/m^2. At Beishawo, one of the most heavily infested areas in Jiminay County, approximately 20,000 ha of grazing land was infested with locusts and the highest density reached more than 10,000 individuals/m^2 for early-instar nymphs. Locust swarms also attacked farming and grazing lands. During the period from 17 July to 6 August 2004, the locust density was monitored seven times. During that period, it was 110 locusts/m^2 on average, with the highest density (3000 adults/m^2) recorded on 18 July.

The management of locusts requires a rapid response and a coordinated effort so that locust populations of increasing density can be found and controlled before they cause serious damage. One of the problems in controlling locusts in the Xinjiang Autonomous Region is that the area is huge and any spraying over a wide area by aircraft can cause serious harm to the livestock grazing there. Thus, spotting areas with high locust density becomes important, although it is practically impossible to spot all swarms and dense populations. Once spotted, however, such areas may be sprayed manually or by using vehicles.

4.3 Better Grazing Management: a Dream or an Economic and Ecological Imperative?

The ecological aspects of a grazed ecosystem are functionally constant, regardless of the sociocultural aspects of the human population interacting with it. Grazing management involves regulation of the process by which animals consume plants to acquire energy and nutrients, primarily through the manipulation of livestock, to meet specific, predetermined production goals. Both the grazing process and associated managerial activities occur within ecological systems and are therefore subject to an identical set of ecological principles which govern system function. These ecological principles impose an upper limit on animal production which cannot be overcome by management. The fact that both the grazing process and efforts to manage it are influenced by a common set of ecological principles justifies the evaluation of grazing management in an ecological context. Thus, regardless of how sophisticated the society, manipulation of temporal and spatial distribution and kinds and numbers of grazing animals is the only means by which humans can manage grazing land to achieve their desired goals.

However, a rapidly expanding human population, escalating degradation of natural resources and increasing socio-economic pressures have all increased the complexity associated with the management of grazed systems. In the past decades, the number of grazing animals in the pastoral regions of China has increased dramatically and has exceeded the carrying capacity of the rangeland ecosystem. As a result, rangeland degradation induced by overstocking has become a serious problem in most of the pastoral areas. In response to public concern about rangeland degradation, the Chinese government has made great efforts to solve the problems related to grassland degradation in the western region over the past decade. In 1999, the government launched the Western Region Development Programme, focusing on reducing economic gaps between the western and other regions and ensuring sustainable natural resources management. In line with these objectives, the government has made great investments into rangeland improvement programmes involving fencing, aerial sowing, rodent and pest control, forage seed production, livestock housing, settlement of the nomads, water supply and farming machinery. In 2003, the significant Grazing Ban Programme, involving the exclusion of grazing in certain months or year-round, was launched by the national government and has been undertaken in several north-western provinces where rangeland degradation is most severe.

Despite the positive impacts of these policies and technical interventions on ecological improvement of grasslands, the improvements have been observed sometimes to be at the cost of the socio-economic well-being of affected communities. Adverse impacts on affected communities led to further degradation of the grasslands because the herding communities were left with very few options except to continue to overexploit the depleted grasslands. Therefore, there is a need to understand better the ecological and socio-economic context of grassland ecosystems. On the one hand, policy and technical interventions have to respond to farmers'/herders' demands and to aim at empowering the capacity of the herders to enable innovations in their production systems based on their local farming and social resources and not be dependent on unsustainable subsidy and infrastructure. On the other hand, there is a need to limit the number of people who rely on grazing for their livelihoods if a balance between herders' livelihoods and rangeland health is to be achieved. This is not easy, as the minority nationalities are exempted from the one-child policy. Outmigration is one of the options, but its success depends much on alternative employment opportunities and the skills of the people affected. Perhaps one of the ultimate solutions to the problem is to increase the investment in education in the pastoral areas so that the younger generation of the herding families will have the opportunity to start a new career elsewhere, instead of continuing their father's job as herders (Li *et al.*, 2008).

References

Aguiar, M.R. and Sala, O.E. (1994) Competition, facilitation, seed distribution and the origin of patches in a Patagonian steppe. *Oikos* 70, 26–34.

Asner, G.P., Elmore, A.J., Olander, L.P., Martin, R.E. and Harris, A.T. (2004) Grazing systems, ecosystem responses, and global change. *Annual Review of Environment and Resources* 29, 261–299.

Cao, S.X. (2008) Why large-scale afforestation efforts in China have failed to solve the desertification problem. *Environmental Science and Technology* 42(6),1826–1831.

Cao, S., Li, C., Xu, C. and Liu, Z. (2007) Impact of three soil types on afforestation in China's Loess Plateau: growth and survival of six tree species and their effects on soil properties. *Landscape and Urban Planning* 83, 208–217.

Cingolani, A.M., Cabido, M.R., Renison, D. and Solis, N.V. (2003) Combined effects of environment and grazing on vegetation structure in Argentine granite grasslands. *Journal of Vegetation Science* 14, 223–232.

Collins, S.L. (2000) Disturbance frequency and community stability in native tallgrass prairie. *The American Naturalist* 155, 311–325.

Conner, J.R., Hamilton, W.T., Kreuter, U.P., Sheehy, D.P., Simpson, J.R. and Stuth, J.W. (1996) *Environmental Impact Assessment of Livestock Production in Grassland and Mixed Rainfed Systems in Temperate, Humid and Subhumid Tropic and Subtropic Zones (except Africa)*, Volume 1. FAO, Rome (www.fao.org/WAIRDOCS/LEAD/X6117E/X6117E00.HTM, accessed 12 May 2008).

Ding, J.Q. and Wang, R. (1998) Impact of exotic species on biodiversity in China. In: *Report on National Condition in Biodiversity in China*. Environmental Science Press, Beijing, pp. 58–61.

Ding, J.Q. and Xie, Y. (1996) The mechanism of biological invasion and the management strategy. In: Schei, P.J., Sung, W. and Yan, X. (eds) *Conserving China's Biodiversity (II)*. China Environmental Science Press, Beijing, pp. 125–156.

Ding, Y., Ren, G., Zhao, Z., Xu, Y., Luo, Y., Li, Q. and Zhang, J. (2007) Detection, causes and projection of climate change over China: an overview of recent progress. *Advances in Atmospheric Sciences* 24, 954–971 (in Chinese).

Dong, Q., Ma, Y., Li, Q., Zhao, X., Wang, Q. and Shi, J. (2005) Effects of stocking rates of yak on community composition and plant diversity in *Kobresia parva* alpine meadow warm-season pasture. *Acta Botanica Boreali-Occidentalia Sinica* 25, 94–102 (in Chinese).

Friedel, M.H., Sparrow, A.D., Kinloch, J.E. and Tongway, D.J. (2003) Degradation and recovery processes in arid grazing lands of central Australia. Part 2: Vegetation. *Journal of Arid Environments* 55, 327–348.

Guo, Z.G., Wang, S.M., Liang, T.G. and Zhang, Z.H. (2004) Preliminary probe into the classification management for grassland resources. *Acta Prataculturae Sinica* 13, 1–6 (in Chinese).

Guo, Z., Liang, T., Liu, X. and Niu, F. (2006) A new approach to grassland management for the arid Aletai region in Northern China. *The Rangeland Journal* 28, 97–104.

Hochstrasser, T. and Peters, D.P.C. (2004) Subdominant species distribution in microsites around two life forms at a desert grassland–shrubland transition zone. *Journal of Vegetation Science* 15, 615–622.

Holzapfel, C. and Mahall, B.E. (1999) Bidirectional facilitation and interference between shrubs and annuals in the Mohave Desert. *Ecology* 80, 1747–1761.

Hu, Z.Z. (1997) *Panorama on Grassland Classification*. China Agricultural Press, Beijing (in Chinese).

Hu, Z. and Zhang, D. (2001) *Country Pasture/Forage Resource Profiles: China*. FAO, Rome (www.fao.org/ag/AGP/AGPC/doc/Counprof/china/china1.htm, accessed 12 May 2008).

Huang, X.Z., Hao, Y.B., Wang, Y.F., Zhou, X.Q., Han, X. and He, J.J. (2006) Impact of extreme drought on net ecosystem exchange from *Leymus chinensis* steppe in Xilin River Basin, China. *Journal of Plant Ecology* 30, 894–900 (in Chinese).

Jiang, X., Zhang, W. and Yang, Z. (2003) The influence of disturbance on community structure and plant diversity of alpine meadow. *Acta Botanica Boreali-Occidentalia Sinica* 23(9), 1479–1485 (in Chinese).

Jiao, S., Han, G., Li, Y. and Dou, H. (2006) Effects of different stocking rates on the structures and functional group productivity of the communities in desert steppe. *Acta Botanica Boreali-Occidentalia Sinica* 26, 0564–0571 (in Chinese).

Jin, R., Li, X., Che, T., Wu, L. and Mool, P. (2005) Glacier area changes in the Pumqu river basin, Tibetan Plateau, between the 1970s and 2001. *Journal of Glaciology* 51(175), 607–610.

Li, J. (1993) Locoweed poisoning. In: Wang, J. and Li, J. (eds) *Poisoning Disease and Toxicology in Animals*. Tian Ze Press, Shaanxi, China, pp. 80–83 (in Chinese).

Li, J. (2003) The present situation and prospect of studies on locoweed in China. *Scientia Agricultura Sinica* 36, 1091–1099.
Li, W. (2004) Degradation and restoration of forest ecosystems in China. *Forest Ecology and Management* 201, 33–41.
Li, X., He, F. and Wan, L. (2007) A review of China's institutional arrangements for rangeland management. In: Li, X., Wilks, A. and Yan, Z. (eds) *Rangeland Co-management. Proceedings of International Workshop, Diqing, Yunnan, China, 13–15 May 2006.* China Agricultural Science and Technology Press, Beijing (in Chinese).
Li, X.L., Yuan, Q.H., Wan, L.Q. and He, F. (2008) Perspectives on livestock production systems in China. *The Rangeland Journal* 30, 211–220.
Li, X.R. (2005) Influence of variation of soil spatial heterogeneity on vegetation restoration. *Science in China* 48(11), 2020–2031.
Li, X.R., Jia, X.H. and Dong, G.R. (2006) Influence of desertification on vegetation pattern variations in the cold semi-arid grasslands of Qinghai–Tibet Plateau, north-west China. *Journal of Arid Environments* 64(3), 505–522.
Liao, G.F. and Jia, Y.L. (1996) *China Grassland Resources.* China Science and Technology Press, Beijing (in Chinese).
Liu, X. (2008) Comment on 'Why large-scale afforestation efforts in China have failed to solve the desertification problem'. *Environmental Science and Technology,* ASAP Article, 10.1021/es801718q, Web release date: 23 September 2008.
Liu, Z.G. and Li, Z.Q. (2006) Plant biodiversity of *Artemisia frigida* communities on degraded grasslands under different grazing intensities after thirteen year enclosure. *Acta Ecologica Sinica* 26(2), 475–482 (in Chinese).
Long, R.J. (2003) Alpine rangeland ecosystems and their management in the Qinghai–Tibetan Plateau. In: Wiener, G., Jianlin, H. and Riuijin, L. (eds) *The Yak,* 2nd edn. Regional Office for Asia and the Pacific, Food and Agriculture Organization (FAO) of the United Nations, Bangkok.
Ludwig, J.A., Wilcox, B.P., Breshears, D.D., Tongway, D.J. and Imeson, A.C. (2005) Vegetation patches and runoff-erosion as interacting ecohydrological processes in semi-arid landscapes. *Ecology* 86(2), 288–297.
Miller, D.J. and Craig, S.R. (1997) *Rangelands and Pastoral Development in the Hindu Kush-Himalayas.* International Centre for Integrated Mountain Development (ICIMOD), Kathmandu.
Pang, H.X., He, Y. and Zhang, N. (2007) Correspondence: Accelerating glacier retreat on Yulong mountain, Tibetan Plateau, since the late 1990s. *Journal of Glaciology* 53(181), 317–319.
Qian, W.H. and Zhu, Y.F. (2001) Climate change in China from 1880 to 1998 and its impact on the environmental condition. *Climatic Change* 50(4), 419–444 (in Chinese).
Qiang, S. and Chao, W.Z. (2000) Survey and analysis of exotic weed. *Acta of Plant Resource and Environment* 9(4), 34–38.
Ren, J., Wang, Q., Mou, X., Hu, Z., Fu, Y. and Sun, J. (1978). The production flow of the grassland and seasonal livestock farming. *Scientia Agricultura Sinica* 2, 87–92 (in Chinese).
SFA (State Forestry Administration of China) (2006) *China Forestry Yearbook.* China Forestry Press, Beijing.
Shi, Y., Shen, Y., Kang, E., Li, D., Ding, Y., Zhang, G. and Hu, R. (2006) Recent and future climate change in northwest China. *Climatic Change* 80, 379–393.
Smith, A.T. and Foggin, J.M. (1999) The plateau pika (*Ochotona curzoniae*) is a keystone species for biodiversity on the Tibetan plateau. *Animal Conservation* 2, 235–240.
Sun, H. and Shi, Y. (1994) *Agricultural Natural Resources and Regional Development of China.* Jiangsu Science and Technology Press, Nanjing, China (in Chinese).
Tanaka, S. and Zhu, D.H. (2005) Outbreaks of the migratory locust *Locusta migratoria* (Orthoptera: Acrididae) and control in China. *Applied Entomology and Zoology* 40(2), 257–263.
Tielborger, K. and Kadmon, R. (2000) Temporal environmental variation tips the balance between facilitation and interference in desert plants. *Ecology* 81, 1544–1553.
Valone, T.J. and Sauter, P. (2005) Effects of long-term cattle exclosure on vegetation and rodents at a desertified arid grassland site. *Journal of Arid Environments* 61(1), 161–170.
Vitousek, P.M., Mooney, H.A., Lubchenco, J. and Melillo, J.M. (1997) Human domination of earth's ecosystems. *Science* 277, 494–499.
Wang, G., Liu, Q. and Zhou, S. (2003) Research advance of dried soil layer on Loess Plateau. *Journal of Soil and Water Conservation* 17(6), 156–169 (in Chinese).

Wang, M., Li, Y., Huang, R. and Li, Y. (2005) The effects of climate warming on the alpine vegetation of the Qinghai–Tibetan Plateau hinterland. *Acta Ecologica Sinica* 25, 1275–1281 (in Chinese).

Wang, S.P. and Li, Y.H. (1999) Degradation mechanism of typical grassland in Inner Mongolia. *Chinese Journal of Applied Ecology* 10(4), 437–441 (in Chinese).

Wang, W. (2000a) Analysis of the plant individual behavior during the degradation and restoring succession in steppe community. *Acta Phytoecologica Sinica* 24(3), 268–274 (in Chinese).

Wang, W. (2000b) Mechanism of degradation succession in *Leymus Chinensis* + *Stipa Grandis* steppe community. *Acta Phytoecologica Sinica* 24(4), 468–472.

Wardle, D.A., Bonner, K.I., Barker, G.M., Yeates, G.W., Nicholson, K.S., Bardgett, R.D., Watson, R.N. and Ghani, A. (1999) Plant removals in perennial grassland: vegetation dynamics, decomposers, soil biodiversity, and ecosystem properties. *Ecological Monographs* 69(4), 535–568.

Wei, X., Li, S., Yang, P. and Cheng, H. (2007) Soil erosion and vegetation succession in alpine *Kobresia* steppe meadow caused by plateau pika – a case study of Nagqu County, Tibet. *Chinese Geographical Science* 17, 75–81.

Winter, H., Seawright, A.A., Hrdlicka, J., Tshewang, U. and Gurung, B.J. (1992) Pyrrolizidine alkaloid poisoning of yaks (*Bos grunniens*) and confirmation by recovery of pyloric metabolites from formalin-fixed liver tissue. *Research in Veterinary Science* 52, 187–194.

Winter, H., Seawright, A.A., Noltie, H.J., Mattocks, A.R., Jukes, R., Wangdi, K. and Gurung, J.B. (1994) Pyrrolizidine alkaloid poisoning of yaks: identification of the plants involved. *The Veterinary Record* 134, 135–139.

Wu, S., Dai, E., Huang, M., Shao, X., Li, S. and Tao, B. (2007) Ecosystem vulnerability of China under B2 climate scenario in the 21st century. *Chinese Science Bulletin* 52, 1379–1386.

Xie, G.D., Zhang, Y.L. and Lu, C.X. (2001) Study on the valuation of grassland ecosystem services of China. *Journal of Natural Resources* 16, 47–53 (in Chinese).

Xie, Y., Li, Z.Y., Gregg, W.P. and Li, D.M. (2000) Invasive species in China – an overview. *Biodiversity and Conservation* 10(8), 1317–1341.

Xu, C., Sui, P., Xie, G. and Gao, W. (2006) Soil water effect and productivity in poplar and wheat–corn agroforestry systems. *Scientia Agricultura Sinica* 39, 758–763 (in Chinese).

Yan, Z. (2007) Concepts and procedures in co-management of rangeland resources. In: Li, X., Wilks, A. and Yan, Z. (eds) *Rangeland Co-management. Proceedings of International Workshop.* Diqing, Yunnan, China, 13–15 May 2006. China Agricultural Science and Technology Press, Beijing, pp. 5–17 (in Chinese).

Yang, L.M., Han, M. and Li, J.D. (2001) Plant diversity change in grassland communities along a grazing disturbance gradient in the north-east China transect. *Acta Phytoecologica Sinica* 25(1), 110–114 (in Chinese).

Yang, X., Zhang, K., Jia, B. and Ci, L. (2005) Desertification assessment in China: an overview. *Journal of Arid Environments* 63, 517–531.

Zhan, X., Li, L., Li, X. and Chen, W. (2005) Effects of grazing on the soil seed bank of a *Stipa krylovii* steppe community. *Acta Phytoecologica Sinica* 29, 747–752 (in Chinese).

Zhao, H., Zhao, X., Zhang, T. and Zhou, R. (2003) Study on damaging process of the vegetation under grazing stress in sandy grassland. *Acta Ecologica Sinica* 23, 1505–1511 (in Chinese).

5 The Mechanisms of Soil Erosion Processes by Wind and Water in Chinese Rangelands

Zhi-yu Zhou[1] and Bin Ma[2]

[1]*Lanzhou University, Lanzhou, China;* [2]*Zhejiang University, Hangzhou, China*

Synopsis

Erosion is a natural geological process; however, the development of pastoral production, the increase in population and especially the intemperate use of rangeland have accelerated soil erosion, which has become one of the environmental security problems confronting human beings, and has stimulated the degradation of rangeland soil, thus decreasing pastoral production. This chapter concerns mainly the mechanisms of soil erosion by wind and water, including natural and other factors influencing soil erosion processes, especially human activities. Finally, a review of available desertification control technologies in north China, including biological, engineering and chemical methods, is introduced.

Keywords: erosion control; rangeland conversion; rehabilitation methodology; recovery pathways; succession; environmental security; salinity; grazing impacts; bulk density; nutrients; soil organic matter

5.1 Introduction

Soil erosion indicates that soils, as well as their materials, are destroyed, dispersed, transported and precipitated by the influence of natural and artificial factors, such as water, wind, frost and gravity. But other factors are involved, such as a decrease of fertility, decline of physical and chemical properties and effects on land use and on ecosystem processes and functioning. Soil erosion is one of the most concerning problems worldwide, but it is especially worrying in China, where the area of eroded soil is 4.92 million km^2, or 51.2% of the total land area of China (Table 5.1).

Because of the long-term increase of population and the further development of livestock husbandry leading to overuse of rangelands, accelerated rangeland soil erosion has become a major environmental problem. Soil erosion also destroys the balance of the ecosystem and restricts sustained development of the economy; therefore, the problem of soil erosion is of major concern.

5.2 The Status of Rangeland Soil Erosion by Wind and Water

5.2.1 Soil erosion by wind

In China, because of the diversity of the soil, the complexity of the landscape and the instability of the climate, as well as the rapid development of livestock production, about 90% of rangeland has been degraded, especially by wind. For instance, a huge number of high-quality rangelands in Inner

Table 5.1. China's soil erosion area classified according to type and severity.

Degree of erosion	Total area		Water erosion		Wind erosion		Freeze/ thaw erosion	
	Area	%	Area	%	Area	%	Area	%
Light	254.03	51.59	91.91	51.23	94.11	50.16	68.01	54.23
Medium	135.05	27.30	49.78	27.75	27.87	14.86	57.40	45.77
Strong	47.62	9.67	24.46	13.63	23.17	12.35		
Very strong	25.76	5.23	9.14	5.09	16.62	8.86		
Extreme	29.96	6.08	4.12	2.30	25.84	13.77		
Upwards medium	238.41	48.41	87.51	48.77	93.50	49.84	57.40	45.77
Upwards light	492.44	100	179.42	100	187.61	100	125.41	100

Mongolia, Qinghai, Gansu and Sichuan Provinces have been degraded during the period 1970–2000 (see also the case studies in Chapters 7–14).

5.2.2 Soil erosion by water

Soil erosion by water is widespread in China, even in the most arid areas. Storm rains can always induce soil erosion by water in mountainous areas, highlands and on all sloping land where vegetation is degraded. At present, soil erosion by water in eastern and central China is being brought under control, but it is worse in western China. Some badly degraded rangelands, which are eroded by runoff from spring thaws, are of concern, especially in the cold arid regions such as the Qinghai–Tibet Plateau (Chapter 13).

5.3 The Mechanism of Soil Erosion

Soil erosion by both water and wind is a product of climatic factors, especially rainfall, which generally is considered a crucial factor. Geology and physiognomy determine the status of the erosion and deposition, local hydrological conditions and soil moisture influence the type and severity of the erosion and vegetation cover and artificial factors affect the process and rate of the soil erosion.

The processes of soil erosion impact on plant productivity in a variety of ways:

- Stripping of the surface soil to expose a less permeable and less fertile subsoil results in lower infiltration rates. Sheet erosion transports water, nutrients and organic matter out of the system more rapidly.
- Loss of soil from profiles in which nutrients are concentrated strongly in the surface results in a lower potential for plant growth.
- Invasion by less palatable (often toxic) plants, particularly woody plants, results in lower surface cover of grasses, reduced infiltration and greater loss of water and nutrients.

Wind erosion has been a major component of the degradation in each of the eight case studies. Dust and sandstorms provide the most spectacular, readily observed evidence of episodes of land degradation in terms of the transport of soil particles, the 'sandblasting' of vegetation and the burying of buildings and infrastructure The main effect of wind is insidious because there is 'winnowing' of the soil and any easily transported material (organic matter, clay, silt and fine sand) is removed.

The loss of available water (rainfall) and nutrients through increased runoff and associated soil loss amplifies the severity of drought and grazing, resulting in more bare ground and/or more ephemeral species. As the litter from ephemerals breaks down rapidly, the impact of subsequent droughts on animal nutrition is likely to be amplified, resulting in livestock mortalities or substantial forced destocking during these drought periods. The loss of high-infiltration microsites and the barriers to overland flow further reduce the productivity of the resource. Excessive surface soil loss often exposes less fertile and less permeable subsoil and sandblasting from windborne sands destroys remaining perennial plants.

5.3.1 Climate

Climate affects soil erosion primarily through rainfall. Soil erosion by water is prevalent in humid areas and soil erosion by wind is prevalent in arid areas. In China, the ecotone of soil erosion by wind and water is correlated with the regions where the annual precipitation is between 200 and 400 mm. Susceptibility to soil erosion changes with the season; for example, water erosion is common in summer when rainfall is more abundant and wind erosion is dominant in winter when rainfall is uncommon and wind velocity increases. Climate conditions are key factors influencing soil salinization, in which precipitation and soil evaporation have a close relation to soil salinization. The majority of halosols are distributed in arid and semi-arid areas and the sea shore area in the northern part of China, which coordinate with the climate conditions.

The majority of regions north of the Yangtze River are subhumid, semi-arid and arid areas, where the ratio of precipitation and evaporation is less than 1. The general trend of the movement of soil water is upwards to the soil surface. Soluble salts are also brought upwards and the water then evaporates, leaving the salts on the soil surface; the long-term accumulation and condensation of salts form salinized soil. In summary, the more arid is the climate, the stronger is the evaporation and the more severe is the salt accumulation. In the arid north-western area of China, evaporation can be more than ten times the precipitation; soil evaporation is absolutely dominant and hence large areas of inland salinized soil are formed.

5.3.2 Geology and physiognomy

The influence of landscape and soil properties (e.g. texture and structure) on the rate of soil erosion varies across China's vast rangelands. Water erosion is particularly serious on the loess plateau, where the soil texture is loose, and in arid or semi-arid sandy rangelands. Along a south-east to north-west axis that stretches over several degrees of latitude, there is a marked shift in the major types of soil erosion. Water is the primary agent in the south-east, but wind dominates in the north-west, while a wind–water ecotone is located in the transitional region.

5.3.3 Precipitation and thaw

The soil erosion caused by rainfall and snowfall, two forms of precipitation, is entirely distinct. Rainfall disaggregates soils by raindrop impact, eroding and conveying the soils in the runoff. Snowfall erodes and conveys soils by water from melting snow. Furthermore, the melting process induces changes in soil attributes that influence the resistance to erosion. Rainfall erosion can take place in most degraded rangelands, but thaw erosion is restricted to mountainous areas in western China and to degraded rangeland in the north-east. For instance, on some degraded alpine meadowland of the Qinghai–Tibet Plateau, many rills and small gullies can appear on slopes where the gradient is greater than 10°. These result from the action of thaw water. The average length and depth of these rills and gullies can reach 13.32 cm and 6.57 cm, respectively, but the deepest can be more than 1 m. These rills expose subsoil and become larger through erosion by wind. The lost soil takes with it organic matter and many nutrient elements, making recovery difficult. In early spring, because the temperature fluctuates around freezing point, freezing and thawing can occur over and over again. These processes can change the physical property of the soil, e.g. the stability of aggregates, water conduction and resistance to erosion and shearing force, thus affecting soil erosivity.

5.3.4 Human action

In historical times, the pastoral regions of China were excellent rangelands, with fertile soils and flourishing vegetation. But, because of the high stress imposed by the incursion of crop agriculture, including the creation of artificial oases, the sharp increases in population of both humans and livestock and industrial development on the rangelands, the soils in the rangelands have degraded. Soil structure and fertility levels have been changed significantly by wind and water. The texture of surface soils, composed primarily of sandy components, provides a physical basis for desertification. Based on these potential factors, any human activities and excessive grazing by livestock can accelerate changes in the physical status of the soil,

Box 5.1. Sandy blowouts – a phenomenon of sand land in north China.

A sandy grassland blowout consists of a depression (blowout in a narrow sense of the word) formed by wind erosion of the topsoil, and the pre-existing sand deposit underlying it, and its adjoining aeolian deposition of sand derived from the depression. A blowout is oval or somewhat fan-shaped, with its long axis consistent with the direction of the prevailing wind. The rims of the sidewalls of a blowout depression are vertical ruptures, which are formed by the topsoil layer along the pre-existing sand wedges breaking and falling off, due to basal undermining. The remaining sidewalls form slopes. The bottom of a blowout is usually part of the surface of an ellipsoid, which becomes flat when downward erosion by wind is restricted by moist sand or by silt soil or a clay interlayer in the underlying sands. Sand from the depression is transported leeward and deposited around the depression, covering grassland vegetation and forming a low parabolic dune with an area up to eight times that of the depression.

inducing the development of soil erosion by wind and water (Zhao, 2004).

The dynamics of dust emission and deposition in the grasslands of Inner Mongolia were investigated by Hoffmann *et al.* (2008) in both grazed and ungrazed plots. Both processes are determined by the intensity of grazing, but dust deposition rates are modified additionally by the topography. Evidence of dust emission was found at all grazed sites (up to 0.8 g/m^2/day) while ungrazed sites seemed well protected. The dust deposition rates on grazed and ungrazed sites were on average 1.3 and 2.4 g/m^2/day, respectively. Leeward slopes had 29–33% higher deposition rates than windward slopes, summits and plain positions.

Rangeland conversion and wind erosion

Wind speed affects soil erosion so that, when the speed exceeds 5–6 m/s (Beaufort scale 4), small particles (0.05–0.1 mm in diameter) start to move. When the wind speed reaches 9 m/s (grade 6), small particles are lifted and become airborne and large particles (0.5–2 mm in diameter) begin to glide. When small particles are blown away, sand is left and the soil becomes sandy, especially in the rangeland converted to cropland, in which there is an abundance of fine particles that would be blown into the air by hurricane-force winds (grade 8). As a result, there are two pathways to degradation of the soil. First, the dunes at the edge of the desert expand. Secondly, the fine particles in non-sandy soils are blown into the air and sand is left.

When rangeland is changed to cropland, soil can be eroded easily, especially after harvest. In spring, because of the increase in temperature differentials between the soil surface and the air, lack of precipitation, strong winds and the freezing/thawing processes, the soil is extremely dry and loose. Such soil is eroded easily, especially in Shanxi, Ningxia, Xinjiang and Inner Mongolia, which are arid or semi-arid regions. Sandy blowouts (wind-eroded holes) are a peculiar feature of sandy land in north-east China (Box 5.1).

Zhang *et al.*'s study (2007) showed that, of 187 blowouts in Inner Mongolia, 87% were induced by the human activities of growing plantations (35.8%) and building roads (34.8%) and housing (16.0%).

Erosion processes in fixed and semi-fixed dunes become activated when livestock move across the area and disturb the thin soil layer. Even on flat land, trampling by livestock can activate wind erosion. Furthermore, the digging up of medicinal plants also induces soil erosion by wind; for example, the holes that result from digging up *Glycyrrhiza guralensia* in Ningxia's desert are about 1 m^2 and are easily eroded by wind.

CRACKING OF THE SOIL LAYER. In arid regions, when the root layer is destroyed by natural erosion and artificial actions such as digging by both humans and animals and impaction, the land surface can form bare patches because the unprotected fine particles are blown away by strong winds. The sand content increases significantly as the finer particles are lost under the sifting action of wind. The soil that remains is more vulnerable to erosion. Then loose spots such as mouse pits can crack, baring non-felting sand. In addition to human activities and the erosion of runoffs, more and more cracks seem to be appearing.

BASAL SAPPING. The loose sands bared by wind erosion at the edges of cracks are blown out when the wind meets the crack edge, creating a low-pressure effect when the airflow comes into contact with the deeper sand layers, because the substrate layer with loose sands is eroded much more easily by wind than the upper layer with a clay, plant root and calcified layer. The substrate sands are eroded by basal sapping, causing the upper layer to lose its support and collapse (Fig. 5.1).

FORMATION OF WIND-ERODED PITS. When the substrate is eroded by the basal sapping effect, the surface layer with grasses cracks and falls under the action of gravity. Wind-eroded pits are formed when the fine particles are blown away and the sands are transferred.

The effects of basal sapping (Fig. 5.1) and gravity have accelerated wind-eroded cracking. As the depth of the pits grows, blowouts develop rapidly as the face area expands quickly and the amount and intensity of the sandy wind increases significantly, intensifying the erosion rate of the wind-eroded pits.

HORIZONTAL DEVELOPMENT OF WIND-ERODED PITS. The datum plane of active wind-erosion pits is the water table or clay interlayer, which is the limit of erosion. Because the datum plane prevents vertical erosion, wind erosion develops only on the horizontal level.

Rangeland conversion and water erosion

The infiltration rates of dryland, grassland and forest are 15–50 mm/h, 50–130 mm/h and 20–700 mm/h, respectively. Plants can prevent the erosion of soil and can transport water to the subsoil; leaves can stop or reduce the impact of raindrops crashing on to the soil, stems can prevent water flowing and roots can hold soils.

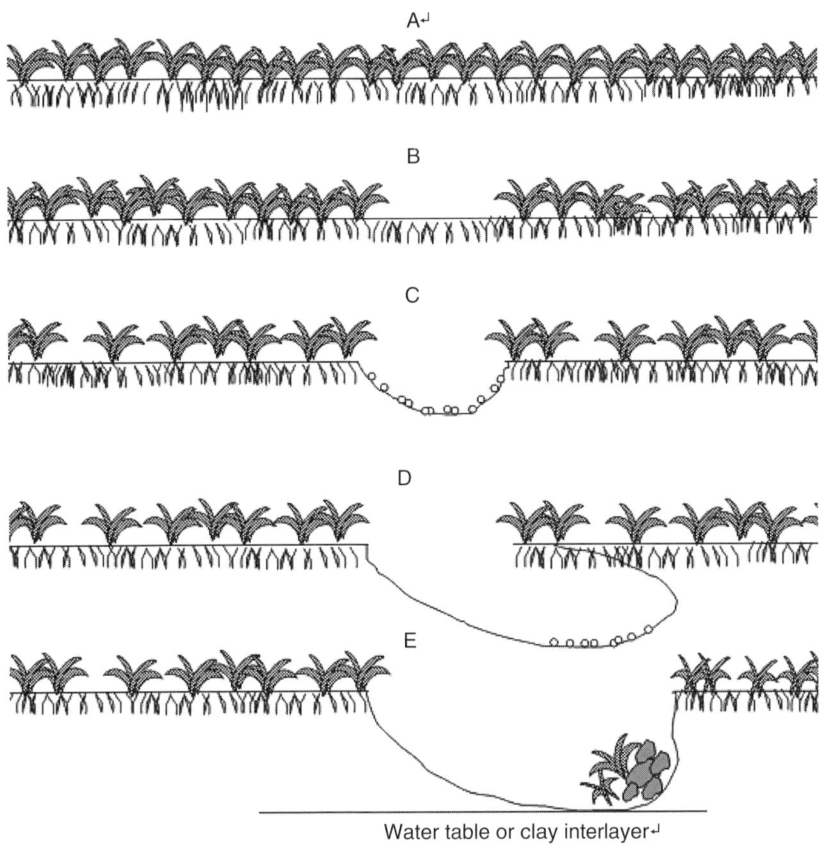

Fig. 5.1. The developing process of wind-erosion pits: A–C, cracking of the soil layer; D, basal sapping; E, the formation of the wind-eroded pit.

If we assign a value of 1 to the soil erosion of bare land, the value in a flourishing pasture is just 0.007; the soil erosion in cropland planted to soybean or maize, however, is 0.75. This indicates that retention or planting of grass can prevent soil erosion. Precipitation in pastoral regions in northern China, most of which is in the form of high-intensity storms, is concentrated from July to September. The inevitable runoff from overgrazed and poorly protected rangelands washes nutrients away and erodes the land surface into small gullies, which can grow into ravines. After land conversion, the effect is much worse.

Taking Suide, Anshai, Tianshui and Xifeng (Fig. 5.2) as examples, we can analyse the soil erosion in typical pastures on the loess plateau and gain a better understanding of the processes and consequences of soil erosion.

The annual eroding rainfall in Suide County, accounting for 25.7% of the rainfall during the flood season, is 129 mm. The eroding rainfall, with storms lasting for 1–6 h, delivers more than 20 mm of rain, with an intensity of greater than 5–12 mm/h. When the slope is less than 10°, the erosion is general non-point erosion; when it is greater than 10°, soil erosion increases with the slope, until it reaches 28°, the maximal erosion slope. Above 28°, erosion decreases with the slope. There is a positive correlation between the length of the slope and soil erosion.

Erosion on 20–60 m slopes, which are the most eroded regions, develop various erosion forms resulting from a mixture of the effects of hydraulics and gravity. Soil erosion by water develops from non-point erosion to rills and gullies. The land surface is chopped up, fragmented, cliffy and rugged. Soil erosion from the watershed to the bottom of a valley is distributed so that the soil erosion at the top of the hill is primarily non-point erosion; however, in the rills at the base of a hillside, it is mainly gully and gravity erosion.

The eroding rainfall events in Suide County during June to September, which account for 89.5% of the annual rainfall, induces 95.7% of the annual soil erosion. Because the time of the eroding rainfall is less but its intensity is strong, soil erosion is caused mostly by summer storms; for example, the soil erosion caused by a storm in August 1998 accounted for 99.3% of the annual soil erosion and 44.4% of the total soil erosion during 1995–1999. Soil erosion is correlated closely

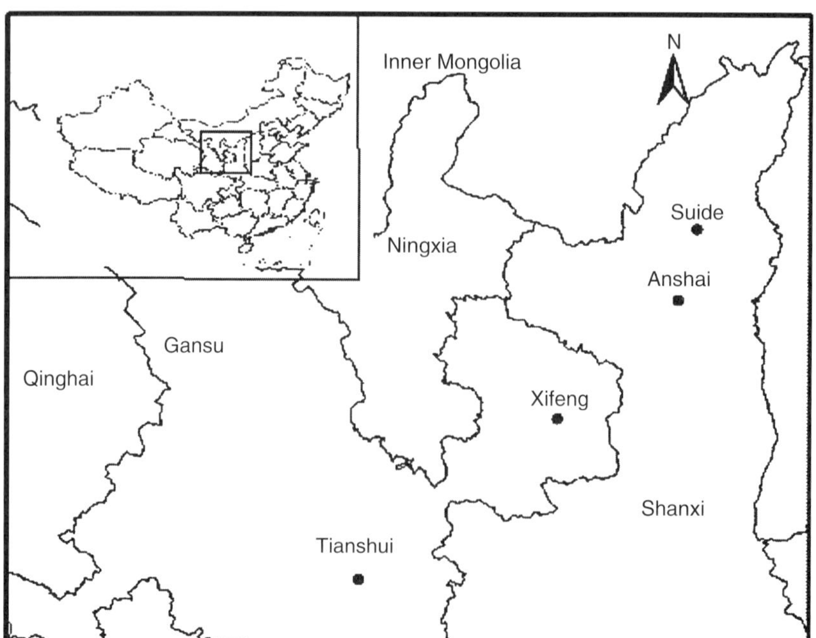

Fig. 5.2. The locations of Suide, Anshai, Tianshui and Xifeng.

Table 5.2. The correlation between slope and rate of erosion.

Slope	0–3°	3–5°	5–15°	15–25°	25–35°	>35°
Erosion rate (t/km^2)	0–1000	1000–1500	2000–3000	3000–8000	5000–10,000	>10,000

with slope; for instance, in the Zhifang River watershed in Zhejiang Province, the soil erosion rate varies with the degree of slope (Table 5.2).

Soil erosion of cropland is greater than that of grassland under the same conditions; for example, the soil erosion of cropland is 67.5% greater than that of wasteland, 92.5% greater than that of pasture and 97.9% greater than shrub-pasture. Rill erosion arises on sites where the slope is greater than 10°, being concentrated at the non-point regions. The width, depth and distance of rills are 12 cm, 8 cm and 87 cm, respectively, but the deepest rills can reach to 100 cm. Soil erosion increases with the length and gradient of the slope, unless the slope is greater than 20°. The erosion of rills accounts for 50–70% of the total soil erosion. Shallow grooves are distributed at the bottom of a slope, accounting for 75% of the total area of a valley. A shallow groove occurs on the surface of a 10–35° slope, especially on a 22–31° slope. The critical gradient of a slope which can give rise to shallow grooves is 18° and the critical length is 40 m. Shallow grooves account for 35% of the total erosion area. There is a logarithmic correlation between the erosion of a gully and the catchment area.

Gullies arise at the base of a valley where the slope is greater than 35° and the annual erosion intensity is 8373 t/km^2. The greatest annual erosion is 18,000 t/km^2 in rangeland where the slope is greater than 25° and the coverage is less than 10%. This land type accounts for 15% of the total area in the county. The second greatest annual erosion is 15,000 t/km^2 in the pastoral-forest zone where the slope is 15–25° and the coverage is 10–30%, accounting for 28% of the total area. The third greatest annual erosion is 8000 t/km^2 in pastoral-forest where the coverage is 30–50%, accounting for 20% of the total area. The soil erosion in other types of landscape is generally less than 4000 t/km^2, accounting for 37% of the total area.

The landscape in Tianshui County in Gansu is mainly ravines and highland, including 15% of tectonic hills and 65% of eroding hills. The soil types in this area include Spodosols, black loose and grey-cinnamon soils. The erosion resistance of the soil of Tianshui is the weakest. The natural vegetation has disappeared almost entirely, with a coverage of just 10%, and 60–70% of the rainfall is concentrated during May and September; sometimes, one rainfall event can deliver 40% of the annual precipitation. In pasture, soil erosion by wind at the top of hills and the stationary front of red soil is slight; forms of soil erosion by water are mainly leprose erosion (rough to the touch, covered with scales), layer erosion, rill erosion and sheet erosion. A study of the Luoyu watershed in Gansu showed that layer erosion accounted for 43% of the total erosion, 20% of erosion being in the watershed. Rill erosions and shallow grooves are also important erosion forms, accounting for 40% of the erosion of a slope. Gravity erosion, represented as landslips, happens mostly on the crags of ravines.

Soil erosion in the period from June to September accounts for 70–85% of erosion in any one year. The vertical distribution of soil erosion is: (i) slight erosion on the top of hills, including splashing down, layers, rills and shallow grooves, accounts for 55% of the total area of the watershed and 46% of the total erosion in the watershed; (ii) erosion on the middle slope, including tracing to the source, cutting, expanding and gravity erosion such as landslip, accounts for 34% of the total area of the watershed and 42% of the total erosion in the watershed; (iii) erosion at the base of grooves, presented as washout and deposition, accounts for 11% of the total area of the watershed and 12% of the total erosion in the watershed.

The eroding rainfall in Xifeng, Gansu (200–276 mm), accounts for 45–63% of the annual rainfall and 54–67% of the rainfall during June and September. Eroding rainfall occurs on average 11 times a year. The study of ^{137}Cs in the South River watershed indicated that:

- the most eroded regions were in the middle slope;
- erosion decreased with the length of the slope; and
- there was deposition at the base of the slope.

The landscape in this region is tableland, which is distinct from highland in that the stream is on the tableland but the bed load comes from the tributary gullies. There is 30% of tableland area, which contributes 67% of the flux.

5.3.5 The chemical mechanism of the erosion process by wind

The stability of soil aggregates

Soils form aggregates from the effects of organic and inorganic bonds. The stability of mineral soil aggregations is influenced by physical, chemical and biological factors. The types and concentrations of clayey mineral, as well as the contents and components of electrolytes, affect the stability of soil aggregations, which decrease as soil aggregations tend to decompose when intumescent soil touches water.

The effects on soil water conduction

A large number of studies show that the water conduction rate of soil decreases with the content of soil solution. High concentration of solution can reduce the adverse effects of sodium (Na) on soil water conduction capacity; the decreasing degree, however, is dependent on the relative concentration of sodium. Sodium ion is the main dispersing ion. Even if the proportion of Na^+ ions is very low among all the absorbed ions, the soil can still disperse and decrease the water conduction capacity.

Fine grains block the capillaries by moving with the water, decreasing the soil's water conduction capacity. In contrast, the distensibility of clay, caused by hydration, is continuous and can also decrease the soil's water conduction capacity. Accordingly, the wet–dry process of soil and a change of soil water content can also induce a reduction of the soil's water conduction capacity, increasing runoff on the surface. Moreover, the distensibility of clay increases with the decrease of electrolyte concentration.

Soil infiltration

The soil leakage rate indicates the volume of water entering the soil per unit area over a given time. When it is raining, the mechanical impact of raindrops on soil aggregates can lead to movement and sedimentation of the soil, forming a 2–3 mm compacted layer, which decreases the permeability and conduction of the soil, increases surface streams and ventilation of the soil and weakens the capacity to grow forage.

There are two mechanisms leading to the formation of a soil crust. (i) Surface soils are disintegrated and compacted by raindrops, which leads to surface sealing. The thickness of the surface seal layer, generally 0.1 mm, depends on the energy of the drops and is characterized as greater bulk density, anti-shear strength, low permeability and conduction. (ii) The physical–chemical dispersion of clay transports it to deep soil by the effusion of water, plugging up the capillaries, which leads to greater bulk density and lower infiltration rates. The chemical factor is dominant during this process. The more a surface seal layer is formed, the greater the intensity of the anti-shear strength of the soil.

Surface runoff and soil erosion

Surface stream and infiltration are two contrary processes: the lower the infiltration, the greater the surface stream. All the factors influencing infiltration and transport can affect the surface stream to some degree, but the primary factors are the form of crust on the soil surface and the characteristics of the rainfall.

There are two contrary effects of soil crust on soil erosion: (i) the surface seal layer increases the resistance capacity for shear strength and decreases the decomposing capacity of surface soil particles, increasing the resistance to erosion; (ii) the surface seal layer decreases infiltration and increases runoff, increasing erosion on the surface.

Soil salinity is one of the main factors influencing soil erosion. Soil erosion is correlated with clay mineral, exchangeable sodium percentage (ESP) and soil solution. Soil erosion increases with soil ESP; for example, when soil ESP increases from 4.6 to 19.3, soil erosion can rise 6.3-fold. Increasing the salinity of soil, by the application of gypsum, for example, can decrease soil erosion significantly. Calcium (Ca^+) from the gypsum replaces exchangeable Na^+, which is leached away. This increases aggregate stability and prevents deflocculation of clay particles.

The effect of magnesium

From the 1960s, magnesium (Mg) was thought to favour stabilization of the soil structure in the same way as calcium. Later studies, however, refute this conclusion because Mg^+ can also bear the characteristics of Na^+ under a low electrolyte content condition, bringing an adverse influence on the stability of the soil structure. Mg^+–Na^+ saturated soil aggregate is easier to disperse than Ca^+–Na^+ soil aggregate, resulting in lower soil water conduction and infiltration rates and greater runoff and erosion rates. The clay dispersion caused by exchanging Mg^+ is more than 5% of the influence of exchanging Na^+, but the effects of exchanging Mg^+ on calcic soil is not obvious. This may be related to the nature of the solution of calcic mineral (Keren, 1991; Curtin *et al.*, 1994).

There are two adverse influences of Ca^+ ion on the stability of soil. The first is a direct influence. The hydration of Mg is 50% greater than Ca^+. In the Mg^+ saturated double electron layer of clay particles, the Stern layer is thicker than the Ca^+ saturated system. The clay particles agglomerate by the charge of the outer surface, so the agglomeration in the Mg^+ system is easier to clash than in the Ca^+ system. The second is the indirect influence of the effects of Mg^+ on the accumulation of the exchangeable Na^+ ions. The absorption of clay to Ca^+ is stronger than to Mg^+; as a result, Mg^+ can induce more exchangeable ions adsorbed on to the clay particles. The influence of Mg^+ on the accumulation of exchangeable Na^+ is dependent on the type of clay mineral and soil organic matter content.

5.3.6 The effects of soil erosion by wind on nutrient elements

According to the second national soil erosion remote-sensing survey in 2000, the area affected by wind erosion was 1.91 million km^2, accounting for 20% of the total land area in China. This area is expanding quickly, as the incidence of heavy dust storms has increased greatly over the past five decades, mainly as a result of intensified soil cultivation. The economic and ecological damage caused by wind erosion is considerable. Heavily affected areas show a loss of nutrients and organic carbon in soils, and heavily degraded soils are much less productive.

Wind erosion is the primary power for nutrient transport. Wind erosion decreases the residual portion of fine components, organic matter and nutrition. The amount of soil moved by wind decreases with height above the terrain, but the proportion of K, Na, Ca, Mg and organic matter and the cation exchange capacity (CEC) of particles in soil moved by wind increases with the height but not with the intensity of the wind. Aeolian sandy soil in desert regions has been sorted and eroded by wind for a long time, so the organic matter, N, P, K and fine particles are lacking, but other mineral and trace elements components remain stable. Airborne dust from eroding surfaces is abundant in nutrients and wind deposition is more fertile than its parent material. The nutrients in airborne dust increase in proportion with the distance of transport. Furthermore, the soluble matter in airborne dust can affect the chemical properties of rain. If the dust deposition is greater than erosion, the nutrients increase in the soil, and vice versa.

5.3.7 The influence of water erosion on soil nutrients

Runoff has an important influence on the redistribution of soil nutrients in all rangelands. The runoff and eroded nutrients are greater when the coverage is low. Water erosion denudes surface land, taking away the surface layer with its abundant nutrients. The influence of water erosion on soil fertility is particularly strong in the agropastoral ecotone, for example, the erosion in Bashang is about 10 mm, and nutrient loss in the runoff in shrubby land is greater than that in grassland, for example, the N loss is 0.33 and 0.15 kg/Mha, respectively. The nutrient loss caused by the runoff in China equals 40 million t fertilizer. Sediment in the Yellow River has been washed down from the arid and semi-arid regions in the upstream and middle reaches of the river.

There is increased awareness of the environmental impacts of soil carbon (C) and nitrogen (N) losses through wind erosion, especially in areas heavily affected by dust storm erosion. Wang *et al.* (2006) have reviewed the recent literature concerning dust storm-related soil erosion and its impact on soil C and N losses in northern China. Compared with non-degraded soil, the C and N contents in degraded soils have declined by 66% and 73%, respectively. The estimated annual losses per cm top layer of soil C and N by dust storm erosion in northern China range from 53 to 1044 kg/ha and 5 to 90 kg/ha, respectively.

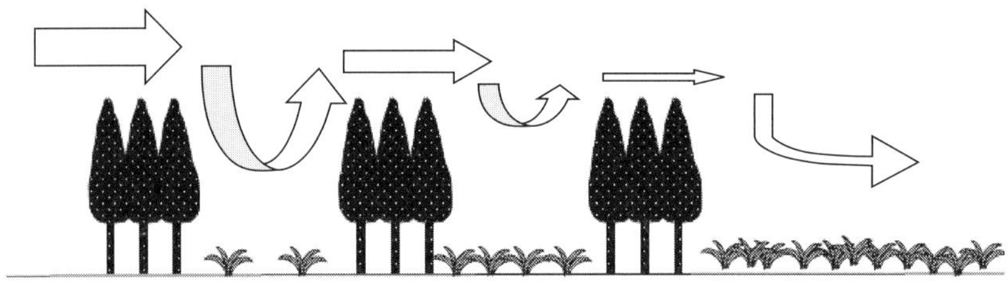

Fig. 5.3. The effect of windbreaks on decreasing the wind speed.

5.4 Control of Soil Erosion

5.4.1 Biological methods

Vegetation can protect the soil surface from wind erosion by covering the surface, slowing down the wind and blocking the sand. The main forms include planting windbreaks of trees and grass, increasing the vegetation and biological crust to protect the surface (Fig. 5.3).

Windbreaks

There are two mechanisms by which windbreaks decrease wind speed. The first mechanism is that windbreak forests drive up the airstream and decrease the surface wind speed. The second mechanism is by increasing the resistance. The distance, height, tendency, density and the arrangement of shrub and grass can influence the effect of the windbreak forest. Woodruff and Siddoway (1965) compared windbreak forests and found that the best effect came from windbreak forests of two to three rows. Zhu *et al.* (1986) and Zhu (1994) found that the effective downwind distance of windbreak forests was 25 times the height of the trees. Fullen and Mitchell (1994) have summarized much of the early work on the re-vegetation of sandy lands, sand dune fixation and other measures.

Coverage of vegetation

It has long been known that vegetation cover has a role in the prevention of erosion by wind and water. The vegetation can protect the soil surface by covering the surface, increasing the height of the coarse layer, decreasing the wind speed and blocking sand (Fig. 5.4). Dong (1999) studied the erosion of aeolian sand soil in a wind tunnel and found that soil erosion decreased significantly with the increase of vegetation; for example, 30% of coverage could reduce erosion sharply and 60% of coverage could prevent soil erosion by wind entirely. Similarly, the results of Huang and Niu (2001), based on a quantitative correlation model of vegetation cover and soil erosion by wind, indicated that 40–50% of vegetation coverage could reduce soil erosion effectively and 60% of vegetation coverage could prevent soil erosion by wind entirely. The challenge, though, is to achieve this level of coverage in nutrient-poor soils with poor texture and structure.

Biological crust

Biological soil crusts increase the resistance of soil to wind by concreting soil, increasing coarseness and coverage (Li *et al.*, 2003). Biological crust is a complex layer formed by soil and organisms, e.g. bacteria, fungi, algae, lichen and moss. The thickness of a biological crust is generally 1–10 cm. The interaction of the rhizoids of moss and lichen, the mycelia of fungi and the protonema of blue algae can fix the soil particles in the underground part of the crust. The biological crust increases the resistance to wind erosion, the height of the coarse layer and the coverage of the eroding particles through fixing the soil particles. This factor is generally included in most wind erosion models and forecasting systems as an important variable.

The structure of the physical crust is destroyed much more easily by collision with windborne particles carried by sandy wind and fine soil is eroded by the wind. The biological crust, however, is resistant to wind erosion. Studies of sandy soil show that the critical wind speed for removing sand in undisturbed biological crust is much greater than that on bare land. The

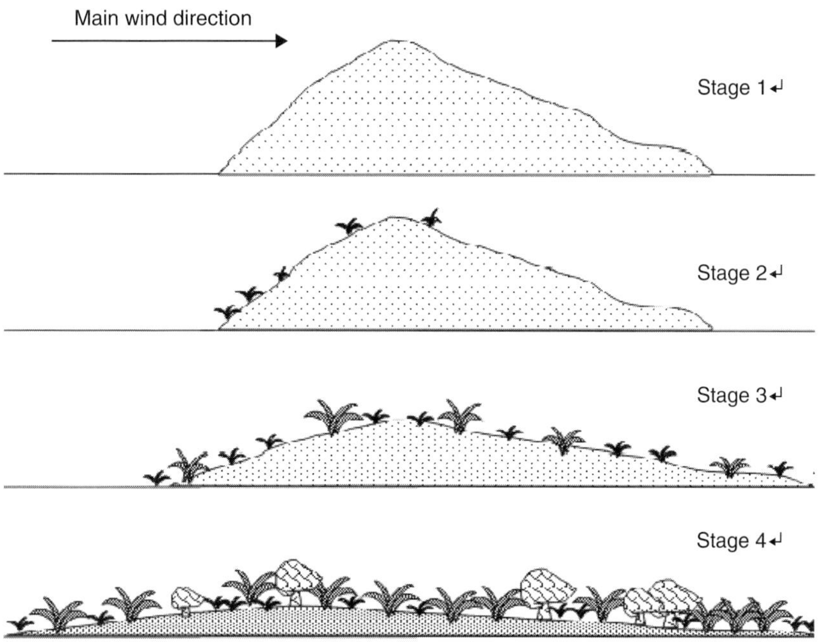

Fig. 5.4. The development of vegetation coverage on shifting dunes.

wind tunnel study of Wang *et al.* (2006) in the Guerbantongte Desert in Xinjiang shows that the influence of biological crust on wind erosion is strong, varying with the crust type and the degree of degradation. The threshold wind speed of moss crust is the highest, the next is lichen crust and the lowest are algae crust and algae–lichen crust. The wind erosion rates of the four crusts at the same wind speed are only 9% of bare land. The flexibility and intensity of crust depend on the resistance capacity to wind erosion but the biological crust is extremely weak in most pastoral rangelands, due to overgrazing and trampling. The resistance to wind erosion by destroyed crust drops sharply, generally nearly 100-fold, compared with undisturbed biological crust. The formation of biological crust is a long-term process, generally taking from several months to many years.

Cryptogamic crusts have long been regarded as important components of dryland ecosystems. In order to reduce and combat the hazards of sandstorms and desertification, it is critical to conserve cryptogamic crusts in arid desert and semi-arid regions. From studies in the Shapatou Research Centre in Ningxia, it can be concluded that algal cover and species richness are correlated positively with soil pH, contents of silt and clay, concentrations of HCO_3, Cl^-, SO_4^{2-}, Mg^{2+}, soil organic carbon and N contents. The number of species and cover of mosses were correlated positively with soluble K^+ and Na^+, but no other relationships were apparent. The percentage of sand in the composition of soil particle sizes and the soil bulk density were correlated negatively to species number and cover for both cryptogam organisms (Li *et al.*, 2003).

5.4.2 Chemical methods

Chemical fixation is an engineering measure involving the spreading of chemical reagents on to the soil surface for fixing mobile sand. The treated soil surface, which forms a protected or fixed layer, protects the soil surface by: (i) preventing erosion of the soil surface; and (ii) transferring sand along the surface, thus preventing its accumulation. Common materials include petroleum products, macromolecule materials and inorganic materials. The results of Dong (1999) showed that there was no erosion, even when the wind speed was as high as 24 m/s.

The functional principle of chemical sand stabilization is based on providing a stable but permeable surface on the sand by spraying diluted cohesive chemical materials on the shifting sand surface. After treatment, runoff and rainwater can infiltrate into the sand and soil layer quickly. However, these cohesive chemical materials will be kept in the sand soil layers and sand particles will be gelatinized together to form a hard preservation crust on the sand surface. Thus, air currents are separated from any loose sand surface and, consequently, the sand surface is protected from wind action. This method is designed to stabilize the shifting sands on semi-fixed dunes. It does not work on holding down and fixing the sand particles in wind-sand flows that are crossing the sand surface. The sand chemical stabilization techniques are popularized mainly to control severe sand disasters along communication lines and transport facilities in the seriously mobile sand areas and around military bases and mineral and industrial sites inside sandy deserts. They are rarely used in rangelands.

5.4.3 Engineering methods

Engineering methods, which increase the resistance of wind erosion through reducing wind speed, changing the structure of the soil and increasing the critical wind speed for sand moving, are a broadly effective approach to preventing wind erosion. This method is not restricted by natural conditions and is easy to implement. There are a huge number of methods and materials (Department of Science and Technology, 2002), but different methods and materials have different effects. Engineering methods also protect the soil by increasing the height of the coarse layer (surface roughness), decreasing the wind speed, reducing the flow of the sand, covering the surface and preventing the exposure of bare soil.

5.4.4 Soil cover using local materials

Covering the surface of the soil by locally available materials is an approach for avoiding further wind erosion. For instance, gravels are the most commonly used materials in north-west China. A study in Dunhuang, Gansu, showed that different gravels should be used on different sands and that the coverage should be greater than 65%.

Another popular approach is the grass chequerboard method, which was developed to protect the Baotou–Lanzhou railway in Shapotou (Fig. 5.5). The studies in Shapotou showed that the effect of the grass/straw chequerboard was best when the height of the grass/straw was 10–20 cm. The cost of the method is low as it uses local materials and labour. Chequerboards reduce by 95% the intensity of sand blown by the wind. The chequerboard stabilizes the dunes through accelerating soil formation, increasing the nutrient content, by the retention of fine particles and by forming a crust on the dune surface. However, the chequerboards need a lot of labour and need refreshing every 3–5 years (Table 5.3). The raw material is abundant everywhere and the ecological effect is significant in not only reducing wind erosion, but also increasing the nutrient content in the soil.

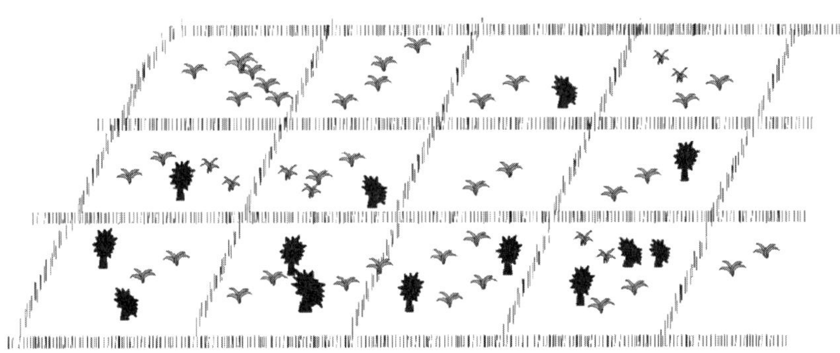

Fig. 5.5. The layout of a grass/straw chequerboard used to stabilize shifting sand. Pioneer plants can be established inside each grid square after the wind velocity is reduced by the 10–15 cm high barriers of straw.

Table 5.3. A review of the available desertification control technologies in north China.

No.	Technique/methods	Sites where applicable	Limitations/benefits	Relative cost effectiveness	Overall rating[a]
Biological methods					
1	Shelterbelt networks to protect farmland	Within farmland Along canal banks	– Only a few tree species suitable – Long-horned beetle damaged – High consumption of water – Good protection results – Making microclimate for crops – Supplying timber	– Relatively expensive – Simple management – Results in yield reduction in the marginal field	4 Effectiveness 4 Durability 4 Maintenance
2	Sand fixation forest for fixing mobile sand dunes	2/3 of leeward side of mobile dunes from bottom	– Hard for shrubs to survive – Labour demanding – Long life (20–40 years) – Fixing sand dunes	– Cheap – Relatively easy to maintain	4 Effectiveness 4 Durability 3 Maintenance
3	Windbreak forest	Between farmland and sand dunes	– Labour demanding – High consumption of water – Good ecological and economic benefits	– Relatively cheap – More effort to maintain	4 Effectiveness 4 Durability 2 Maintenance
4	Enclosure for grazing land and forest (grazing bans)	Desert grassland Forest area	– Increasing biodiversity – Few labour demands	– Cheap – Easy to maintain	4 Effectiveness 4 Maintenance
5	Aerial sowing for grazing land and afforestation	Loess plateau Desert grazing land	– Must have aircraft – Relatively high concentration of rainfall – Efficient for making grazing land and afforestation	– Cheap over large areas – Low labour cost	4 Effectiveness 4 Durability 3 Maintenance
6	Blocking in front and pulling from behind	Dune chains	– Labour demanding – Reduces sand blowing off dunes – Stabilizing mobile dunes	– Relatively expensive	4 Effectiveness 4 Durability 3 Maintenance
7	Grass kulum[b] to block wind and sand and to create pasture	Pasture land	– Labour demanding	– More effort to maintain	4 Effectiveness 3 Durability 2 Maintenance
8	Integrated management of small watershed with planting	Loess plateau	– Labour demanding – Can cause blowout – Long life – High social value as it provides cash for local people	– Relatively expensive – More effort to maintain	4 Effectiveness 4 Durability 2 Maintenance
9	Combating soil secondary salinization with vegetation	Mismanaged irrigation areas Lower reach of river	– Labour demanding – Few species – Improving soil	– Costly – More effort to maintain	2 Effectiveness 4 Durability 2 Maintenance
10	Combating industrial mining-induced desertification with vegetation	Mining area	– Labour demanding – High consumption of water – Good ecological and economic benefits	– More effort to maintain – Costly	2 Effectiveness 4 Durability 2 Maintenance

Continued

Table 5.3. Continued

No.	Technique/methods	Sites where applicable	Limitations/benefits	Relative cost effectiveness	Overall rating[a]
Engineering methods					
11	Clay sand barriers	2/3 of leeward side of mobile dunes from bottom	– Must have clay – Labour demanding – Preventing rainwater from infiltration (crust on surface) – Long life	– Costly	4 Effectiveness 4 Durability 4 Maintenance
12	Straw chequerboard	2/3 of leeward side of mobile dunes from bottom	– Must have local supply of straw – Labour demanding – Short life (2–4 years)	– Cheap – Low labour cost because of low opportunity cost of rural labour	4 Effectiveness 2 Durability 3 Maintenance
13	Straw or clay sand barriers combining with vegetation	2/3 of leeward side of mobile dunes from bottom	– Must have local supply – Labour demanding	– Relatively cheap – Easy to maintain	5 Effectiveness 5 Durability 4 Maintenance
Engineering in combination					
14	Building farmland by levelling sand dune with water	Sand dune	– Must have water – Less labour demanding – Good results	– Cheap – Easy to maintain	5 Effectiveness 4 Durability 4 Maintenance
15	Building water conservation project, reclaiming barren land and improving soil to form new oases	Intermountain basins surrounded by snow-capped peaks	– Must have water – Labour demanding – Long life – High social value as it provides cash for local people	– Relatively expensive – Low labour cost because of low opportunity cost of rural labour	4 Effectiveness 4 Durability 3 Maintenance
Chemical methods					
16	Covering sand dune with pitch or making sand barren with asphalt	Sand dune	– Must have chemical materials – Labour demanding – Changing soil surface – Long life	– Expensive – Easy to maintain	4 Effectiveness 4 Durability 4 Maintenance
17	Using some chemical materials (such as plastic film, dry water or soil moisture protector) to protect or supply water for afforestation	Arid areas	– Must have chemical materials – Labour demanding – Short life – Good results	– Expensive	4 Effectiveness 1 Durability 2 Maintenance

[a] The rating is on an arbitrary scale of 1 (poor) to 5 (excellent).
[b] Kulum is a Mongolian word to describe plantings in a dune enclosure, natural meadows or plots between dunes where the water and soil are suitable.

References

Curtin, D., Steppuhn, H. and Selles, F. (1994) Effects of magnesium on cation selectivity and structural stability of sodic soils. *Science Society of America Journal* 58, 730–737.

Department of Science and Technology, State Forestry Administration (2002) *Technologies for Controlling Sands and Rehabilitating Sand Lands*. Forestry Publishing House, Beijing.

Dong, Z.B. (1999) A review on the forecast of soil erosion by wind. *Soil and Water Conservation in China* 6, 17–19 (in Chinese).

Fullen, M.A. and Mitchell, D.J. (1994) Desertification and reclamation in North Central China. *Ambio* 23(2), 131–135.

Hoffmann, C., Funk, R., Wieland, R., Li, Y. and Sommer, M. (2008) Effects of grazing and topography on dust flux and deposition in the Xilingole grassland, Inner Mongolia. *Journal of Arid Environments* 78(5), 792–807.

Huang, F.X. and Niu, H.S. (2001) The quantitative analysis of sand transport rate of wind erosion and vegetation coverage in Maowusu desert. *Acta Geographica Sinica* 56, 700–710.

Keren, R. (1991) Specific effect of magnesium on soil erosion and water infiltration. *Soil Science Society of America Journal* 55, 783–787.

Li, X.R., Zhou, H.Y., Wang, X.P., Zhu, Y.G. and O'Conner, P.J. (2003) The effects of sand stabilization and revegetation on cryptogam species diversity and soil fertility in the Tengger Desert, Northern China. *Plant and Soil* 251(2), 237–245.

Wang, X., Oenema, O., Hoogmoed, W.B., Perdok, U.D. and Cai, D. (2006) Dust storm erosion and its impact on soil carbon and nitrogen losses in northern China. *Catena* 66, 221–227.

Woodruff, N.P. and Siddoway, F.H. (1965) A wind erosion equation. *Soil Science Society of America Journal* 29, 602–608.

Zhang, M.D., Wang, X.K., Sun, H.W. and Feng, Z.W. (2007) HulunBuir sandy grassland blowouts influence of human activities. *Journal of Desert Research* 27(2), 214–219.

Zhao, L. (2004) A study on the process of desertification due to grassland reclamation. *Grassland of China* 2004(2), 68–69.

Zhu, Z.D. (1994) The condition and prospect of desertification in China. *Geographical Research* 13, 104–113.

Zhu, Z., Liu, S., Wu, Z. and Di, X. (1986) *Deserts in China*. Institute of Desert Research, Lanzhou, China, 135 pp.

6 Processes in Rangeland Degradation, Rehabilitation and Recovery

Victor R. Squires
University of Adelaide, Australia

Synopsis
The mechanisms and processes of degradation and rehabilitation are analysed. The distinction is made between restoration and rehabilitation at the landscape level and the implications for whole rangeland ecosystems are considered. The recovery phase is considered and the keys to successful recovery including technical interventions are analysed. A set of guiding principles for artificial rangeland improvement (re-seeding) is presented.

Keywords: definitions; restoration ecology; re-seeding; re-vegetation; site selection; economics; cost effectiveness; constraints to large-scale rehabilitation; guiding principles

6.1 Introduction

Sustainability is the long-term maintenance of the resources on which livestock and forage production depends. Degradation represents an undesirable change from sustainability. It is useful, when assessing rangeland degradation, to distinguish between 'deteriorated', which is judged to be reversible, and 'degraded', which is not reversible economically. However, it should be noted that the definitive discrimination in terms of 'reversible' and 'non-reversible' is apparent only with the benefit of hindsight. Rigid attempts to define 'degradation' obscure the fact that degradation and the associated loss of productivity occur on timescales from years to decades (and beyond). The term 'degradation' is used here, in the general sense, to embrace both reversible and non-reversible aspects of resource damage.

Deterioration and degradation both describe a more fundamental change in rangelands, namely a loss of landscape function (Ludwig *et al.*, 1997). Deteriorated and degraded rangeland environments are characterized by a reduced capacity to absorb rainfall and by increased runoff, greater surface disturbance and greater patchiness, loss of surface soil nutrients and overall poorer nutrient availability. Any recovery must depend on arresting and reversing these losses. An understanding of the processes that lead to land degradation is an important first step.

6.2 Mechanisms of Degradation in Pastoral Rangelands

The core desirable component of rangelands used for grazing is the palatable perennial plant species (grasses, forbs and shrubs). These species provide: (i) animal nutrition, especially in periods of extended drought; and (ii) soil surface protection from wind- and water-driven erosion. The major loss of desirable perennial plant species generally occurs under the combination of heavy use and drought. Rangelands also contain annual ephemeral species that are usually of high nutritional value when moisture is available but which break down rapidly once conditions are dry. Hence, ephemeral species do not usually contribute greatly to forage supply in drought or to cover for soil protection.

Perennial grasslands are the main understorey species which have a predominant summer rainfall component. Perennial grass plant 'density' is usually measured as grass basal cover (GBC). GBC fluctuates greatly with climate conditions, with severe drought causing substantial decline. Prolonged drought can cause high mortality of perennial grasses, irrespective of grazing pressure, but the combination of drought and heavy grazing results in lower GBC than would occur with drought alone. Thus, the combination of drought and heavy grazing substantially accelerates the decline in per cent GBC. Continued heavy grazing after the drought is over can continue the decline in GBC, resulting in a near-complete loss of 'desirable' perennial grass species, causing a loss of productivity and/or vegetation change. Several mechanisms contribute to the rapid decline of perennial grasses under drought and grazing. Perennial grasses depend on substantial root systems to survive periodic drought and so allocate more photosynthate to roots with the onset of dry conditions. However, grazing or defoliation reduces the photosynthate available for partitioning to root growth and hence reduces root biomass, resulting in a lower chance of survival under severe water stress. Some perennial grasses have synchronous release of meristems (buds/growth points) following drought. Heavy grazing at this time can remove all meristems, resulting in the death of plants. For perennial grasses with low viable seed production or transient seed stores, recovery after drought is likely to be slow. For these reasons, perennial grasses are likely to disappear under heavy grazing after the drought is over.

This knowledge of the susceptibility of desired perennial grasses to the combination of drought and heavy grazing is generally missing for the key species in China's pastoral rangelands. The loss of GBC of perennial grasses contributes to an amplification of degradation processes in several ways:

- decreased infiltration, increased runoff and soil loss through water erosion;
- increased soil loss through wind erosion; and
- decreased nutrient cycling and lower soil microbial activity.

Perennial shrubs and subshrubs are important components of pastoral rangelands. Where edible shrubs are found in high enough densities, they provide adequate and stable nutrition for animal production during dry periods and also, in some places, at other times! Under the light stocking rates of the past, green grasses and forbs, dry herbage (mainly grasses), were grazed in preference to shrubs. This dietary preference hierarchy resulted in an 'inbuilt rotational grazing system' with concentration on shrubs only in the dry times and a rapid switch back to herbaceous species once effective rainfall had occurred. This release of grazing pressure allowed shrubs to recover. The shrubs played a key role as a stable element providing long-lived resistant structures in the landscape. Loss of edible shrubs through heavy use in recent decades has led to a build-up (or invasion) of less palatable species, or to erosion of the soil surface.

There are generally few available data on the longevity of the shrubs in China's pastoral lands or on their demography under different intensities of grazing pressure, but evidence from Australia and South Africa is that some are relatively short-lived (10 years) and that others live 70–100 years (Watson *et al.*, 1997a,b).

Soil erosion is an important consequence of overgrazing and drought. The mechanisms are explained in Chapter 5.

6.3 Recovery and Rehabilitation Defined

Restoration ecology is a field of study that provides a conceptual framework for efforts to improve, repair, rehabilitate and restore damaged land (Jordan *et al.*, 1987; Hobbs and Norton, 1996). In reality, there is a continuum of input that encompasses three distinct sets of actions or activities. It is about a broad set of activities (enhancing, repairing or reconstructing) on degraded ecosystems.

Restoration refers to the reinstatement of the original ecosystem in all its structural and functional aspects. Restoration can be thought of as resetting the ecological clock. Restoration is about the reassembling of species into communities that have a chance to grow, develop and rebuild local biodiversity.

Rehabilitation is the process by which the impacts of degradation are repaired. Rehabilitation is the term used for the progression towards reinstatement of the original ecosystem (Bradshaw, 1990). In agricultural lands (cropland as well as grazing land), it usually involves (re)developing conditions conducive to the establishment and retention of appropriate surface vegetation to reverse resource degradation processes. Creating appropriate conditions may range from relatively simple actions, such as reducing grazing pressure or cropping frequency, to complete landscape reconstruction. Selection of vegetation type and species to rectify resource degradation will invariably be driven by local circumstances and budget constraints. Increasingly, it is being recognized that within this setting, 'appropriate surface vegetation' commonly requires the establishment of deep-rooted perennial vegetation. Trees in particular have been promoted as an essential component in developing sustainable agricultural landscapes (see Chapters 10 and 11 for an outline of efforts to stabilize sand, protect oases and improve rangeland).

Reclamation describes the general process whereby the land surface is returned to some form of beneficial use. Reclamation is judged to be successful if it restores the natural capital of the flora and fauna and the productivity of the land that has previously been seriously degraded. Where reclamation is guided by ecological principles and promotes the recovery of ecological integrity, the term restoration has been used.

Re-vegetation seeks to change a plant community having undesirable characteristics to one with desirable characteristics. It may involve reseeding with the existing suite of perennials, or complete replacement with a preadapted species from another place. The goal of many rangeland re-vegetation projects is to re-establish native species and restore natural community functions. Re-vegetation normally involves changes in community composition, plant cover and density, and reduction in competition from undesirable species. Perenniality is usually emphasized because degradation has often been the consequence of replacement of the original perennial vegetation with annual crops and rangeland species. Perennial vegetation offers greater groundcover and soil protection. The introduction of perenniality into dryland landscapes necessitates alternative production systems such as alley cropping, phase rotations and agroforestry.

Rangeland improvement is concerned with the increase of the grazing value of any piece of rangeland. The grazing value is measured through the output of the rangeland in terms of animal production, or by the measured increase of primary production (biomass) in qualitative and quantitative terms. Any improvement, however, is based on an assessment of the past and present situations of the rangeland, i.e. a relatively accurate evaluation of the resource and its evolution with time. That means the availability of a baseline study and its subsequent monitoring.

The improvement methods used are just as diverse as the rangelands themselves; they may be fully natural or highly artificial, depending on the weight of the human action and investment involved in the improvement process. Improvement may involve just restoring the balance between these two entities by, for example, adjusting the stocking rate to the carrying capacity, i.e. the number of animal this vegetation may sustain in the long run.

The increase of the grazing value of natural vegetation may be achieved in various ways using interventions either singly or in combination. Natural vegetation includes plant species that are palatable to both livestock and game and are preferred by them, but also other species that are ignored by large herbivores. The second group of plants increases in importance as the intensity of the grazing pressure mounts. The proportion of both entities in a given site is an indication of the grazing value of the vegetation under consideration. Improvement of grazing lands is thus the increasing of the first entity at the expense of the second, opposite to the usual overgrazing practice.

6.3.1 Rehabilitation of degraded land in China

Rehabilitation of degraded rangeland productivity and restoring its ecological function have become one of the key tasks for combating desertification (Bradshaw, 1984; Werner, 1990). In China, there are 186.038 million ha (Mha) of rangeland in desertification-prone areas and 105.237 Mha of rangeland (about 56.6% of the total area) has suffered from degradation, to a greater or lesser extent (Wang, 2006).

The topic of land degradation in China's pastoral rangelands has received growing attention in recent years from many concerned agencies and individuals. The field is intrinsically complex, involving as it does the consequences of decisions taken by literally millions of people relating to the management and custody of the land resource. It is not surprising, then, that opinions of what should be done about these matters vary greatly in their nature and conclusions.

Scientists from institutes and universities and the local grassland bureaux have made numerous attempts to rehabilitate degraded rangelands and reverse the trend towards domination by less desirable plants (often shrubby species – although sometimes toxic plants). In severe cases, sand encroachment and other forms of desertification have been common. Measures to reverse desertification and rehabilitate affected areas have also been a focus of much work (see Chapter 11).

Although remediation work has been widespread throughout the 'Three Norths' region, even from the 1960s and onwards, generally it has been difficult to find out precisely what was done and where and what the specific objectives of the treatments were. Experiments were poorly documented because several different agencies were involved and different aspects were handled by specific bureaux without lodgement of results and data sets in a common file. Although county-level technicians organized and supplied labour for the remediation treatments, few records were kept on the scientific objectives of the various activities, and most records that were kept have been lost.

Most early references made to treatments are of a very general nature. Compounding this lack of specificity is the possible disposal of old records in periodic clean-up activities. Much of the specific documentation that is needed to pinpoint the exact nature of the remediation was either lost or never recorded to begin with. As a result, many treatments were never documented properly and many treatments were never evaluated properly or assessed with regard to effectiveness. Some attempts have been made (Fullen and Mitchell, 1994; Mitchell *et al.*, 1996; Xin *et al.*, 2003) to draw together documentation of treatments and evaluate the outcomes for particular sites, e.g. Shapatou in Ningxia. Despite this work, a significant number of treated areas remain undocumented and the impact and effectiveness of the various treatments are not well known.

There is one conclusion that can be drawn; there are few measures that are likely to be cost-effective and sustainable in pastoral lands where rainfall is low and unreliable and the land has limited production capabilities. Also, the cost-effectiveness of the remediation is very poorly documented.

6.4 Constraints to Rehabilitation Using Artificial Establishment Techniques

6.4.1 Immense areas of degraded rangeland of low economic value

Pastoral rangelands include extensive and diverse acreage. Most of the problems associated with the word 'degradation' in rangelands are, in reality, manifestations of the fact that the natural resources of the drylands yield little or no economic rent. There is a gradient in economic rent from the arid desert margin to the subhumid. At the extreme extensive arid margin, the economic value of the land and its related natural resources is so low as not to justify any management intervention on purely economic criteria alone (Squires and Andrew, 1998). That is, the per-unit cost exceeds the social value. Such lands and their associated resources would be under a regime of open access. Therefore, any use, no matter how destructive, is less wasteful of resources than the high cost of trying to rehabilitate them.

The enormous size of this area simply precludes comprehensive treatment of all seriously depleted sites. Few sites now support a desirable vegetative cover. Many sites support less productive and undesirable weedy (even toxic) species and unsatisfactory watershed conditions. However, the cost of correcting these problems may not always justify extensive artificial treatments. Site improvement may be attained better through careful management. Numerous sites on steep, inaccessible slopes cannot be treated with existing equipment. Topographical and vegetative conditions are usually very diverse within most areas, and site preparation and planting equipment are not always versatile enough to treat all circumstances. Consequently, some areas cannot be treated properly (see below).

6.4.2 Climatic conditions

Many arid or semi-arid rangelands that occupy extensive areas within the pastoral areas of north and north-west China cannot be treated satisfactorily using current re-vegetation and restoration measures. Arid conditions and irregular moisture patterns may not be conducive to seedling establishment.

It is simplistic to focus on the lack of rainfall as the major cause of rangeland degradation. Some commentators, even today, view climate change as the sole cause. However, drought on its own does not cause degradation of the scale described in the case studies in this book (Chapters 7–14). Drought has been a feature of China's landscapes for tens or even hundreds of thousands of years. China's arid rangelands have adapted to recurrent drought and have probably weathered droughts far worse than those encountered in the past 60 years. The climate has fluctuated significantly at least three times around a dry and probably cool regime over the past 5000–8000 years (Chun *et al.*, 2002; Feng *et al.*, 2006).

There will always be periods of deterioration resulting from seasonal and unpredictable annual fluctuations of climate. Differing levels of resilience in the resource to perturbations in different systems lead to different manifestations of ecosystem instability (Stocking, 2005).

Large areas are normally treated and seeded only once. Uniform stands may not develop, yet replanting is costly and impracticable. Regions receiving less than 200 mm of annual precipitation are the most difficult to treat – this condition applies throughout much of western Inner Mongolia, western Gansu and much of Xinjiang. Even if suitable species were found and if there were sufficient seed available, appropriate planting techniques for successful planting of these species may not be available. Many semi-arid rangelands need improvement, but changes can often be attained more easily through proper long-term management (see below) than through artificial re-vegetation. Many species that occupy arid sites are extremely valuable plants and should be retained or enhanced. However, these plants are not easily cultured and are not well suited to artificial planting. Suitable substitute species that could be used in their place are not known. Consequently, many arid and semi-arid sites must be managed carefully to minimize abuse and stimulate natural recovery.

6.5 Artificial Re-vegetation Considerations

Similar factors must be considered in determining if management or re-vegetation should be employed to improve a degraded rangeland. However, certain factors must be looked upon quite differently, depending on which approach is used. For example, the size of an area requiring restoration or rehabilitation is a major factor to be considered. A large area may be difficult to manage due to differences in topography, access or season of use. Improvements may not be achieved easily. Similarly, the area may be so diverse that artificial re-vegetation may be difficult to achieve using a single method or closely related methods of site preparation and seeding.

The following are some factors to consider in determining the applicability or practicality of artificial re-vegetation. The list is not considered all-inclusive. Other issues may also be important, particularly in specific areas. However, the factors discussed below must be considered before developing improvement measures.

6.5.1 Site suitability

USDA research (2004) emphasized the importance of correctly discerning the capabilities of a site prior to treatment. Too often, attempts are made to convert a vegetative community to a complex of desirable but unadapted species. The site must be capable of sustaining the selected species. In addition, species included in the seed mixture must be compatible with one another and with the existing native species.

Some attempts have been made to improve shrublands by seeding grasses or by introducing other shrub species. In many cases, treatments have failed and less productive plants have invaded. Failure to recognize the suitability or capability of these sites has resulted in the loss of the adapted native plants.

Sufficient information is not always available to determine the adaptability of many introduced and native species. Some species are difficult to establish through artificial seeding and the desired complex of adapted species is not always achieved. However, it is not advisable to seed or plant substitute species that are adapted marginally but established easily. A site may be capable of sustaining a complex array of species. However, initial attempts to re-establish certain species may be unsuccessful. Soil crusting and high salt content in the soil surface often limit seedling establishment of species on some sites. Rodent foraging may seriously limit seedling survival. Livestock selectively graze some species, particularly broadleaf herbs, limiting their survival even when planted under favourable climatic and soil conditions. Animals tend to concentrate on seeding project sites unless they are well protected, especially if the adjacent rangelands are devoid of an adequate forage cover.

Weed infestation and slow or erratic seedling growth of many seeded species often diminish their success. Artificial plantings or natural seedlings are often not successful and attempts to restore large areas from a single planting cannot always be achieved. These factors significantly influence site suitability for improvement by either management or artificial re-vegetation. The current status of the plant community must also be considered when designing a re-vegetation programme. Newly developed or introduced plant materials must be able to establish, persist and reproduce. If they are unable to reproduce satisfactorily, stands ultimately deteriorate. Many productive and palatable forage plants have been established successfully, but they have been short-lived and unable to reproduce by natural seeding, and stands have slowly disappeared.

Various introduced (exotic) herbs and shrubs perform favourably from initial plantings on degraded rangeland sites. However, some have failed to survive when insect outbreaks and other unusual stress events occur. Similar situations have been encountered when highly desirable native species have been planted on sites where the species does not normally exist, even when such sites are quite similar to the origin of collection. Some ecotypes of a particular species demonstrate specific site adaptability; unadapted ecotypes may then be sorted out quite rapidly. Other ecotypes may be equally sensitive, but climatic or biological events that affect their survival may not occur frequently. Consequently, these ecotypes may persist for an extended period before being eliminated.

Perhaps the most critical issue to be considered in re-vegetating semi-arid and arid sites is the availability of soil moisture for seedling establishment. Attempting to seed areas that receive erratic amounts of moisture is extremely hazardous. Seeds of many species require periods of cold-moist stratification to initiate germination. In addition, developing seedlings must receive sufficient moisture to assure establishment. Attempting to plant in areas dominated by weeds, or during periods when soil moisture is unfavourable for growth, is ill advised. The seeding of species with different germination and growth characteristics can be successful if the moisture requirements of all species are met. Problem sites may be capable of supporting a specific array of species, but current planting techniques are not satisfactory for planting many sites. Consequently, the site must be suitable for: (i) maintaining the planted species; and (ii) applying currently available methods of treatment.

6.5.2 Status of soil and watershed conditions

Sites that have been degraded and subjected to erosion are normally the most critical areas requiring artificial restoration. Protection must be provided for on-site and downstream resources. However, barren and eroding soil surfaces are not normally satisfactory seedbeds. Recovery of natural re-vegetation is often prevented because of unstable surface conditions and a limited soil seed bank. Artificial seeding, including site preparation, is difficult and costly to achieve on unstable watersheds. Areas should not be allowed to deteriorate to the point that rehabilitation or other costly measures are necessary to re-establish a plant cover.

Soil conditions must be surveyed carefully to assure that a satisfactory seedbed can be created. Soil conditions may need to be improved. Too often, herbaceous understorey species have been lost. The change in plant composition reduces soil protection. Problem areas may be ranked depending on their values and the severity of the disturbance. The most critical areas may then be selected for treatment. The feasibility of treating the candidate sites must be considered in developing rehabilitation plans.

6.5.3 Appraisal of resource values

Restoration or rehabilitation projects have been completed on various sites to improve forage production without carefully determining the best specific locations where these resources are to be found. Large areas are often treated assuming 'the more hectares treated, the more forage provided'. This assumption is sometimes incorrect. Re-vegetation projects should be designed to provide cover, forage and protection on sites where the greatest benefit can be derived. It is obvious that treatments must be carried out efficiently. Large tracts of land can be treated more easily than isolated sites, for example, by aerial sowing. However, treatments should be designed to accomplish the goals of the project and the needs of targeted land users.

6.5.4 Selective treatment and impacts on associated areas

Artificial treatments can be designed to restore critical areas indirectly. Artificial re-vegetation can, and does, benefit both the treated area and adjacent sites. Consequently, areas having good access and highly productive soils can often be treated, leaving adjacent sites to recover naturally. However, the untreated sites must be able to recover. Highly palatable species, or plants that provide seasonal forage, can be seeded on to specific sites to attract and hold grazing animals on adjacent areas. Treating an area of sufficient size is necessary to disperse animal use and allow the seeded species and untreated areas a chance to develop. Not all untreated sites respond favourably. Areas that are nearly devoid of desirable species or are dominated by weedy plants do not generally respond to a reduction of grazing.

Selective treatment, an important practice, can be used to promote successional changes and supplement improved habitat, seasonal availability of herbage and forage quality. Adding an appropriate shrub (e.g. *Caragana*) or herb (e.g. lucerne) to the existing vegetation can enhance forage resources, restore specific species and control weeds. Interseeding selected species into existing stands is an important technique to improve large areas without excessive costs.

6.5.5 Management and control of access

Treated sites must be managed to retain species composition, plant vigour and productiveness. Treated sites may require special protection that cannot be provided. If this occurs, the value of the project is lost. Treated areas must be of appropriate size to accommodate seasonal use during the time of plant establishment and over a long-term maintenance period. Areas must be of sufficient size and diversity to respond to climatic conditions and associated biotic factors that influence plant succession. Some treated areas may be heavily grazed to such an extent that weeds are able to invade during stressful periods. The treated sites must be able to accommodate all forms of use, including somewhat abnormal events such as insect attacks and drought.

Treated sites should be managed or used as intended initially. Too often, areas are seeded or treated to provide 'special purpose' pastures designed to fill a feed gap or provide a protein supplement at a critical time, but are then used as general grazing for livestock, despite the fact that the areas may not be designed to accommodate these constant high levels of use. Treated sites regress if not managed properly. Improper use, particularly during the period of seedling establishment, can eliminate certain species and decrease the overall success of the project.

6.5.6 Availability of adapted plant materials

Rehabilitating ranges usually requires the inclusion of various locally adapted native species in the seeding. Restoration projects require seeding diverse mixtures of native species. Seeds of many native species are not always available and substitute species are frequently planted (or seeds of the correct species are obtained from irrigated 'seed increase' sites far from the local area). The lack of adapted ecotypes of many species limits re-seeding opportunities. The use of introduced grasses has facilitated many rehabilitation projects. However, the more commonly available grasses and broadleaf herbs do not satisfy all resource needs. Seed sources must be found and/or developed to assure the use of desirable and adapted native plants.

6.5.7 Site improvement costs

The costs incurred in restoration and rehabilitation ultimately determine the site treatment and seeding practices to be employed. However, it is difficult to determine the value of stable plant communities; not only for forage production, but also for soil protection, biodiversity protection and watershed protection. Benefits cannot be calculated wholly on the increased production of forage. All benefits must be considered over the entire life of the project (Le Houérou, 2000).

Improvement of vegetative and edaphic conditions on some sites can be achieved through proper management, as well as by manipulative plantings. Sites that have been subjected to serious abuse, or that lack needed cover or forage resources, can be improved by various methods. Prior to the development of any site improvement programme, land managers must first discern the resource needs and suitability of an area for treatment (see the ten principles below). Then appropriate methods and techniques can be developed through management schemes or artificial measures, or both. Factors that influence site improvement through management are discussed first. Factors that are of special concern when considering restoration or rehabilitation are presented next. Factors that influence management decisions are also important considerations in developing planting programmes. Unless an adequate representation of desirable species that are capable of recovery and natural spread remains, artificial seeding is unnecessary. If managed properly, plants that have been weakened by excessive grazing and browsing can normally recover and begin producing seed within a few years. Plants growing in arid environments may require a longer period to recover.

Even grazing ban areas may require many years to recover following heavy grazing. Some disturbed areas have remained in an almost static condition for more than 15 years, even with protection from grazing. However, in one case in Inner Mongolia, considerable improvement resulted following 3 unusually wet years in succession. It is often this co-occurrence of favourable events (soil moisture and favourable temperature) that determines the success of the recruitment of perennial plants to the population.

Woody species that exist in mountain communities normally have the capacity to recover and spread quickly when managed correctly. Woody species growing at lower elevations are usually exposed to more adverse climatic conditions and many are less capable of natural spread. Thus, recovery in salt desert shrublands and shrubs on low foothills is slow.

Many native communities are capable of self-regeneration by natural seeding or sprouting. However, replacing individuals that die naturally is an entirely different situation from repopulating a broad area where most species have been depleted by grazing. A disturbed site may still support some species, but not others. This is quite common on most overgrazed rangelands. The more desirable forage plants are often lost by selective grazing. Other remaining, but less desirable, species may be capable of recovery, but the important forage species may not reappear without some means of artificial seeding. This should result in an increase in total herbage production. However, the recovery of important broadleaf herbs frequently does not occur. Some broadleaf species usually occurring on specific microsites may not dominate a community, but they are important as seasonal forage. Unfortunately, these same species are often eliminated by grazing and do not persist in sufficient numbers to recover, even when protected for extended periods. If desirable species are not present, improvement by natural means may be unattainable.

6.5.8 Status of soil conditions

Soil and watershed conditions are critical resources that cannot be allowed to deteriorate.

If disturbance has progressed to the extent that soil loss is serious, rehabilitation measures must be implemented. If adequate protection of the soil and watershed through management is not realized within a satisfactory period, artificial re-vegetation measures will be required. A long recovery can be accepted if the soil and watershed resources do not deteriorate appreciably during the initial stages of natural recovery. However, both the physical and chemical condition of the soil affects seedling establishment and growth. Soil surfaces must be conducive to seedling establishment if the vegetation is to recover. An open, but stable, surface may exist, but surface crusting or freezing may prevent seedling establishment. In addition, lowering of the water table through down-cutting of the stream channel can, and does, influence areas adjacent to the drainage. Wind erosion and lack of surface organic matter are highly detrimental to seedbed conditions. These and other features must be considered when assessing soil and watershed conditions.

Protecting the soil resource may be necessary before attempts are made to improve habitat or forage conditions. This has been a major concern in many circumstances. The vegetation in these areas can often recover satisfactorily through protection, but eroding areas may respond more slowly. In addition, the occurrence of intense summer storms and other climatic events can be expected and can have devastating and long-lasting impacts.

6.5.9 Management strategy

Rangeland sites in fair condition are usually able to recover through natural processes. However, providing protection from human-induced changes is often difficult. Winter, spring and autumn ranges may constitute small, but important, portions of a broad geographical area. Attempts to restrict the use of a broad area for sufficient time to allow recovery of these seasonal ranges may not be practical. Continued livestock use on these broad areas may not be compatible with natural recovery. A well-designed management system to improve conditions may require a long-term commitment. Management strategies must ensure that the following conditions are created:

1. Development of suitable seed banks.
2. Creation and protection of adequate seedbeds.
3. Protection of plants for sufficient time to provide an acceptable composition of most critical areas.

6.5.10 Impacts on other resources

Few areas can be managed to support just one use, yet treatment practices are often developed to enhance a single primary resource. In these cases, attention must be given to the expected impacts on other resources. For example, the value and impact of management schemes must be determined for other uses. In addition, management strategies that are used to regulate animal distribution, population numbers and seasonal use must be developed as part of the rehabilitation programme. The decision to treat an area artificially is normally based on the value of numerous resources. For example, a site essential in maintaining biodiversity that may also be an important watershed area might receive treatment priority.

6.5.11 Management considerations – status and condition of existing vegetation

Restoration or rehabilitation projects are not usually contemplated unless the native communities have been severely disturbed, resulting in adverse watershed conditions and loss of desirable vegetation. If an area is capable of recovery and natural spread (thickening up), artificial seeding is unnecessary. If managed properly, plants that have been weakened by excessive grazing and browsing can normally recover and begin producing seed within a few years. Plants growing in arid environments may require longer to recover.

Proper management is the key to the improvement or maintenance of acceptable plant cover and soil stability. Successful re-vegetation may change plant and watershed conditions dramatically. Yet, without appropriate management, improvements can be lost. The following are some factors that influence decisions on whether to attempt to improve a specific site.

6.6 Ten Principles of Rangeland Re-vegetation (see USDA (2004) for more detail)

Principle 1: The proposed changes to the plant community must be necessary and ecologically attainable

The general goal of most re-vegetation projects is to change a plant community having undesirable characteristics to one with desirable characteristics.

Principle 2: The terrain and soil must support the desired changes

The potential productivity of a site must be considered when planning re-vegetation projects. Various site characteristics affect productivity significantly. The most important features are:

- depth of the soil surface and subsurface horizons;
- soil texture and the amount of salt in surface and subsurface horizons; and
- occurrence and location of hardpans or restrictive layers in the soil profile.

Principle 3: Precipitation must be adequate to assure establishment and survival of indigenous and planted species

Water is often the most critical factor affecting seedling survival and plant establishment in semi-arid and arid regions. Generally, re-vegetation efforts should not be initiated in areas receiving less than 230 mm of annual precipitation.

Principle 4: Competition must be controlled to ensure that planted species can establish and persist

Young seedlings of most species are usually unable to compete with established vegetation. Undesirable, highly competitive species must be removed or reduced in density to allow seedling establishment of the planted species. Individual methods do not usually eliminate all plants completely but can reduce competition sufficiently to allow seeds of the planted species to germinate and establish. Treatments can often be difficult to select and implement where retention of existing and desirable species is wanted.

Principle 5: Plant and manage site-adapted species, subspecies and varieties

Factors important in determining which plant materials should be selected for seeding are:

- use of site-adapted species and populations;
- presence, density and composition of indigenous plants;
- availability of seed or planting stock; and
- project objectives.

Principle 6: A multispecies seed mixture should be planted

Many early re-vegetation projects emphasized the use of a limited number of species. For most rangeland re-vegetation projects today, however, there are many reasons for seed mixtures rather than single species:

- Restoration of native plant communities usually requires the reintroduction of a variety of species to provide community structure and function.
- A combination of species is normally required to initiate natural successional processes.
- A variety of species that are adapted to the diverse microsites occurring within major sowings should be planted.
- Mixtures reverse the loss of plant diversity and enhance natural recovery processes following natural impacts from insects, disease organisms and adverse climatic events.
- Chances for successful seeding are often improved when mixtures are planted.
- Mixtures can provide improved groundcover and watershed stability.
- Mixtures produce communities that provide greater potential for reducing weed invasion and for providing for a balance in the use of all resources.
- Combinations of species can provide a better-quality habitat including cover and seasonal forage.
- Total forage production and seasonal succulence can be increased with mixtures.
- Mixtures are generally more aesthetically pleasing and match natural conditions.
- Mixtures provide diverse habitats required to sustain wildlife species.

Seeded mixtures should include the various growth forms, that is: grasses, forbs, shrubs and trees that existed prior to disturbance. Seeded and indigenous species must be compatible and able to establish and develop together. Successional changes must occur that will result in the ultimate development of a desirable plant community. A few special situations such as providing immediate groundcover to stabilize erosion may occasionally dictate the seeding of only one or a few species. Because some shrubs establish and grow much more slowly than many herbs, planting individual woody species with plants having similar establishment and growth characteristics is recommended. Selectively planting different species in separate rows or spots is sometimes required.

Principle 7: Sufficient seed of acceptable purity and viability should be planted

It is important to calculate seeding rates carefully. Planting excessive seed is unnecessarily expensive and increases competition among seedlings and indigenous species. Low seeding rates, on the other hand, may jeopardize stand establishment. It is essential that seeding rates be determined on a pure live seed (PLS) basis. The number of PLS per unit of weight varies greatly among species and seed batches. Seed must be tested for purity and germination and tagged properly with the current results to enable the operator to calculate seeding rates.

Principle 8: Proper seed coverage is essential for successful germination and seedling establishment

Depth of planting is generally determined by seed size. However, it is also influenced by the special requirements of individual species. As a general rule, seeds should not be covered more than three times the thickness of the cleaned seed. Seed of certain species are best seeded on a disturbed surface with shallow soil coverage. But some species do better if planted deep (5–7.6 cm). Soil type and surface conditions also influence seeding depth. Most species benefit from firm seedbeds, but some do well in loose soils. Heavy soils may crust and prevent emergence. Light-textured soils are less likely to crust or become compact; however, they dry rapidly and thus deeper planting depths are recommended.

Principle 9: Plant during the season that provides the most favourable conditions for establishment

Late autumn and winter sowings have been most successful in some areas because:

- The inherent seed dormancy of many species is released by overwinter stratification.
- Seeds are in place in early spring when soil moisture is most likely to be available for germination, seedling emergence and growth. Early emerging seedlings are better able to resist high summer temperatures and drought.
- Seed predation by small mammals and birds is less likely to occur if seeds are planted when these animals are less active. Seeding too early in autumn may result in precocious germination following unseasonably warm periods coupled with any autumn rains. Seed losses to mammals and birds can also be high during this period. Transplanting should be completed in early spring when the soil is wet and before active growth of the transplant stock or the native vegetation has begun. Autumn transplanting is not generally recommended unless soils are moist and are likely to remain moist until they freeze.

Principle 10: Newly seeded areas must be managed properly

As a general rule, newly seeded areas should not be grazed for at least two or three growing seasons following planting. Poor sites and slow-growing species may require a much longer period of non-use. When grazing does occur, it should be regulated carefully.

6.7 Natural Means of Rangeland Rehabilitation

The main task of rangeland restoration seems to be one of repair, or reassembly, of damaged landscapes and biota, but, in fact, managers and scientists must assemble entirely new communities of plants and animals. The goals of particular restoration projects vary greatly, although they often contain the same set of potentially incompatible qualities; that is, the new community may be required to be self-sustaining, stable and minimally disruptive to native biota, and yet produce a high yield of introduced animals.

Recovery of natural re-vegetation is often prevented because of unstable surface conditions and a limited soil seed bank. Too often, herbaceous understorey species have been lost. Improvement of vegetative and edaphic conditions on some sites can be achieved through proper management. Areas that are nearly devoid of desirable species or dominated by weedy plants do not generally respond to a reduction of grazing. Sites that have been subjected to serious abuse, or that lack needed cover or forage resources, can be improved by various methods. If managed properly, plants that have been weakened by excessive grazing and browsing can normally recover and begin producing seed within a few years. Plants growing in arid environments may require a longer period to recover. Unless an adequate representation of desirable species that are capable of recovery and natural spread remains, artificial seeding is necessary.

Even grazing ban areas may require many years to recover following heavy grazing. Some disturbed areas have remained in an almost static condition for more than 15 years, even with protection from grazing. However, in one case in Inner Mongolia, considerable improvement resulted following 3 unusually wet years in succession. It is often this co-occurrence of favourable events (soil moisture and favourable temperature) that determines the success of the recruitment of perennial plants to the population.

Woody species that exist in mountain communities normally have the capacity to recover and spread quickly when managed correctly. Woody species growing at lower elevations are usually exposed to more adverse climatic conditions and many are less capable of natural spread. Thus, recovery in salt desert shrublands and shrubs on low foothills is slow.

Many native communities are capable of self-regeneration by natural seeding or sprouting. However, replacing individuals that die naturally is an entirely different situation from repopulating a broad area where most species have been depleted by grazing. A disturbed site may still support some species, but not others. This is quite common on most overgrazed rangelands. The more desirable forage plants are often lost by selective grazing. Other remaining, but less desirable, species may be capable of recovery, but the important forage species may not reappear without some means of artificial seeding. This should result in an increase in total herbage production. However, the recovery of important broadleaf herbs frequently does not occur. Some broadleaf species usually occurring on specific microsites may not dominate a community, but they are important as seasonal forage. Unfortunately, these same species are often eliminated by grazing and do not persist in sufficient numbers to recover, even when protected for extended periods. If desirable species are not present, improvement by natural means may be unattainable.

Natural recovery processes must be considered in predicting secondary successional changes. Although some desirable species may not be present on a disturbed site, their re-entry may depend on factors other than the adverse effects of grazing; for example, some shade-dependent plants are not able to survive if overstorey species are not present. The shade-tolerant species will not appear until overstorey plants have become established, assuming a viable seed bank remains.

The recovery capabilities of individual species must be evaluated correctly to decide on methods of improvement. Some plants spread well from seed, even under stressful situations. Others are rarely seen, even though abundant seed crops are produced most years. Some species are site specific, existing as pure stands but intermixed with other communities. If these stands are eliminated or seriously diminished, natural recovery is extremely slow. Recovery is affected by limited seed sources, low plant density and poor distribution of parent plants.

Although more time may be required to achieve natural recovery, this may be the most practical approach. However, land managers must understand that during the period of recovery, the vegetation may not furnish the desired forage and cover. Until a complete recovery of all species is attained, all resource values may not be provided.

Exclusion of grazing in many situations allows natural regeneration and is the preferred approach on extensive low-value rangelands and has been applied globally (Noy-Meir *et al.*, 1989; Pucheta *et al.*, 1998). Experience from China indicates that exclusion of livestock can increase productivity of degraded rangeland (Bao and Chen, 1997; Wu and Ci, 2002; Wu, 2003; Wang *et al.*, 2004). More detail about this approach is set out in the eight case studies in this book (Chapters 7–14).

References

Bao, Y.T. and Chen, M. (1997) The study of changes of plant diversity on degenerated steppe in enclosed process. *Acta Scientiarum Naturalium Universitatis NeiMongol* 28(1), 87–91.
Bradshaw, A.D. (1984) Land restoration now and in the future. *Proceedings of the Royal Society, London. Series B* 223, 1–23.
Bradshaw, A.D. (1990) The reclamation of derelict land and the ecology of ecosystems. In: Jordan, W.R., Gilpin, M.E. and Aber, J.D. (eds) *Restoration Ecology: A Synthetic Approach to Ecological Research.* Cambridge University Press, Cambridge, UK, pp. 53–74.
Chun, C.H., Pang, J. and Li, P. (2002) Abruptly increased climatic aridity and its social impact on the Loess Plateau of China at 3100 BP. *Journal of Arid Environments* 52(1), 87–99.
Feng, Z.D., An, C.B. and Wang, H.B. (2006) Holocene climatic and environmental changes in the arid and semi-arid areas of China: a review. *The Holocene* 16(1), 119–130.
Fullen, M.A. and Mitchell, D.J. (1994) Desertification and reclamation in North Central China. *Ambio* 23(2), 131–135.
Hobbs, R.J. and Norton, D.A. (1996) Towards a conceptual framework for restoration ecology. *Restoration Ecology* 4, 93–100.
Jordan, W.R. III, Gilpin, M.E. and Aber, J.D. (eds) (1987) *Restoration Ecology.* Cambridge University Press, Cambridge, UK, pp. 257–270.
Le Houérou, H.N. (2000) Restoration and rehabilitation of arid and semi-arid Mediterranean ecosystems in North Africa and West Asia: a review. *Arid Soil Research and Rehabilitation* 14, 3–14.
Ludwig, J., Tongway, D.G., Freudenberger, D., Noble, J.C. and Hodgkinson, K.C. (eds) (1997) *Landscape Ecology, Function and Management Principles from Australia's Rangelands.* CSIRO Publishing, Collingwood, Australia.
Mitchell, D.J., Fearnehough, W., Fullen, M.A. and Trueman, I.C. (1996) Ningxia desertification, development and reclamation. *China Review* 5 Autumn/Winter, 27–31.
Noy-Meir, I., Gutman, M. and Kaplan, Y. (1989) Responses of Mediterranean grassland plants to grazing and protection. *The Journal of Ecology* 77(1), 290–310.
Pucheta, E., Cabido, M., Diaz, S. and Funes, G. (1998) Floristic composition and above ground net plant production in grazed and protected sites in a mountain rangeland in central Argentina. *Acta Oecologica* 19(2), 97–105.
Squires, V.R. and Andrew, M.H. (1998) Management interventions: are they feasible in arid zone livestock production systems? *Annals of Arid Zone* 37(3), 205–214.
Stocking, M.A. (2005) Integrated ecosystem management: its evolution as an approach for managing natural resources. In: Jiang, Z. (ed.) *Integrated Ecosystem Management.* Proceedings of a Conference, Beijing, November 2004. China Forestry Publishing House, Beijing, pp. 23–39.
USDA (2004) *Restoring Western Ranges and Wildlands.* General Technical Report RMRS-GTR-136-Vol. 1. US Forest Service, Rocky Mountain Research Station, Fort Collin, Colorado.
Wang, T. (2006) *Deserts and Desertification in China.* China Science Press/Longmen Book Co., Beijing (in Chinese), 560 pp.
Wang, T., Wu, W., Xue, X., Sun, Q.W., Zhang, W.M. and Han, Z.W. (2004) Spatial–temporal changes of sandy desertified land during last 5 decades in northern China. *Acta Geographica Sinica* 59, 203–212.
Watson, I.W., Westoby, M. and Holm, A.McR. (1997a) Demography of two shrub species from an arid grazed ecosystem in Western Australia 1983–1993. *Journal of Ecology* 85, 815–832.
Watson, I.W., Westoby, M. and Holm, A.McR. (1997b) Continuous and episodic components of published demographic change in arid zone shrubs: models of two *Eremophila* species from Western Australia compared with published data from other species. *Journal of Ecology* 85, 833–846.
Werner, P.A. (1990) Principles of restoration ecology relevant to degraded rangelands. *The Rangeland Journal* 12(1), 34–39.
Wu, B. and Ci, L. (2002) Landscape change and desertification development in the Mu Us Sandland, Northern China. *Journal of Arid Environments* 50, 429–444.
Wu, W. (2003) Dynamic monitoring to evolvement of sandy desertified land in Horqin Region for the last 5 decades, China. *Journal of Desert Research* 23, 646–651.
Xin, R.L., Feng, Y.M., Hong, L.X., Xin, P., Wang, G. and Ke, C.K. (2003) Long-term effects of revegetation on soil water content of sand dunes of Northern China. *Journal of Arid Environments* 57(1), 1–16.

Part III

Case Studies of Degradation and Recovery

The purpose of describing these case studies is to gain an understanding of what causes land degradation and what actions and information sources are needed to prevent further degradation episodes. This book is not intended as a history, but uses previous histories and documentation to interpret the causes of degradation and recovery.

Because recovery sometimes occurs decades after the degradation episode, it has not been possible to quantify the extent to which initial productivity and resource conditions have been restored. In the case study areas where there has been a considerable loss of soil, irreversible change may well have occurred and a return to initial productivity is unlikely to happen.

7 Case Study 1: Hulunbeier Grassland, Inner Mongolia

Lu Xinshi,[1] Ai Lin[1] and Lv Shihai[2]
[1]*Beijing Forestry University, Beijing, China;* [2]*Chinese Environmental Science Academy, Beijing, China*

Synopsis

This is an examination and analysis of animal husbandry development and its role in the local economy of part of the vast Hulunbeier Grasslands. The interactions between climate, land degradation and changes in population density of humans and their livestock are based on a review of data for the past 30–50 years. The extent to which accelerated land degradation has occurred and the relative contribution of anthropogenic factors and climate change are considered. The impact of social and economic development within the region on the exploitation of land and water resources has been a major contributor to land degradation. The role of the responsibility system and the allocation of grazing user rights are examined and some proposals about future actions are presented.

Keywords: primary productivity; carrying capacity; stocking rate; grazing ban; ecological migration; land tenure; grazing user rights; land conversion; policy issues; socio-economics; feed balance; re-seeding; aerial sowing; grazing system; pen feeding; fodder crop; artificial pastures

7.1 Brief Statement of the Problem

Hulunbeier sandy grasslands (Fig. 7.1) in the north-east of the Inner Mongolia Autonomous Region (IMAR) are now at the stage of evolution from sandy grasslands to sandlands. One-third of the rangeland in the Hulunbeier League is severely degraded. Vegetation coverage in the rangeland has decreased by 30–80%. The number of plant species in the rangelands has reduced from 130 to 30. Many palatable plants have disappeared from the rangeland.

The area of degraded grasslands has more than doubled since the late 1980s and the area converted to cropland has increased dramatically (see Section 7.4 below).

Soil and water erosion has become obvious and desertification processes have accelerated. The area of shifting sand dunes has expanded. Under increased human disturbances in the form of animal production, depletive utilization of rangeland resources, land conversion for monoculture and dryland cultivation operations and the influence of climate changes (Chapter 3), the Hulunbeier rangeland is undergoing a fast degradation process.

For the rangeland, ever-increasing grazing pressure and excessive trampling, in particular during the spring season, cause large areas of rangeland degradation. Plant growth and vegetation recovery were checked at the critical time for plants which were beginning spring growth. This stimulated a vicious cycle for the grassland deterioration, i.e. grazing pressure was heavy, herbage growth became less and herbage regrowth was slow, so grazing pressure became even heavier.

Land degradation is caused by a combination of natural factors (infestation by rodents and insects and changing climatic factors – such as timing and intensity of rainfall) and human

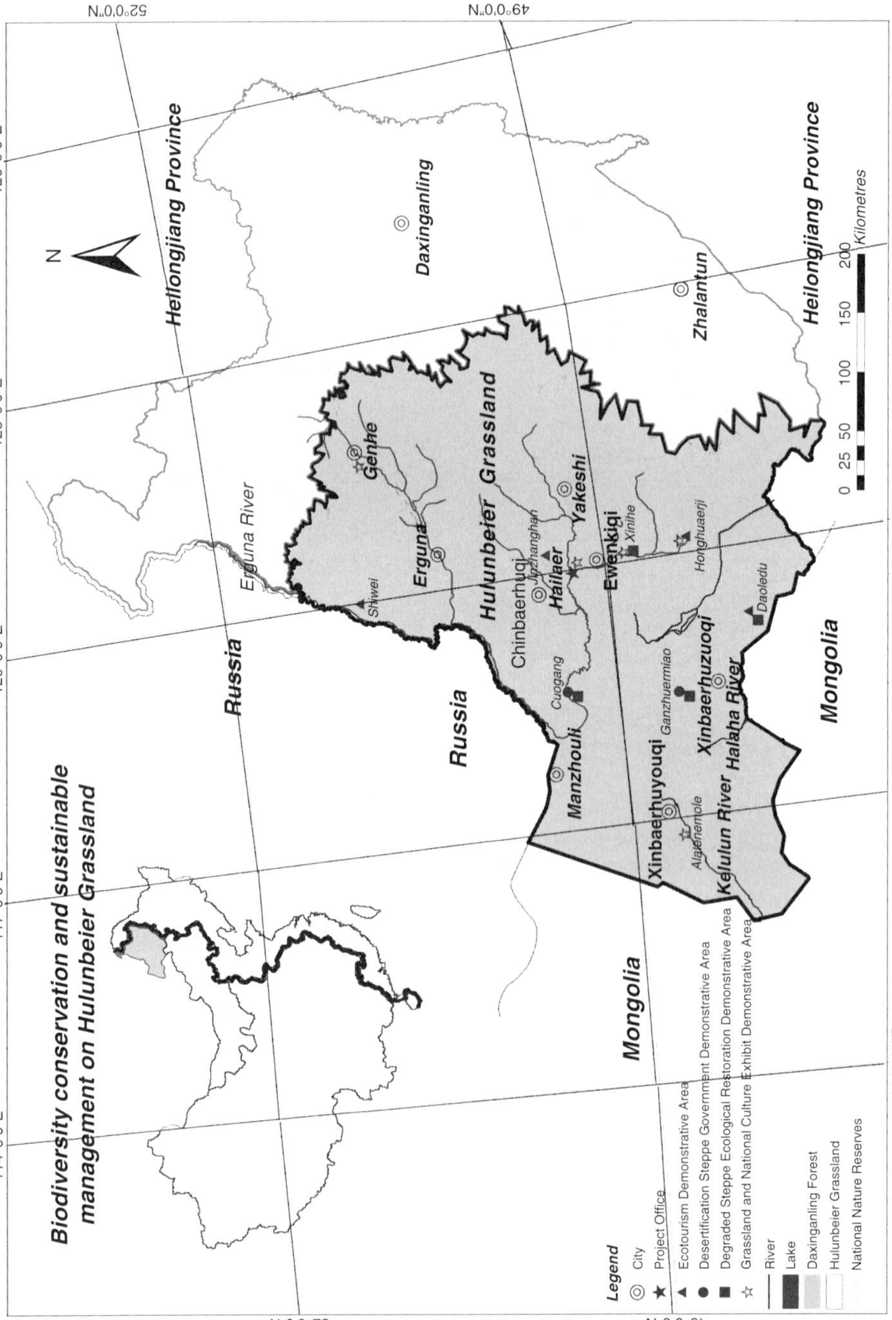

Fig. 7.1. Map showing the location of the Hulunbeier Grassland with place names.

factors such as inappropriate land-use policies, inadequate rangeland management and overharvesting of rangeland products. The human-induced factors are exacerbating: (i) overall poor understanding of the functioning and resilience of ecosystems; and (ii) lack of awareness by various levels of government officials of the medium- and long-term environmental impact of intervention technologies such as stall feeding.

7.2 Natural Resources and Environmental Features

The climate is semi-arid or subhumid in the North Temperate Zone in China. It is drought prone, windy and has a long severe winter and spring, with a warm and damp summer and autumn. Annual average temperature is −2.0° to 0°C, with an absolute minimum temperature of −49°C. Annual accumulated temperature ≥10°C is within the range of 1800°–2200°C and the frost-free season is about 90–110 days. Yearly precipitation is 230–380 mm, 80% of which falls in June, July and August, and the aridity index is 1.2–1.5. Annual average wind speed is 3.5–4.5 m/s; the maximum wind velocity is 26 m/s, with about 20 days every year above force 8 on the Beaufort scale. Annual wind-sand days vary from 140 to 220, depending on location.

There are many rivers, lakes and swamps in the Hulunbeier rangelands where there are relatively favourable water resource conditions. More than 400 sand lakes differing in size meet the herders' drinking water demands all year round. The larger rivers are Hailaer and its five tributaries. Groundwater is available at a depth of about 4 m.

Under the influence of geography, climate, soil and other natural conditions, the Hulunbeier Grassland has developed complicated vegetation types with abundant species, especially in the transition from temperate steppe to marshy grassland, which can be seen in the eastern part of Hulunbeier. There are over 1220 species of vascular plants from 108 families and 468 genera, accounting for 84.4%, 70.3% and 53.5%, respectively, of the entire Inner Mongolian vascular plants. Zonal vegetation is comprised of temperate marshy grassland, temperate steppe and five other grassland types (Table 7.1).

7.3 General Status of Grassland Degradation

Hulunbeier Grassland was, until recently, a relatively well-preserved pastoral area in the north of China, but, regrettably, there has been much species loss. The number of plant species in the rangelands has reduced from 130 to 30. Many palatable plants have disappeared from the rangeland.

Recently, affected by many factors such as the changing climatic environment (Chapter 3), excessive deforestation in watersheds, overgrazing, uncontrolled firewood collection and the

Table 7.1. Areal distribution and ranks of grassland subtypes in Hulunbeier.

Grassland types (subtype)	No.	Total grassland area (ha)	Total grassland area (%)
Grassland total area		9,950,780	100.00
Plain and hill grassland	1	4,376,220	43.98
Low wetland meadow	2	2,520,160	25.33
Swampy lowland meadow	3	744,800	7.48
Upland meadow	4	739,526	7.43
Plain and hill grassland meadow	5	586,287	5.89
Low and middle mountainous meadow	6	307,447	3.09
Sandy grassland	7	292,640	2.94
Salinized lowland meadow	8	278,420	2.80
Marsh grassland	9	95,120	0.96
Sandy grassland meadow	10	16,160	0.10

implementation of the Grassland Law (Zhang *et al.*, 2007), the trends of accelerated land degradation, vegetation degeneration and dune activation have become serious. Sand movements have become obvious in the Hailaer River valley and in three other sandy strips in the Yimin River valley. These have had an impact on the local animal husbandry and on the herdsmen's daily life.

As revealed by the 1:100,000 scale satellite-borne Thematic Mapper (TM) image in the 1990s, the grassland area undergoing wind erosion in Hulunbeier was 4,316,200 million ha (Mha), accounting for 43.4% of the total grassland area, among which the grassland area with serious desertification was 558,000 Mha, or about 12.9%, and the steppe area with sand on the surface was 1,013,000 Mha, or around 23.5% of the total area.

In the past 50 years, the percentage of desertified grassland has increased. In 1950, it was 0.18%, while by 2000 it had risen to 18.48% – a tenfold increase (Fig. 7.2). The total area of various degraded grassland in Hulunbeier was 328,000 Mha at the end of the 1980s, about 2.9% of the IMAR total grassland area, but at the end of the 1990s, it was 558,000 Mha, about 12.9% (Table 7.2). Over the 10 years, the net increase was 230,000 Mha, at an annual average growth rate of 7.0%. The annual growth rate of lightly degraded grassland (LDG) was the highest (7.8%), which was five times faster than the national average. At the current rate of expansion, the Hulunbeier Grassland will become desert after 40 or 50 years, directly threatening the security of local industry and agriculture, and people's lives and property in the downstream region will also be threatened.

A 5-year study on sandy land in the northern part of the Hulunbeier desertified grassland on 107 sites showed that the botanical composition

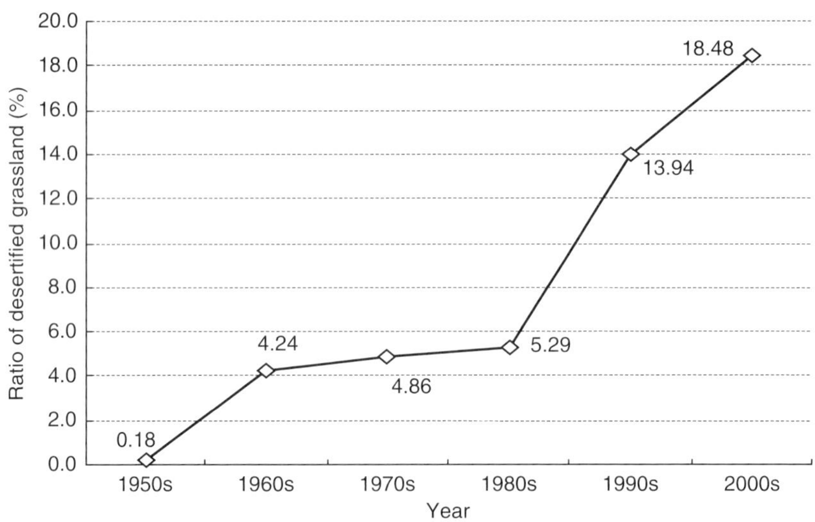

Fig. 7.2. Percentage of degraded rangeland since 1950.

Table 7.2. Expansion of desertified grassland in Hulunbeier from the 1980s to the 1990s.

Degradation category	End of 1980s (Mha)	End of 1990s (Mha)	Total increase (Mha)	Growth rate (%)	Yearly average growth rate (%)
SDG	5,695	9,595	3,900	68.5	6.9
LDG	14,161	25,163	11,002	77.7	7.8
PDG	308,303	522,920	214,617	69.6	7.0
Total	328,159	557,677	229,518	69.9	7.0

SDG, serious desertification; LDG, light desertification; PDG, potential desertification.

was gradually becoming less diverse, with herbaceous perennials being reduced sharply. Vegetation was obviously trending towards more xeric forms and annual plants became dominant. Community structure and diversity have been lost little by little. Compared with typical grassland, the height, coverage and biomass of slightly degraded grassland separately dropped 56.6%, 80.8% and 74.2%, respectively. The abundance, variety and homogeneity of species reduced 76.4%, 56.7% and 32.0%, respectively. Ecological dominance rose 2.1 times. The β diversity of the plant community increased 14.6 times. Community similarity was at the lowest level. The grassland habitat changed completely and the system lost its stability. In addition, the proportion of many primary species, such as *Stipa krylovii*, *Aneurolepidium chinense* and *Agropyron desertorum* fell by 38.0%, 48.9% and 89.9%, respectively, and the ecological niche overlap diminished greatly. All these demonstrated that the availability of resources (water, nutrients) declined because of grassland degradation and the sociability of perennial primary species in the community was reduced. In short, the vegetation was in the process of regressive succession. The grassland survey in the 1980s showed that the proportion of the quality forage species (Gramineae and Leguminosae) was, respectively, 40–75% and 14–41%, but declined to 32–50% and 5–23% in the early 21st century.

7.4 Rangeland Conversion – a Major Factor in Land Degradation in Hulunbeier

The sandy desertified land areas of the Hulunbeier Grasslands increased from 8065 km^2 in 1989 to 20,893 km^2 in 2000, a leap of 159%. Meanwhile, 3613 km^2 of grasslands were converted to cropland from 1986 to 1996, which made an increase in the total cultivated land of 34.8% and the centre in gravity of farmland moved about 33 km north-westwards, approaching the central part of the Hulunbeier Grasslands. About 80% of the newly developed farmlands are from grasslands in south Xin Barag Left Banner and south-east Ewenki Autonomous Banner, which are mostly sandy grasslands and very vulnerable to desertification. Now, a new round of project-initiated grassland cultivation of forage field construction is under way. The grassland ecology and geological environment of the Hulunbeier Grassland is under threat from a new wave of increasing human-induced pressure. For example, three large-scale, state-owned farms were established in quick succession in Chinbaerhu Qi when the Barhu grassland conversion was started at the end of the 1950s. The whole agricultural acreage was 68,000 ha in 1959 and soon reached 4.7 million ha (Mha); after that, rangeland conversion gradually slowed down due to the contradiction between farming and herding, but remained

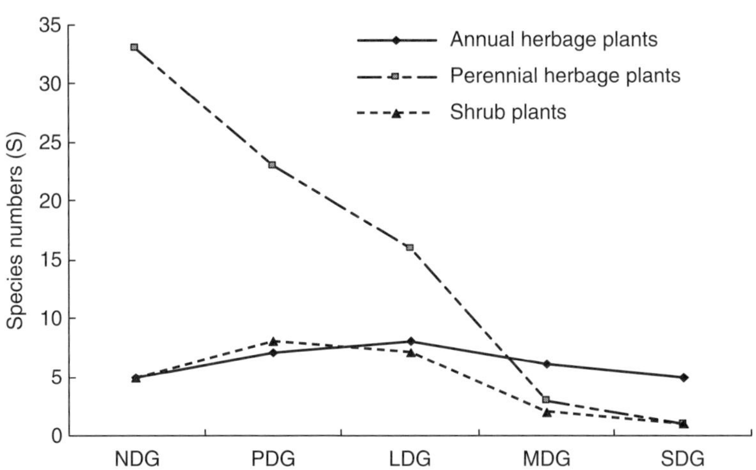

Fig. 7.3. Changes of species life forms in different stages of desertification (data from Lv Shihai). NDG, no desertification; PDG, potential desertification; LDG, light desertification; MDG, medium desertification; SDG, serious desertification.

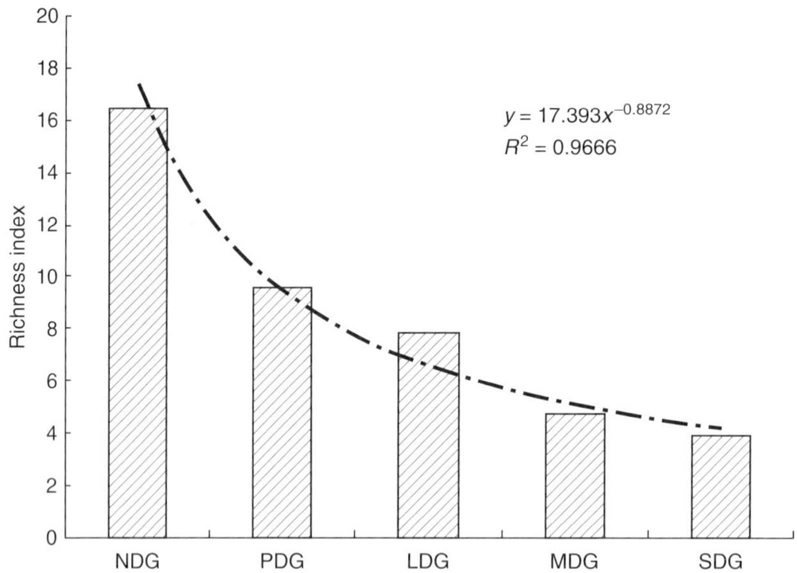

Fig. 7.4. Changes of species richness of communities in different stages of land degradation (data from Lv Shihai). NDG, no desertification; PDG, potential desertification; LDG, light desertification; MDG, medium desertification;SDG, serious desertification.

35,000–40,000 ha/year. In the late 1980s, on Barhu grassland, a second wave of rangeland conversion began because of the nationwide demand for food grains (Fig. 7.5).

At the same time that there was a shrinking area of grasslands, livestock (sheep units) increased from 2.71 million to 4.8 million from 1989 to 1999.[1] Dramatic shifts in the pattern of use have occurred over the past 50 years (Fig. 7.6).

7.5 Grassland Policy and Regulations

7.5.1 Confirmation of the right and contract system of the grassland

Since the 1980s, there has been a transformation in grassland ownership from long-term single national ownership to national ownership and collective ownership. These changes have been backed by local legislation. The reform adjusts the production relationship to the development of productivity, bringing the farmers/herdsmen more autonomy and providing opportunities for further reform of the grassland. In 1984, the contract responsibility system concerning both grass and livestock was carried out, which included 'the socialization of the grassland, the contracting management, the evaluation of the livestock and raising the livestock by each household' (Chapters 2 and 15). The practice of the system has played a part in protecting, constructing and utilizing the rangeland, but it did not increase the herdsman's income. In order to solve this problem, the autonomous regional government carried out the policy of 'Two Rights and One System', namely, to put the grassland's ownership and usufruct and the contracting responsibility system into effect. In 1996, the Autonomous Regional Government issued 'Regulations about Further Implementing the Two Rights and One System'. According to the unified arrangement of the autonomous region, local administration began to instigate with each household the contract system for the grassland and pasture, ensuring a policy of 30 years without any change. This is an important reform with regard to production in the pasture areas after implementation of the household contract system, realizing the important link between labour force and grassland – the fundamental

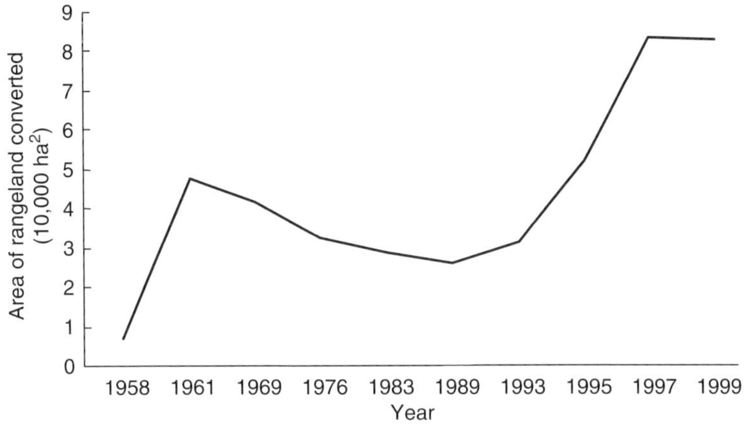

Fig. 7.5. The area of rangeland converted to cropland in Chinbaerhu from the 1950s to the 1990s.

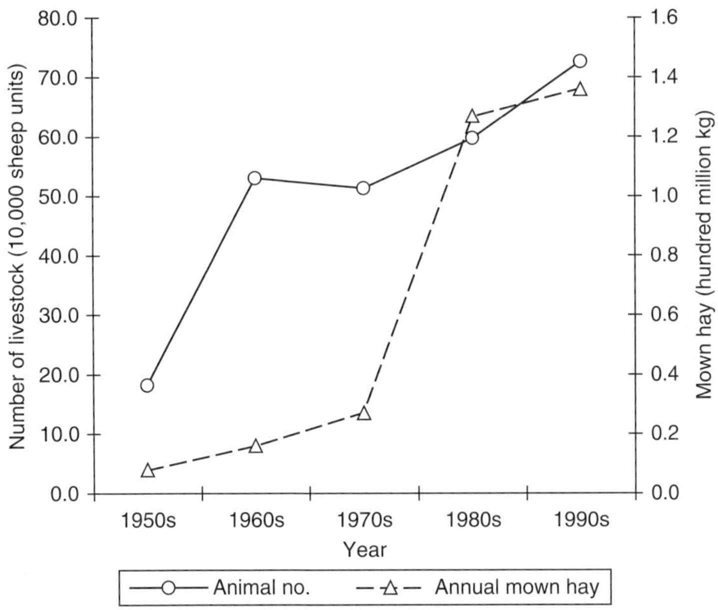

Fig. 7.6. Large-scale rangeland conversion and rapidly rising livestock populations were a feature in Chinbaerhu from the 1950s to the 1990s.

production material in the pasture area. As a result, the raising of livestock developed considerably. The size of the pasture contracted to each household was determined by the number of their family and livestock, with the preservation of some collective grassland to be used as alternative grassland, summer encampment, winter encampment and so on. The remaining pasture was divided into two parts, with seven people and three livestock (or six people and four livestock) as a unit. The contracted pasture size was determined according to the calculated figure of each household. The contracted period was valid for 30 years. If the household raised no objection, the contract would be signed. At present, the grassland has been distributed largely to each household. In addition, part of the pasture that belongs to the grassland construction project has been enclosed in recent years, which has effectively prevented an excessive pasturage, restored the coverage of the vegetation and increased the yield of grass. Each household has an appropriate size of pasture land. The problem is that the enclosed pasture area belongs to the collective, so anyone can herd there; as a result, some areas are prone to overgrazing, which will degenerate the pasture land severely.

7.5.2 Balance system between grassland and livestock

To strengthen grassland protection, construction and rational use, and to promote the sustainable development of stockbreeding, the local government has formulated 'The Management of Balancing the Grassland and Livestock'. Calculation about the forage grass has been conducted quantitatively so as to determine its viable stock carrying capacity. The appropriate carrying capacity of the grassland is determined by the 'Calculation Standard of Natural Grassland Resource for Appropriate Stock Capacity' formulated by the autonomous region.

To strengthen the management and protection of the grassland, use the grassland resources rationally, improve the eco-environment, regulate use of the grassland by non-herdsmen, ensure the lawful rights of the herdsmen and increase their income, two orders were issued under the provision of the 'The Grassland Law of PRC', 'The Grassland Management Regulations of Inner Mongolia Autonomous Region', 'Advice about the Liquidation of the Non-Herdsmen's Occupation of Pasture Land and Regulations of the Usufruct Circulation of the Grassland under the Law' and 'Decision About Enhancing the Grassland Protection and Construction' issued by the Hulunbeier's People's Government. They are 'Temporary Measures of the Non-Herdsmen's Utilization of Grassland Management' and 'Temporary Measures of the Viable Grassland Management'. Each September, the contracted grassland user will sign an obligation document, which includes a report of the current condition of the grassland, the kinds and number of livestock and general feed balance. Measurements dealing with overloading include: (i) planting and purchasing forage grass, increasing the supply of forage grass; (ii) confining forage grass, ceasing herding periodically or grazing alternately in certain areas; (iii) optimizing the composition of livestock herds and increasing the slaughter rate; and (iv) large stock-keepers leasing their grassland to non-stock-keepers or small stock-keepers (i.e. annually subcontracting the grassland). By so doing, these measures aim to achieve a balance between grassland and livestock. In order to promote this balance, the grassland user should use alternative management of the grassland under the provision of the 'Method of Grassland Contract Management Right in Inner Mongolia Autonomous Region'.

7.5.3 Grassland supervision and management system

The Grassland Supervision Office is authorized to investigate and punish people who violate 'The Grassland Law'. Its general task is to strengthen the protection, management, construction and rational use of the grassland to protect and improve the eco-environment and to develop modern stockbreeding. This Grassland Workstation–Grassland Supervision Office is composed of four agencies, namely, the Grassland Construction Office, the Grassland Supervision Office, the Comprehensive Office and the Financial Office. The Grassland Supervision Office regularly sends workers to the villages, patrols the grassland, establishes checkpoints, takes fire prevention measures and examines the effects of the construction project already carried out.

7.6 Recovery – Technology Systems for Recovering Desertified Rangeland

China has many effective scientific achievements in grassland improvement projects arising from the study of the processes and origin of grassland degradation and the most effective methods for vegetation restoration on degraded sites. Similarly, improved grazing management (rest rotation, deferred grazing), the assessment of reasonable usage based on feed balance and the promotion of pen/yard feeding of livestock have contributed to reducing the pressure on the rangelands. The most effective approaches are fencing to regulate livestock numbers and season of use, long-term enclosure to allow natural regeneration of plants, grazing bans, rangeland re-seeding, grassland establishment of artificial (sown) pastures and livestock yard-feeding technology.

7.6.1 Fencing and exclosure as management tools

Exclosure and grazing ban experiments were conducted in typical grassland. Results showed that over the 17 years from 1982 to 1999, grass output of the populations of *S. krylovii,* the populations of *S. grandis* and the populations of *Thymus serpyllum* increased by about 8–9 times, 3–8 times or by about 3 times, respectively (Table 7.3). The height of grass layer increased nearly 5–10 times, 3–6 times and 3 times, respectively. Vegetation cover improved 3–4 times, 2–3 times and 3 times, respectively. Plant species biodiversity increased two- to threefold, three- to fourfold and twofold, respectively. So exclosure is the main measure to recover vegetation and stabilize the rangelands (Zhao *et al.*, 1994).

Research by one of the authors (Lv Shihai) indicates that exclosure is one of the most economic and effective measures for restoring rangeland. In the absence of grazing and trampling, the environment improves step by step. Some 'locally extinct' species intrude little by little and gradually occupy a dominant position. The content of organic matter and overall soil fertility improvement speed up as cover increases and forward succession occurs. Biomass output also improves (Table 7.4).

7.6.2 Rotational grazing on summer pasture

A comparison of rotational grazing and free grazing was made in Baolige and Erjihzuoersumu in Dongwe Banner. Usually, the summer grazing there lasts for about 100 days from 1 June until 11 September. The comparison of rotational grazing and free grazing showed that grass height increased 7–15 cm compared with free grazing. It was 5–10 cm before rotational grazing began. Grass cover reached 50–55% – more than 16–18 percentage points higher than free-grazing grassland – herbage biomass was 113.8 g/m^2 – 23.8 g/m^2 more than free-grazing grassland.

7.6.3 Shallow tillage as a rehabilitation technique

Under some suitable conditions, this approach performs well. For example, soil layer thickness was >20 cm and the existing degraded vegetation consisted mainly of rootstock grass, or stoloniferous grasses. A tyned implement (small ripper), mounted or drawn by tractor,

Table 7.3. Impact of exclosure on different plant communities in typical grassland.

Community types		*Stipa krylovii* populations				*Stipa grandis* populations				*Thymus serpyllum* populations	
Slope direction		NE		SE		N		S			
Growing year		1982	1999	1982	1999	1982	1999	1982	1999	1982	1999
Exclosure	Plant species (no./m^2)	6	15	5	13	5	19	3	9	8	18
	Cover (%)	25	95	20	80	20	65	20	35	20	55
	Grass layer height (cm)	10	54	8	55	8	52	5	18	3	9
	Biomass (kg/Mha)	750	6050	550	4500	550	4200	550	1550	750	2255

Table 7.4. Effect of exclosure on plant community composition and structure.

Years of exclosure	Foliage cover (%)	Grass layer height (cm)	Density (plant/m²)	Biomass (DM g/m²)	30 cm root (DM g/m²)	Litter (DM g/m²)
1 year	11.3	15.8	33.0	37.4	40.1	0
4 years	24.5	20.3	96.9	232.6	251.7	41.3
7 years	32.6	25.3	355.2	168.9	348.3	84.6
11 years	44.5	28.2	654.7	116.7	310.8	92.7
17 years	51.5	27.7	834.1	156.2	325.4	110.2

DM, dry matter.

Table 7.5. Impact of shallow tillage on rangeland vegetation in Hulunbeier.

Place	Height (cm)		Increase compared with zero tillage (%)	Cover (%)		Increase compared with zero tillage (%)	Density (bunches/m²)		Increase compared with zero tillage (%)
	Shallow tillage	Zero tillage		Shallow tillage	Zero tillage		Shallow tillage	Zero tillage	
Dongwu	31.5	22.5	40.0	54	43	18.6	252	109	131.3
RIGSTS[a]	21.9	14.3	53.2	45	39	15.4	277	137	102.2
Abaga Banner	18.0	13.3	35.3	35	30	16.7	204	111	83.8
Cattle farm site	96.5	72.6	32.9	95	70	35.7	743	691	7.5

[a] RIGSTS, Research Institute of Grasslands Science Shallow Tillage.

loosens the soil along the contour to a depth of 10–15 cm, in strips every 30 m. Destruction of existing vegetation is minimized. The work is done in the rainy season in mid- to late June. No harrow is used after ripping. The Research Institute of Grasslands Science, Chinese Academy of Agricultural Sciences, employed this technology in the Hulunbeier Grassland. The community appearance and botanical composition changed markedly between the treated and the untreated (zero till) rangeland. Grass height in the shallow tillage treatment increased 32.9–53.2% more than in the untreated, cover expanded 15.4–35.7% and density was up 7.5–131.3% (Table 7.5).

7.6.4 Artifical grassland (sown pasture) establishment

Artificial grassland plays an important role in ecological recovery because it allows pressure to be shifted from the rangeland. Under suitable conditions (soil and moisture), established artificial grassland can increase forage output and quality, which could also relieve the grazing load pressure on rangeland and play a part in the promotion of animal husbandry development by increasing forage supply and thus increasing incomes (Table 7.6).

7.6.5 Livestock pen/yard feeding technology

Remarkable results can be obtained through many effective methods, such as establishing artificial grassland with high yield and dependable crop, enhancing the storage and processing of forage, e.g. silage, and use of warm pens (or yards) for feeding livestock over winter. Such an approach conserves the winter pastures, improves nutrition of pregnant ewes and cows and improves the utilization ratio of straw from 30% to upwards of 90%. It can also allow time for deferment of grazing for 2–3 weeks in spring after 'green-up'. Pen feeding can be more effective if there is artificially planted grass or a forage crop to augment

Table 7.6. Vegetation conditions in perennial rangeland and artificial pasture and relation to income-generation outcomes.

Grassland type	Grass cover (%)	Grass height (cm)	Grass output (hay t/Mha)	Edible forage proportion (%)	Crude protein output (kg/Mha)	Annual general income/ grassland area (RMB[a]/Mha)	Yearly net income/ grassland area (RMB/Mha)
Sown pasture	>95	87.6	9.14	99	1148.5	6875	5050
Exclosed rangeland	96	27.7	3.93	76.7	462.6	1965	1440
Free-grazing rangeland	69	14.4	2.91	74.1	342.5	2535	775

[a] RMB yuan.

hay. Taking the livestock off the rangeland effectively protects the vegetation during the winter, and especially at the vulnerable period in early spring. The amount of manure that can be collected is also a factor. The flexibility of the system allows changes to mating time to produce lambs or calves at a time that meets market demand, e.g. the spring festival.

7.7 Lessons Learnt: How Can Further Rangeland Degradation be Prevented?

For centuries, the herdsmen in Hulunbeier lived a life through natural stock raising, extensive farming and an undeveloped way of production, which resulted in low yield of individual animal products and slow turnover of herds. Therefore, the number of livestock kept on increasing and the output and economic effects failed to grow in step with each other. The development of stock raising in this area has been impeded due to some economic development in the industrial and urban sectors and the changing social environment, including the inward migration of farmers. The relatively low level of education in pastoral areas makes the extension of science and technology difficult.

Government effort to promote more scientific methods and obtain higher productivity from the land has led to the excision of large tracts of the best rangeland for conversion to cropland. This has increased pressure on the remaining rangelands as human population increases and demand for food grains goes up. Many of the past policy initiatives had unintended consequences, and these lie at the base of today's rangeland degradation problems.

Implementation of the Grassland Law and the various IMAR government regulations has caused hardship to herders and exacerbated the degradation problem. Restricting people and their livestock to fixed areas without providing adequate alternative fodder or grazing resources has not proved workable. While there is some arable land that might be used to grow fodder, or even feed grains, there is the dilemma of how to feed the burgeoning human population and this leads to inevitable conflict over what to grow on the shrinking area of arable cropland.

Lack of proper markets for livestock and livestock products (cashmere, wool, meat, hides) has led to lower than hoped for turn-off and a tendency for herders and farmers to keep unproductive livestock rather than sell on these 'surplus' animals. Proper price incentives for premium-quality livestock products do not yet exist. Market failure is an indirect cause of rangeland degradation, as is the lack of proper rural credit and the opportunity for herders/farmers to borrow modest sums on a long-term basis to invest in technology and better-quality livestock.

Property rights and better land tenure arrangements could go a long way towards getting a commitment from land users to exercise better stewardship of the rangeland resource.

Large-scale rangeland conversion to cropland needs to cease and the abandoned croplands and some marginal areas should be sown to artificial pastures.

Note

[1] *Hulunbeier Yearbook*, Statistics Bureau of Hulunbeier League, Hailar (1999).

References

Statistics Bureau of Hulunbeier League (1999) *Hulunbeier Yearbook*. Statistics Bureau of Hulunbeier League, Hailar, China.

Zhang, M.A., Borjigin, E. and Zhang, H. (2007) Mongolian nomadic culture and ecological culture: on the ecological reconstruction in the agro-pastoral mosaic zone in northern China. *Ecological Economics* 62, 19–26.

Zhao, H., Li, S.G., Zhang, T.H., Okhuro, T. and Zhou, R.L. (1994) Sheep gain and species diversity in sandy grassland, Inner Mongolia. *Rangeland Ecology and Management* 57(2), 187–190.

8 Case Study 2: Horqin Sandy Land, Inner Mongolia

Jiang De-ming, Kou Zhen-wu, Li Xue-hua and Li Ming
Institute of Applied Ecology, Chinese Academy of Sciences, Shenyang, China

Synopsis

The Horqin Sandy Land is one of the major tracts of such land in north-east China. Here, we present an analysis of the changes that have occurred over the past 50–60 years and evaluate the underlying causes of rangeland degradation. Data are presented on climate change and we analyse and discuss the relationship between the changing trend of climate in recent years and the onset and spread of land degradation. Vegetation restoration on sandy land receives special attention. Successional pathways for both the degradation and recovery phases are outlined.

Keywords: re-vegetation; restoration ecology; soil; wind erosion; overgrazing; dust and sandstorm; socio-economics; demography; land-use change; aridity index

8.1 Statement of the Degradation Problem

This vast aeolian sand land (total area 51,750 km^2) was formed on the alluvial plain of the Xiliaon River in north-east Inner Mongolia during the Quaternary period. However, due to population growth in past decades, steppe and arable land were overused. Livestock and human populations rose rapidly, with excessively high stocking rates and, consequently, forest and grass vegetation was destroyed Overgrazing (by cattle, goats, sheep, camels and horses), clearing of land for cropping and overcutting of trees and shrubs in this vulnerable ecosystem have resulted in accelerating land degradation and desert encroachment.

Desertified lands have spread rapidly in the past three decades and new dust and sandstorm sources are expanding. The latest desertification period would thus have occurred during the past 100 years, with a noticeable acceleration in the last 30 years. The total population of the Horqin Sandy Land increased from 936,400 in 1947 to 3,480,200 in 1996, the annual average growth rate was 5.22% and the population density increased from 10.44 persons per km^2 in 1947 to 38.8 persons per km^2 in 1996. The domestic animal population has increased more than sixfold during the same period.

South-western Horqin has reached a stage of severe desertification, while in other areas water and wind erosion continues unabated, causing more serious land degradation year after year. Desertification is progressing in this area and is damaging not only the rangeland ecosystem but also agricultural fields and infrastructures such as railways and roads. Little remains of what was once the natural vegetation in the region. Some relics of poplars (*Populus simonii*), willows (*Salix matsudana*, *S. gordejevii* and other species), wild peach trees (*Prunus armeniaca*) and elm trees (*Ulmus pumila* and other species) can still be found. Many lakes have dried up or become highly mineralized and the water table in the aquifers has fallen drastically.

8.2 Location of the Horqin Sandy Land

8.2.1 Geography and geomorphology

The Horqin Sandy Land (118°35'–123°30'E, 42°41'–45°15'N, 178.5–631.9 m above sea level) is located on the western side of the north-east plain of China. Its principal area is distributed on the alluvial plain of the lower branches of the Western Liaohe River and the upland sandy land is scattered across the alluvial mesa in front of the Great Xing'an Mountains. The district includes 17 banners (counties) in the east of Inner Mongolia, the total area of which is 51,750 km^2, and they are distributed mainly in Tongliao, Chifeng, the west of Jilin Province and the north-west of Liaoning Province (Fig. 8.1).

The main geomorphology in the Horqin Sandy Land includes sand dunes, sand plains, interdune swales and rocky hills. The dominating feature is the mixture of sand dune and interdune swales. The sand dunes can be classified into fixed dune (vegetation coverage >40%), semi-fixed dune (vegetation coverage 20–40%) and shifting dune (vegetation coverage <20%). The relative height of the sand dunes and gently sloping sandy land is generally 10 m and 3–5 m, respectively, and the maximum height is about 20–30 m. The inter-dune lowland can also be classified into wet meadow, alkali meadow and sand meadow.

8.2.2 Climate

The Horqin Sandy Land experiences a continental climate due to the influence of high-pressure air from Mongolia and the distance from the ocean. The climate is characterized by frequent drought and strong winds in spring, torridity and concentrated rainfall in summer, short cool days in autumn and long cold days in winter. Annual mean temperature is *c*.5.2–6.4°C and the accumulated temperature (≥10°C) is *c*.3000–3200°C (Table. 8.1). The total solar radiation of the Horqin Sandy Land is 5200–5400 MJ/m^2, of which the proportion in the growing season (July–September) is 65%. The total solar radiation during that period (daily mean temperature ≥10°C) is 2800 MJ/m^2, which is 50% of the

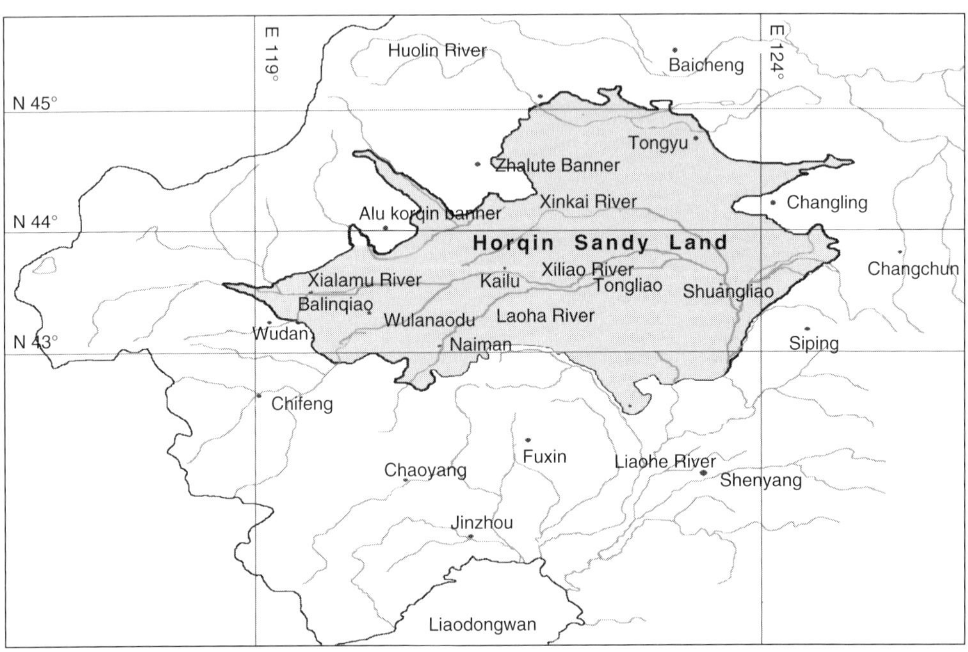

Fig. 8.1. Geographical location of the Horqin Sandy Land.

Table 8.1. Average monthly temperature of the central regions in the Horqin Sandy Land (in °C). Partial data from Zhu (1994).

	Spring			Summer			Autumn			Winter		
Region	3	4	5	6	7	8	9	10	11	12	1	2
Kailu County	−2.6	7.8	16.0	21.0	23.9	22.1	15.7	7.0	−3.5	−11.7	−14.7	−10.8
Zhalute Banner	−2.7	7.5	15.8	20.8	23.4	21.7	15.3	6.5	−3.6	−11.1	−13.3	−10.6
Kulun Banner	−1.2	8.3	16.5	20.9	23.7	22.3	16.4	8.1	−2.0	−10.1	−12.6	−9.3
Naiman Banner	−1.6	8.2	16.5	20.9	23.6	21.7	15.8	7.6	−2.5	−10.7	−13.1	−9.8
Kezuozhong	−2.9	7.2	15.6	20.9	23.7	21.9	15.4	6.5	−4.4	−13.6	−16.2	−12.1
Kezuohou	−2.0	7.5	15.5	20.1	23.2	21.8	15.6	7.2	−3.0	−11.4	−14.0	−10.5
Wengniute Banner	−2.33	8.03	15.7	20.0	22.4	20.6	14.6	6.63	−2.8	−12.5	−9.83	−9.8
Tongliao City	−2.8	7.6	15.9	21.0	23.7	21.9	15.4	6.9	−4.0	−12.9	−15.3	−11.7

annual amount. Annual rainfall is *c.*343–500 mm: the southern and eastern regions receive more and the northern and western regions receive less. Under the influence of the South-east Asia monsoon (tropical maritime air mass), there is an uneven intra-annual distribution of rainfall. The majority of rainfall occurs in summer (June–August), which accounts for 70–75% of the annual amount, but the proportion of rainfall in winter and spring is only 11–16%. The year-by-year rainfall shows great variation, with a maximum of *c.*606.5 mm and a minimum of *c.*136.9 mm. The dryness coefficient is *c.*1.0–1.8.

8.2.3 Main soil types

The main zonal soils in the Horqin Sandy Land are dark brown forest soil, chestnut soil and dark loessial soil, and the non-zonal soils are sandy soil, meadow soil and saline–alkalized soil.

Dark brown forest soil belongs to temperate forest soil and is distributed mainly in the medium-low mountains to the south of the Great Xing'an Mountains. The natural vegetation is mainly coniferous and broadleaved mixed forest, and shrub and herbage grow prolifically under the Mongolian oak and birch forest. The soil parent material is mainly residual and slope deposit.

Chestnut soil is distributed mainly in the Xila Mulun River and to the north of the Xinkai River, which is in the extreme east of the Eurasian continent. The vegetation is mainly steppe, comprised of perennial herbaceous plants.

Dark loessial soil, i.e. grey cinnamon forest soil, is distributed mainly in the series of loess hills and the terraces of Chifeng, Aohan Banner, Kulun Banner and Wengniute Banner. The preserved area of dark loessial soil is very small, due to long-term improper cultivation and erosion.

Sandy soil can be classified into aeolian sandy soil, grassing sandy soil, meadow sandy soil and chestnut sandy soil, of which the distribution area of aeolian sandy soil is the largest. According to the fixing state and vegetation coverage, aeolian sandy soil can be classified into fixed, semi-fixed and shifting. The characteristics of aeolian sandy soil are coarse texture, poor nutrients, low water-holding capacity and low fertility, which are unfavourable for plant growth. The vegetation coverage in aeolian sandy soil is generally restricted to drought- and grazing-tolerant plants.

Meadow soil is semi-hydromorphic soil developed under the effect of meadow vegetation and a high water table and is distributed mainly in the flood plain and in river valley of the Western Liaohe River, the Xila Mulun River, the Laoha River and the Jiaolai River Basin. The parent material is the riverine flood alluvial deposit and lake deposit. The high groundwater level and soil nutrient favour the growth of meadow vegetation. Nowadays, most meadowland has been cultivated into farmland, except for some parts of low wet and saline meadow that are still rangeland.

Saline–alkalized soil, occupying a small area in the Horqin Sandy Land, can be classified into alkalized solonchak and meadow solonchak, which is scattered mainly in meadow or former meadows.

8.2.4 Vegetation

The Horqin Sandy Land is in the transition zone where elements of the Mongolian flora, the Changbai Mountain flora and the north China flora merge. The main zonal vegetation is woodland steppe, which is an ecotone between typical steppe and forest steppe. In the past 100 years, owing to population growth and strong disturbances such as overgrazing and overcutting, the original woodland steppe has been destroyed entirely and substituted by secondary vegetation at different succession stages.

Most of the pioneer vegetation in the shifting sand dune is annual or biennial species, notably *Pennisetum flaecidum*. They are the earliest species that invade bare shifting sandy land, but the community structure is very simple. A few psammophyte species form single species communities. *Chenopodiaceae* such as *Agriophyllum arenarium* are typical pioneer species in the Horqin Sandy Land.

Shrub and semi-shrub vegetation in sandy land forms a stable community structure with strong resilience to disturbance. The main shrub communities include *Caragana microphylla*, *P. humilis* and *Atraphaxis manshurica* and the main semi-shrub communities include *Artemisia halodendron*, *A. frigida* and *Ephedra distachya*.

The herbage plants in the fixed sand dune include mainly *Glycyrrhiza uralensis*, *Diarthron linifolium*, *Cleistogenes chinensis*, *Lespedeza davurica*, *Agropyron cristatum*, *Delphinium grandiflorum*, *Thalictrum squarrosum*, *A. siversiana*, *Chloris virgata* and *Tribulus terrestris*.

8.3 Degradation Causes within the Horqin Sandy Land

8.3.1 Changing trend of modern climate

Taking Wengniute Banner and Kangping County of the Liaoning Province, which are located in the west and the south-east of the Horqin Sandy Land, respectively, as our examples, we analyse and discuss the relationship between the changing trend of climate in recent years and the onset and spread of land degradation.

As shown in Table 8.2, the wind velocity reaches a maximum in spring and winter, while the precipitation is restricted to the low valley, where there is low vegetation coverage and much bare ground. The ground surface and rangeland vegetation are exposed to damage by wind erosion for about half of each year, and the loose sandy soil forms large blowouts and develops

Table 8.2. Monthly changes of climate in Wengniute Banner and Kangping County.

	Temperature (°C)				Precipitation (mm)				Wind velocity (m/s)	
	Wengniute Banner		Kangping County		Wengniute Banner		Kangping County		Wengniute banner	
Month	Monthly average	Season average	Monthly average	Season average	Monthly average	Proportion (%)	Monthly average	Proportion (%)	Monthly average	Season average
3	−2.33		0.06		5.9		13.03		3.875	
4	8.03	7.14	9.71	8.845	10.78	11.31	23.28	13.72	4.2	4.017
5	15.73		16.77		25.7		38.9		3.975	
6	20.03		21.69		56.68		80.18		2.825	
7	22.38	20.99	23.91	22.85	113.6	70.29	153.1	69.71	2.3	2.342
8	20.55		22.95		93.1		149		1.9	
9	14.55		17.045		43.58		53.67		2.575	
10	6.63	6.14	9.19	8.143	14.5	16.62	20.56	15.05	3.025	2.945
11	−2.75		−1.805		4.2		8.3		3.225	
12	−12.48		−8.845		1.6		2.94		3.7	
1	−9.825	−10.7	−12.35	−9.628	2.45	1.37	3.395	1.526	3.425	3.475
2	−9.8		−7.69		1.08		2.03		3.3	
Whole year	5.892		7.553		374.7		548.29		3.194	

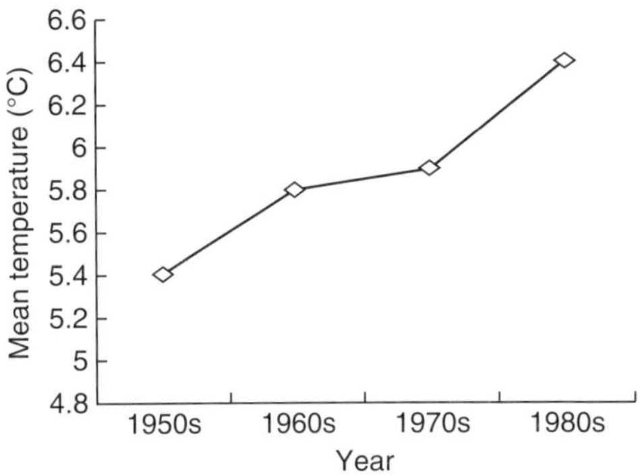

Fig. 8.2. Mean temperature changes over several decades in Wengniute Banner.

further into shifting sand dune. The degradation process develops more intensely in spring than in winter because of the higher temperature and evaporation. Wind and drought happen simultaneously in spring and winter and these are the most important meteorological factors leading to desertification in the Horqin Sandy Land.

There was an increase of 1.0°C in the average temperature between 1950 and 1980 in Wengniute Banner (Fig. 8.2). The average temperature increased by 1.06°C (from 6.88°C in 1960 to 7.94°C in 1990) in the Kangping County of Liaoning Province (Fig. 8.3). The climate shows a warming trend over the past few decades in the Horqin Sandy Land (see also Chapter 3).

A polynomial match was applied to analyse the average temperature and annual precipitation of Wengniute Banner from 1959 to 1990 and of Kangping County from 1959 to 2000. The climate change trend of the two areas was similar, which showed that the temperature

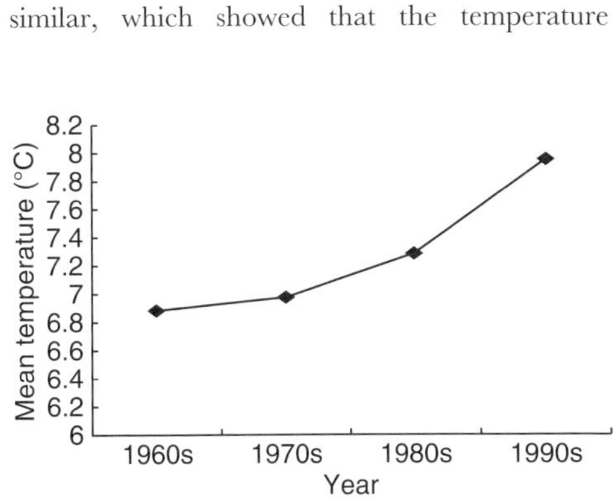

Fig. 8.3. Mean temperature changes over several decades in Kangping County.

increased after the low values recorded from 1950 to 1970 and accelerated in the late 1980s, while the precipitation showed an 'S' curve during this period (Figs 8.4 and 8.5). The temperature and precipitation lines of the Horqin Sandy Land show a tendency towards a warmer and drier climate. The annual average wind velocity is now lower, with an obvious fluctuation occurring in the 1980s (Figs 8.6 and 8.7), suggesting that the climate would be less stable if the temperature increased. The analysis of climate change in Wengniute Banner and Kangping County shows a trend towards warming and drying in the Horqin Sandy Land.

8.3.2 Impact of human activities on desertification

Desertification developed rapidly in line with population growth in the 20th century in the Horqin Sandy Land. Intensive human activities exceeded the limit of the natural carrying capacity, destroyed the ecosystems and brought on accelerated desertification, which reached serious proportions.

Growth of population

The total population of the Horqin Sandy Land increased from 936,400 in 1947 to 3,480,200 in 1996, the annual average growth rate was 5.22% and the population density increased from 10.44 persons per km^2 in 1947 to 38.8 persons per km^2 in 1996. In Wengniute Banner, the total population was 419,400 in 1986 and had increased by a factor of 2.53 times greater than that in 1950. The population density increased from 13.96 persons per km^2 in 1950 to 35.6 persons per km^2 in 1986. The total population of the Zhelimu League was 2.5 million in 1983, with a density of 41.5 persons per km^2, increasing at an annual rate of over 3% during the period 1949–1983. The change of cultivated land and livestock is shown in Table 8.3. UN reports pointed out in 1977 that the population density should not exceed 7 persons per km^2 and 20 persons per km^2 in arid and semi-arid areas, respectively. Under the conditions of unrestrained production, the rapid growth of population and demand for raw materials and food resulted in the expansion

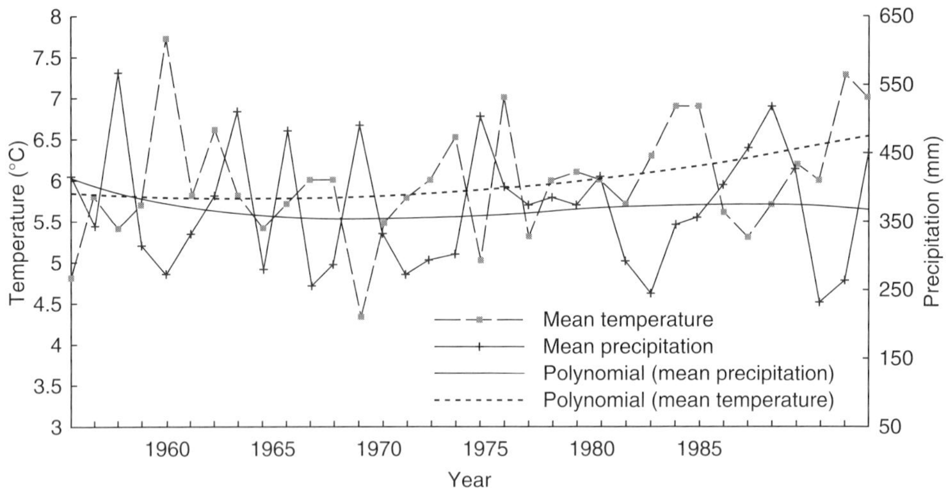

Fig. 8.4. Fitted curves of mean temperature and precipitation over the years in Wengniute Banner.

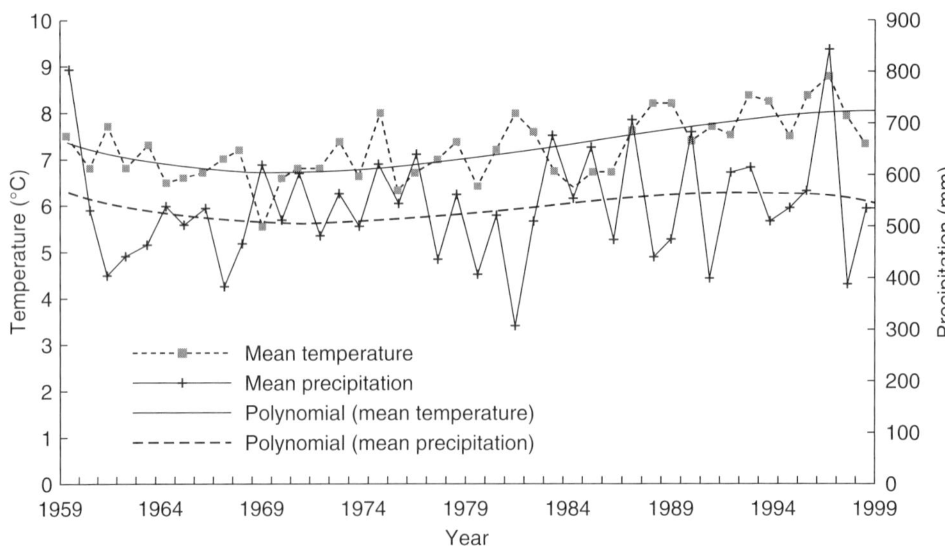

Fig. 8.5. Fitted curves of mean temperature and precipitation over the years in Kangping County.

of cropland area and other land-use changes to cater for the burgeoning migrant population.

For Kailu County, Kezuozhong Banner, Tongliao and Kezuohou Banner, located on the western Liaohe River plain, the population growth rate was comparatively fast and the annual rate was above 4.3% because of the relatively advantageous natural conditions and unprecedented urban expansion. The population growth rate of Zhalute Banner was the highest (22.18%), while those in Kulun Banner and Naiman Banner were lower (3.07% and 3.47%, respectively).

Extensive cultivation

Cultivation of rangelands is the most serious factor that destroys natural vegetation. The

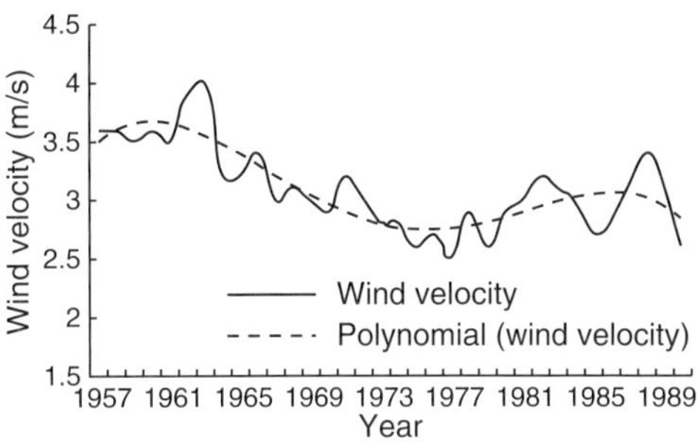

Fig. 8.6. Fitted curves of mean wind velocity over the years in Wengniute Banner.

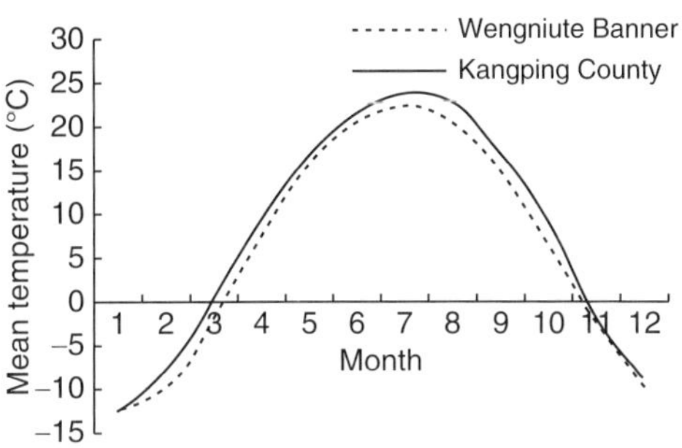

Fig. 8.7. Inter-annual mean temperature change of Wengniute Banner and Kangping County.

destructive effect of cultivation on ground surface and vegetation is very rapid and irreversible in the short term. With the acceleration of erosion and formation of blowouts, the shifting sand begins to spread and form laminar sand flows. By the 1990s, a great deal of low-lying sandy lands and steppes in the east of the Horqin Sandy Land had been cultivated in response to the rising demand for food. For instance, 600 million ha (Mha) of cropland were cultivated around 1998 in the Xinaili and Fuxing villages belonging to the Maolin countryside, Kailu County, Tongliao. Many areas of partial steppes and the interdune swales that were in good condition were cultivated for farmland. The original elm woodland steppe was seriously destroyed by overcultivation. Over the past 50 years, with the successive exploitation and cultivation of sandy land, the total area of cropland in the Horqin Sandy Land reached a peak in the 1960s. The dynamics of farmland in Tongliao, which is a main area of the Horqin Sandy Land, are analysed in Fig. 8.8. The cultivated land area was 634,700 Mha in 1960, an increase of 152,700 Mha (32%) over that in the initial period of New China. There was another cycle of land conversion in Tongliao in 1996 and the area broke through 500,000 Mha (an increase of about 11% over that in 1949) after slipping downward for 20 years.

Overgrazing

Overgrazing is the main cause of rangeland degradation. The main consequences of overloading rangeland are summarized as follows: (i) a decrease in rangeland area with an increase in cultivated land; (ii) a decrease in rangeland area with an increase in livestock number; and (iii) degradation of the rangeland, with a decrease in forage yield and a decline in the carrying capacity because of both natural and the above-mentioned factors (Zhao *et al.*, 1994). Taking five banners located to the north

Table 8.3. Changes of average land area and livestock product per capita of Wengniute Banner.

Item	1950	1957	1965	1978	1980	1985	1986
Total population	165,756	222,999	281,364	380,385	394,076	418,600	419,373
Total land area (km^2)	1,187,600	1,187,600	1,187,600	1,187,600	1,187,600	1,187,600	1,187,600
Land per capita (Mha)	7.12	5.29	4.19	3.11	2.99	2.85	2.83
Arable land per capita (Mha)	0.64	0.55	0.43	0.31	0.29	0.27	–
Livestock	8.5888	16.5576	11.8784	19.1224	19.1430	19.4407	19.7050
Livestock per capita	0.52	0.74	0.78	0.50	0.49	0.47	0.47
Sheep	56,512	254,803	471,975	495,711	469,490	430,855	419,271
Sheep per capita	0.34	1.14	1.68	1.3	1.1	1.03	1.0

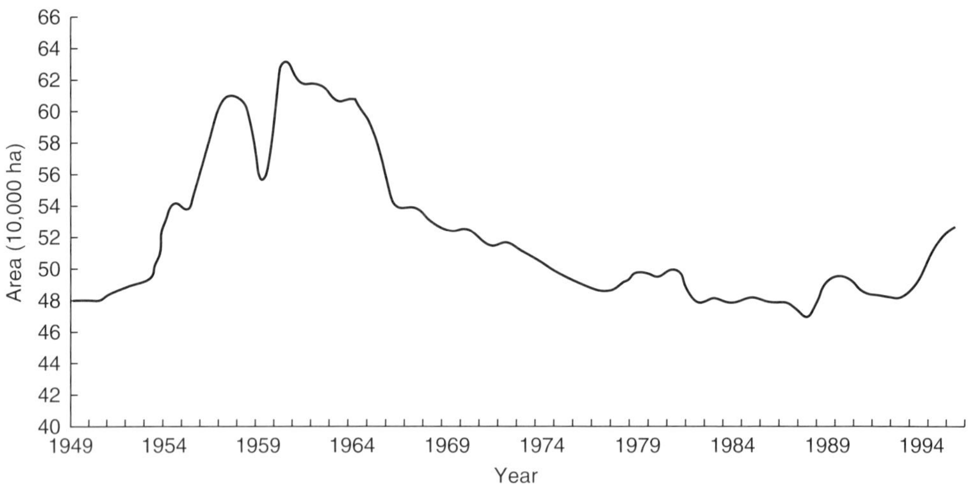

Fig. 8.8. Changes of farmland area in Tongliao in the 50 years to the mid-1990s.

of Chifeng City as examples, all natural rangeland areas in the five banners total 3,595,000 Mha (Table 8.4). The theoretical carrying capacity of the warm season in a year of above-average rainfall, a normal year and a poor year was 5,760,000, 4,610,000 and 3,450,000, respectively, and that of the cool season was 5,283,000, 4,204,000 and 3,126,000, respectively. In 1985, the actual number was 6,722,000 and 5,892,000 in the warm and cool seasons, respectively, which was taken as standard; the overload volume of each banner was 15–72% in a normal year and above 4% even in a year of above-average rainfall. Even in a colder than normal winter/spring, it was 83.19% in Aluhorqin Banner. The rangeland carrying capacity of Wengniute Banner was 79.5, 220.5 and 229.5 per 100 ha in the 1950s, 1960s and 1970s, respectively, and increased to 244.5 per 100 ha in the 1980s. In Naiman Banner, the rangeland area occupied by one sheep was only 0.15 Mha.

Overcutting

With the population growing, the demand for firewood increased. The shrubs and semi-shrubs of the sandy area were the main firewood resource in the Horqin Sandy Land. Most of these have now deteriorated because of overcutting. Before the

Table 8.4. Overloading situation of natural rangeland in the main counties of northern Cheifeng in 1985 (million ovine units).

	Overloading in warm seasons (proportion in theoretical carrying capacity (%))			Overloading in cool seasons (proportion in theoretical carrying capacity (%))		
Banner	Full year[a]	Normal year	Poor year	Full year	Normal year	Poor year
Balinyou	5.17	22.89	40.61	3.58	20.56	37.54
Banner	(5.83)	(32.29)	(76.39)	(4.22)	(30.28)	(73.71)
Balinzuo	22.31	47.77	73.22	16.42	37.93	59.44
Banner	(17.53)	(46.92)	(95.88)	(15.27)	(44.09)	(92.12)
Wengniute	39.19	59.43	79.66	23.02	42.96	62.91
Banner	(38.74)	(73.44)	(131.26)	(23.08)	(53.84)	(105.13)
Aohan	23.53	36.18	48.83	4.17	20.63	37.09
Banner	(37.20)	(71.49)	(128.65)	(5.84)	(37.59)	(96.55)
Aluhorqin	5.17	22.89	84.12	13.71	46.67	79.63
Banner	(2.64)	(14.61)	(71.57)	(83.19)	(35.40)	(80.53)

[a] A year of above-average rainfall.

1980s, it was a particularly common phenomenon every winter to collect grass and shrubs as firewood. In the Kangping County of Liaoning, this was estimated to be 4–5 million kg every year (if every person collected only 100 kg). Furthermore, this activity caused forage roots to be pulled up; the overwinter organs of some monocotyledons and many hemicryptophytes were laid bare and regenerative sprout was damaged, thereby seriously decreasing their ability to overwinter. In some places with dense human population, the cutting of sand-fixing vegetation has caused the fixed sand dunes to reactivate. For instance, 1340 families in the Eleshun countryside of the Kulun Banner of Inner Mongolia could consume 9266.7 Mha of shrub for firewood every year. At the same time, other actions such as cutting sand willows for basket making, etc., digging liquorices and cutting herbal plants such as *Ephedra* spp. can also destroy the original vegetation and reactivate the sand dunes – accelerating land degradation.

8.3.3 Disturbance system of the Horqin Sandy Land

In the Horqin Sandy Land, the main natural disturbance factors include climate, soil and animals; the extent and scale of other factors such as fire and pollution are relatively small.

The climatic disturbance factors include mainly drought, storm, flood, snow, frost and hail, which have influenced biota seriously and have also changed the environment of living systems to some extent. The statistical result of the occurring frequency of the main climatic disturbance factors in the Horqin Sandy Land from 1976 to 1981 is shown in Table 8.5.

The lack of precipitation causes drought and then damages agricultural production and livestock husbandry further. Recurrent and widespread drought is one of the main natural disturbance factors in the Horqin Sandy Land.

The wind, with speeds above 8 on the Beaufort scale (>15 m/s), can be called a disastrous gale. All the monsoons, typhoons, cyclones and local gales induced by severe convective weather can cause damage and enormous loss. There are now even more days with strong wind in the Horqin Sandy Land. According to research, the wind velocity required to entrain soil particles of 0.1–0.25 mm is 4–5 m/s in dried and exposed sandy ground surfaces. In most parts of the Horqin Sandy Land, the sand particle size of the ground surface is mainly 0.5–0.01 mm, the annual average wind velocity is above 4.0 m/s, the wind velocity in spring (3–5 months) is above 6.0 m/s and the maximum is above 20 m/s. There are many days throughout the whole year in which the wind velocity is above 4.5 m/s. The sand particles of the bare ground surface can be moved and form wind-sand currents under the impact of a strong wind, which can then cause a blown sand disaster (farmlands in some regions have to be re-seeded up to five to six times). The dominant climatic disturbance factor in the Horqin Sandy Land is the synergistic effect between drought and the gale-force winds.

Soil disturbance includes mainly aridization, salinization and the loss of nutrients. Soil degradation interacts with other disturbance effects to bring about changes in the vegetation. For example, the elm woodland steppe vegetation, the main vegetation type in the Horqin Sandy Land, has been replaced by secondary sandy vegetation and sparse meadow vegetation. Great numbers of the original plant species have been lost and some

Table 8.5. Statistics of natural disasters in the Horqin Sandy Land.

	1276–1948		1949–1981	
Disaster	No. of times	Frequency (%)	No. of times	Frequency (%)
Flood damage	6	0.89	9	15.6
Drought damage	8	1.19	9	28.1
Storm damage	4	0.59	3	9.4
Snow damage	4	0.59	1	3.1
Frost damage	0	0	8	25.0
Hail damage	3	0.44	6	18.8

wind-sand-enduring and saline–alkaline-enduring plant species have gradually occupied the dominant niches. Many shifting sands, and some stretches of saline–alkali lands, have appeared in many locations throughout the region. The exceedingly serious deterioration of the soil and climate conditions brings many difficulties that impede vegetation restoration. Instability of the soil matrix, lack of nutrients and high levels of soil salinization–alkalization are the dominant soil disturbance factors in the Horqin Sandy Land.

Animal disturbance is reflected mainly in the destructive effect of locally abundant wild animals, such as rodents, and insect pests, such as grasshoppers, on the rangeland ecosystem. Rodent damage is one of the main reasons for rangeland degradation in the Horqin Sandy Land. For example, rodent-damaged pasture accounts for 500,000 Mha every year in Tongliao. The damage to rangeland includes: (i) the large amount of soil that is excavated during the construction of burrows, which forms many mounds of different size – the dry mounds can then be eroded by wind in the spring and winter and cause land degradation; (ii) the rodents harvest the aboveground biomass (30–300 g in a circadian day) – the amount consumed daily when the population density of rodents is high represents a significant loss of pasture; and (iii) the rodents gnaw the roots of pasture, causing damage and even death of the plants.

The quantity of wolf and fox has decreased sharply because of overkilling in order to protect livestock. The population of other rangeland animals and birds, except for rodents, is in sharp decline.

8.4 Ecological Processes of Vegetation Degradation and Restoration

Over the past 50 years or more, with the rapid growth of population, exploitation of land resource and climate warming, the vegetation of the Horqin Sandy Land has been strongly disturbed and the original forest steppe or rangeland vegetation has been destroyed and replaced by sand-tolerant vegetation and salinity-tolerant meadow vegetation. As land degradation has worsened, community structure and species composition, population density, vegetation coverage and aboveground biomass have decreased sharply and synusial structure and life forms have tended to be simplified.

8.4.1 Ecological processes of vegetation degradation

The main plant species in the fixed sand dunes in the Horqin Sandy Land include: *Elymus* woodland, *C. microphylla*, *E. monosperma*, *A. manshurica*, *Hedysarum fruticosum*, *A. halodendron*, etc. The plant species in the semi-fixed sand dunes include: *C. microphylla*, *S. flavida* and *A. halodendron*. The plant species in the semi-shifting sand dunes and shifting sand dunes include: *S. flavida* and *A. halodendron*. *C. microphylla* is the dominant shrub as well as an important foundation species in fixed and semi-fixed sand dunes. However, gradually, it is becoming the dominant species of secondary succession in the shifting and semi-shifting sand dunes and shows scattered distribution in sand dunes (Table 8.6).

The mesophytic *S. flavida* occurs only in fixed and semi-fixed sand dunes and sometimes becomes the dominant species in shifting and semi-shifting sand dunes. Generally, *S. flavida* invades first in those interdune swales with good soil moisture. With the flow of upwind sand dune, the primary interdune swales were gradually buried by sand and most plant species disappeared, but *S. flavida* grew vigorously, developed upwards along the sand dunes with increasing height as the sand piled up and formed thickets in the shifting sand dune. With the aggravation of wind erosion, the top of the shifting sand dune was gradually cut flat and the roots of *S. flavida* were exposed by the wind. *S. flavida* died after the sand dune had moved forward, but colonized the next shifting sand dune.

The vegetation in fixed sand dunes is generally undamaged vegetation or slightly degraded vegetation. Community structure and species composition, population density, vegetation coverage and aboveground biomass are relatively high, but will obviously decrease when desertification occurs. But there is a significant difference in the decreasing amplitude because of the different sensitivity of the various plant communities to desertification. The characteristics of species

Table 8.6. Dominance changes of species in the degradation process of a *Caragana microphylla* community.

Serial number	Species	Fixed sand dune	Semi-fixed sand dune	Semi-shifting sand dune	Shifting sand dune
1	*Pennisetum flaecidum*	18.32	6.123	8.645	0
2	*Agropyron cristatum*	18.08	0	0	0
3	*Setaria viridis*	11.34	19.07	1.358	0
4	*Corispermum thelegium*	10.23	4.762	3.818	0
5	*Lespedeza davurica*	5.641	0	0	0
6	*Salsola collina*	5.476	2.431	0	0
7	*Chenodium acuminatum*	4.340	0	0	0
8	*Artemisia scoparia*	2.630	0	0	0
9	*Trigonella korshinskyi*	2.388	0	0	0
10	*Cleistogenes chinensis*	1.871	0	0	0
11	*Astragalus adsurgens*	1.844	0	0	0
12	*Cynachum sibiricum*	1.828	3.885	0	0
13	*Diarthron linifolium*	1.663	0	0	0
14	*Caragana microphylla*	1.581	44.76	16.25	0
15	*Carex duriuscula*	1.388	0	0	0
16	*Saposhnikovia divaricata*	1.222	0	0	0
17	*Artemisia halodendron*	1.165	5.531	18.75	0
18	*Bassia dasyphylla*	1.013	0	0	0
19	*Chloris virgata*	0.958	0	0	0
20	*Geranium dahuricum*	0.895	0	0	0
21	*Dianthus chinensis*	0.749	0	0	0
22	*Euphorbia humifusa*	0.748	0	0	0
23	*Tribulus terrestris*	0.628	0	0	0
24	*Delphinium grandiflorum*	0.624	0	0	0
25	*Allium senescens*	0.528	0	0	0
26	*Kummerowia striata*	0.498	0	0	0
27	*Polygonum divaricatum*	0.433	0	0	0
28	*Tragus berteronianus*	0.362	0	0	0
29	*Ixeris chinensis*	0.357	0	0	0
30	*Echinops gmelinii*	0.297	0	0	0
31	*Lappula echinata*	0.225	0	0	0
32	*Thalictrum squarrosum*	0.146	0	0	0
33	*Asparagus gilbus*	0.078	0	0	0
34	*Leonurus sibiricus*	0.078	0	0	0
35	*Dragocephalum moldiavicum*	0.0752	0	0	0
36	*Potentilla flagellaris*	0.0750	0	0	0
37	*Euphorbia esula*	0.0743	0	0	0
38	*Xanthium strumarium*	0.073	0	0	0
39	*Artemisia sieversiana*	0.073	0	0	0
40	*Agriophyllum arenarium*	0	10.89	23.68	100
41	*Salix flavida*	0	2.563	27.50	0
	Total amount	39	9	7	1

composition change as degradation becomes more severe are that there are still zonal steppe plants such as *C. songorica* and *A. cristatum* in the slightly degraded fixed sand dune and most of the plants in the semi-fixed and semi-shifting sand dunes are annual psammophilous weeds such as *A. squarrosum*, *Corispermum hyssopifolium* and *Salsola collina* Pall.

The vegetation change in interdune meadow was obviously related to the change of soil water and salt content. Most of the meadows show a tendency of habitat drought, soil salinization, soil aridification and vegetation change. During soil salinization, characterized by an increasing distribution of halophytic vegetation, some drought-enduring and sand-enduring species started to invade the rangeland from the edge of the sand dunes and the dominance of the original fine pasturage gradually decreased.

8.4.2 Process of spontaneous generation of vegetation and restoration

The progressive succession of vegetation in the Horqin Sandy Land means first the fixing process of sand dunes and then the successive process of fixed sand dune to steppe. The fixing process of sand dunes undergoes three stages: (i) shifting sand dune with *A. arenarium* as the dominant species; (ii) semi-fixed sand dune with *S. flavida* and *A. halodendron* as the dominant species; and (iii) fixed sand dune with *A. cristatum*, *C. chinensis* and *C. microphylla* as the dominant species.

The succession series can be classified into four kinds according to the different starting points.

Psammosere

The succession began with bare sand land or shifting sand dune. The succession process of the vegetation community started with the bare sand land or shifting sand dune being fixed, which was the endogenetic succession not influenced by underground water or landform. The pioneer plants invading the shifting sand dune were mainly annual plant communities such as *C. thelegium*, *A. arenarium* and so on, but there was still instability, causing it to remain in the shifting sand dune stage. Later, with invasion by *S. flavida* or *A. halodendron*, the sand dune was fixed to a certain extent and became a semi-fixed sand dune. After that, the coverage of sand dune increased with the development of *C. microphylla* and *Polygonum divaricatum* and a turf layer, which fixed the sand dune. A *P. flaecidum* community began to develop in two directions after the fixed sand dune stage; one was a *C. microphylla* community, or a *P. divaricatum* community, and the other was a *C. chinensis* community. From the perspective of retrogressive succession, a *C. microphylla* or *P. divaricatum* community, after being damaged, would divide into three degradation directions: (i) *A. halodendron* community or *S. flavida* community; (ii) *P. flaecidum* community; and (iii) *C. chinensis* community.

Hydrosere

The succession, dominated by water condition, started from patches of water in the interdune swales and meadow. The initial stage of progressive succession was: *Typha orientalis* community → *Phragmites communis* community → *P. alopecuroides* community or *P. alopecuroides* and *Agrostis matsumurae* community, and then the succession generally developed in two directions, one was that the *S. microstachya* community from interdune swales developed into a *Betula populifolia* community, and another was that the succession from meadow developed into a *Hemarthria compressa* community → mixed *P. alopecuroides* community → *Arundinella anomala* community → *Leymus chinensis* community → *C. chinensis* community. Retrogressive succession was defined by climate and soil condition and included mainly two succession processes: a desertification process and an alkalization process. The *C. chinensis* community, after being damaged, started to degrade to a *P. flaecidum* community through the desertification process; however, the *L. chinensis* community, *A. hirta* community, *P. communis* community, *H. altissima* community, after being damaged, started to degrade to a halophyte community through the alkalization process.

Succession series of abandoned land

After the lands with the *L. chinensis* community, *A. hirta* community, *P. communis* community and *H. altissima* community were cultivated, a *Setaria viridis* community, *C. virgata* community and *Artemisia* community started to form gradually and after several years became an area of 'rhizomatous forages' dominated by an *Artemisia* community, but including *L. chinensis*, *P. communis*, *P. alopecuroides* and *H. altissima*. Then, the land would gradually become abandoned land dominated by a *P. flaecidum* community. This successive process was related to soil water condition and soil texture.

Halosere

The succession, dominated by soil salt content, started from alkaline patches. Few plants, except for halophytic vegetation, can grow on alkaline land because of the total salt quantity exceeding 0.5%. The pioneer vegetation invading sandy land was mainly annuals or biennials, such as *Suaeda glauca*, *Kochia scoparia* and *A. anethifolia*. With the growth of these plants, an *L. chinensis* community started to form and the soil salt content gradually decreased. Generally, the *S. glauca* community, *Eragrostis pilosa* community and *L. chinensis* community showed a concentric distribution around the alkaline patch.

8.4.3 Effects of vegetation restoration on ecological environments

Change of mechanical composition of soil

Soil mechanical composition changed gradually after the shifting sand dune became fixed. With the development of vegetation and the passage of time, the sand dune was fixed. The content of coarse sand (>0.1 mm) tended to decrease, while the content of silt and clay tended to increase (Table 8.7). The establishment of artificially established vegetation improved the sand microenvironments and reduced the probability of the fixed sand being blown off. With the increase of vegetation coverage and the decline of wind velocity in the fixed sand area, much wind-blown material was deposited on the soil surface, which resulted in an increased content of clay and silt.

Although the deeper soil layer was not influenced directly by wind, the aeolian deposit was affected by changes in infiltration rate and water-holding capacity. The belowground parts of plants could also respond to the change in the content of coarse sand, clay and silt in the soil. Rhizosphere organisms and chemical action also influenced the soil mechanical composition, but the effects occurred very slowly. So the deposition of aeolian deposits after the establishment of artificial vegetation would change the soil mechanical composition directly. The effects of different vegetation type and development age on soil mechanical composition were different. As a whole, the effect of artificial vegetation such as *L. bicolor* and *H. fruticosum* is to increase the content of clay and silt.

The change of soil porosity and soil bulk density after the shifting sand dune was fixed was

Table 8.7. Changes of mechanical composition of soil in the process of vegetation restoration. Source: Li and Bai (1984).

			Mechanical composition (mm)		
Species	Method and age	Soil layer	Mid-coarse sand >0.25	Silver sand 0.1–0.25	Clay and silt <0.1
Shifting sand dune	–	0–10	14.67	83.89	1.44
		10–40	14.09	84.33	1.78
		40–60	11.24	87.06	1.70
Caragana microphylla	Seeding 11 years	0–10	6.47	88.73	4.80
		10–40	7.51	89.09	3.20
		40–60	5.22	90.78	4.0
Hedysarum fruticosum	Seeding 8 years	0–10	0.78	86.79	12.43
		10–40	9.29	82.86	6.85
		40–60	6.37	88.36	5.32
Artemisia halodendron	Buried branch 10 years	0–10	8.85	85.30	5.85
		10–40	13.02	85.26	1.72
		40–60	14.76	81.43	1.56
Salix flavida	Inserted branch 10 years	0–10	7.74	85.06	7.20
		10–40	4.02	91.76	4.22
		40–60	4.07	93.34	2.59
Lespedeza bicolor	Planting 9 years	0–10	5.77	88.85	10.38
		10–40	8.93	84.33	6.74
		40–60	2.45	90.07	7.48

Table 8.8. Changes of soil bulk density of different types of artificial vegetation.

Soil layers (cm)	Shifting sand dune	Natural *Caragana microphylla*	*Caragana microphylla* 15 years	*Caragana microphylla* 5 years	*Hedysarum fruticosum* 15 years	*Artemisia halodendron* 15 years	Artificial *Pinus sylvestris* 15 years
0–20	1.665	1.511	1.539	1.593	1.533	1.599	1.555
20–40	1.694	1.695	1.571	1.682	1.654	1.594	1.519
40–60	1.625	1.64	1.56	1.7	1.614	1.693	1.667
60–80	1.623	1.626	1.626	1.656	1.64	–	1.507

Table 8.9. Changes of soil porosity in the restoration process of vegetation.

Soil layers (cm)		0–20	20–40	40–60	60–80
Shifting sand dune	Non-capillary porosity (%)	3.30	3.11	3.46	2.80
	Capillary porosity (%)	33.53	33.10	32.44	33.51
	Total porosity (%)	36.83	36.21	35.89	36.31
Natural *Caragana microphylla*	Non-capillary porosity (%)	5.13	4.98	4.01	3.78
	Capillary porosity (%)	35.86	32.80	33.22	33.27
	Total porosity (%)	40.99	37.78	37.23	37.05
Caragana microphylla 15 years	Non-capillary porosity (%)	4.17	3.864	3.74	3.978
	Capillary porosity (%)	37.38	36.65	36.19	35.95
	Total porosity (%)	41.55	40.51	39.93	39.92
Caragana microphylla 5 years	Non-capillary porosity (%)	3.69	3.95	3.70	2.75
	Capillary porosity (%)	33.79	34.817	33.77	33.17
	Total porosity (%)	37.48	38.76	37.47	35.92
Hedysarum fruticosum 15 years	Non-capillary porosity (%)	3.92	4.89	4.99	4.471
	Capillary porosity (%)	34.59	36.18	37.1406	34.52
	Total porosity (%)	38.50	41.07	42.03	39.99
Artemisia halodendron 15 years	Non-capillary porosity (%)	5.82	4.45	5.1	–
	Capillary porosity (%)	36.81	34.17	34.75	–
	Total porosity (%)	42.63	38.62	39.85	–
Artificial *Pinus sylvestris* 15 years	Non-capillary porosity (%)	6.425	5.764	4.074	7.33
	Capillary porosity (%)	36.63	34.06	34.91	35.83
	Total porosity (%)	43.06	39.82	38.99	43.17

reversed (Tables 8.8 and 8.9). With the establishment and development of artificial vegetation, both non-capillary porosity and capillary porosity gradually increased, but the magnitude of the increase was different in each vegetation type. The effects of artificial *Pinus sylvestris* forest on soil porosity were the most obvious.

Change of soil water in the process of vegetation restoration

With the establishment of artificial vegetation and the gradual growth of individuals, the soil water status gradually deteriorated and became aggravated with plantation age. The loose texture, weak capillary action and dry surface sandy layer of the shifting sand dune could prevent water vaporization. Furthermore, owing to sparse vegetation and low transpiration, the water deposited in sand layers could be preserved and the water condition of the shifting sand dune was relatively better.

With the increase in the individual number and biomass of vegetation, water consumption by transpiration increased rapidly. Soil water stores were depleted, so the water in soil and plants would suffer a serious shortage in drought years, as evidenced by the appearance of a deep dry sand layer that could restrict the

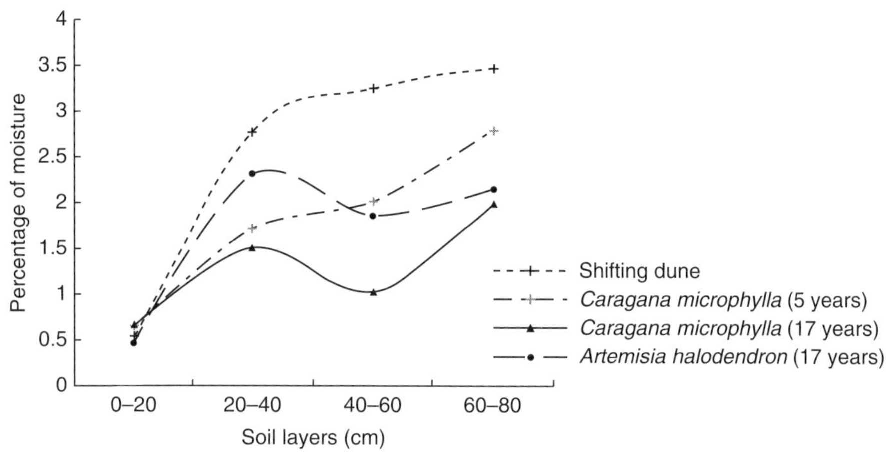

Fig. 8.9. Changes of water content of soil in the process of vegetation restoration.

growth of plants. As shown in Fig. 8.9, the soil water condition gradually deteriorated after planting *C. microphylla* and *A. halodendron* in the shifting sand dune. Due to the different transpiration intensity and aboveground biomass, the soil water consumption intensity of various artificial vegetation types was different. For example, the soil water condition was better in an *A. halodendron* community than that in a *C. microphylla* community.

Change of soil fertility

With the increase of vegetation coverage after the establishment of artificial vegetation, the action of the wind-sand current declined, the dust and granule materials of the air were gradually deposited and much litter was trapped into the soil every year. Under the action of hydrothermal conditions and microorganisms and soil animals, the litter and root debris underwent a series of chemical changes and a crust gradually formed on the ground surface. The forming of the crust and the changes of soil properties intensified the pedogenesis and also created advantageous conditions for sand-loving vegetation. Meanwhile, the natural vegetation under the artificial vegetation gradually developed, the aboveground standing biomass gradually increased and the soil improvement action of plants was also gradually enhanced. Soil organic matter, through the humification process, was transformed into soil humus, which improved the soil nutrient status. The effects of accumulating soil nutrients in various vegetation types were different, but the improvement action gradually increased over time.

The soil nutrient content of the *C. microphylla* sand-fixing community at different ages is shown in Table 8.10. The soil nutrient content has obviously changed with the development of the *C. microphylla* community. Soil organic matter in the soil surface, especially at 0–1 cm depth, gradually increased. The content in the community after 14–19 years was 5.0–8.3 times higher than that in the shifting sand dune, but there was not any obvious change below a depth of 11 cm. This means that the main reason for influencing soil organic matter is to increase litter and atmospheric deposition, while the effect of the plant root system is also very obvious. The content of available N, total N and available P of the soil surface (0–10 cm) was 1.29–1.37 times, 1.5–1.65 times and 1.6–2.86 times, respectively, those in the community founded for 14–19 years than that in moving sand dune. The content of available N was 1.5–1.65 times that in the community after 14–19 years than that in the shifting sand dune to a depth of 11–35 cm, while the content of total N and available P showed no obvious difference.

Influence of artificially planted vegetation on wind-sand current

The influence of artificial vegetation on the wind-sand current was very obvious. At a height of

Table 8.10. Soil nutrient content of sand-fixing community of *Caragana microphylla* at different ages.

Age (years)	Soil layers (cm)	Organic matter (%)	Available (mg/kg)			Total amount (%)		
			N	P	K	N	P	K
Shifting sand dune	0–10	0.03	27.9	4.2	46.2	0.012	0.003	2.34
	10–28	0.02	21.3	5.3	45.6	0.009	0.002	2.30
14	0–1	0.23	54.8	10.5	102.2	0.023	0.012	2.33
	1–12	0.16	35.9	7.8	55.6	0.015	0.019	2.21
	12–35	0.02	32.3	5.5	61.5	0.010	0.001	2.95
17	0–1	0.30	63.2	13.5	151.8	0.026	0.012	2.30
	1–12	0.25	38.2	12.0	107.8	0.022	0.011	2.09
	12–35	0.03	33.3	4.9	53.3	0.008	0.009	4.36
19	0–1	0.68	82.5	15.2	250.8	0.048	0.018	2.15
	1–11	0.15	36.6	6.8	77.0	0.016	0.011	2.50
	11–32	0.02	35.2	4.6	48.9	0.012	0.001	2.26

0–20 cm, the sand particles of the wind-sand current in the shifting sand dune concentrated mainly on the surface layer and the amount in the height of 0–2 cm accounted for 45.7% (Fig. 8.10), but accounted for only 20–25% in the same layer of artificially established vegetation. At a height of 10 cm, the sand content of the wind-sand current in the shifting sand dune and in the *A. halodendron* and *Populus* communities accounted for 86%, 82% and 50%, respectively, of the total amount in the height of 20 cm. With the increase of height, the wind-sand current changed. The extent of change depended on the growth status and the coverage of the artificial vegetation, so the sediment discharge was less in the artificial vegetation than that in the shifting sand dune when it was windy. In windy weather, sand particles of different sizes were distributed at different heights in areas with and without artificial vegetation. Generally, artificial vegetation resulted in the sand dune becoming fixed and coarse sand being confined nearer to the surface of the dune. The higher the proportions of silver sand and clayier soils, the higher the probability of the dune becoming fixed.

8.5 What Can We Learn to Prevent Further Degradation Episodes?

For at least a millennium, this area supported nomadic animal husbandry, but the influx of

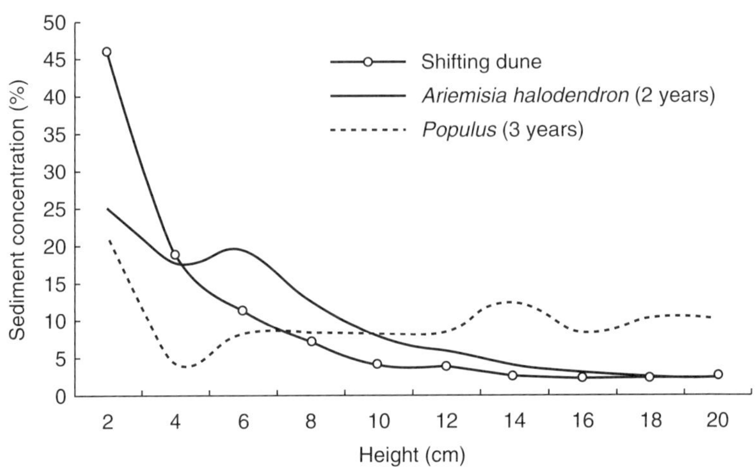

Fig. 8.10. Structure of wind-sand current of different artificial vegetation types.

Han settlers and the expansion of agriculture during the past century, and particularly the past few decades, have destroyed much of the natural vegetation, giving rise to large, unstable dunes. In early times, Horqin was not a semi-desert but rangeland with dispersed trees (savannah-type woodland), in transition between dense forest and the steppe zone. The rolling sand-sheet was deposited during the last glacial period (Würm, 12,000 years BP). During 10,000 years of vegetation growth, thick dark topsoil developed. Since historical times, the region has gone through several cycles of human-induced desertification and subsequent recovery, when human pressure lessened. By the middle of the 10th century, Horqin had developed into a prosperous agricultural and grazing area. At the time of the Jin Dynasty in the 12th century, the forest had reportedly disappeared and the land was covered with sand. By the beginning of the 17th century, Horqin thrived again but, after the middle of the 19th century, with the Qing Dynasty pursuing a policy of 'reclaiming wastelands', forests and rangelands were once again destroyed and bare soil surfaces gave way to wind erosion. Fertile dark topsoil vanished and extensive dune fields gradually built up.

Overgrazing has been the main reason for rangeland degradation over recent decades. The main causes of overloaded rangeland are summarized as follow: (i) a decrease in rangeland area, with an increase in cultivated land; (ii) a decrease in rangeland area with an increase in livestock number; and (iii) degradation of rangeland, with a decrease in forage yield and a decline in the carrying capacity because of natural factors and the changing human and livestock populations.

Grazing restriction has been reported to be effective for the control of desertification in many semi-arid regions. However, economic reasons often make grazing bans (complete exclosure) difficult to carry out. Grazing control has been applied to the Horqin Sandy Lands by means of seasonal exclosure, whereby grazing is allowed from November to April. The harvesting of hay is also allowed once during September–October. This has proved more effective and is also more workable, as it still allows herders some access (Katoh *et al.*, 1998).

Policy is implicated, too, in the accelerated land degradation that has been witnessed since the 1980s. Since decollectivization, government policies have promoted the household responsibility system through the allocation of grazing user rights as the best solution to maximize pastoral productivity and control desert expansion in grassland areas. Fieldwork contradicts this optimism (Williams, 1996). Data and participant observation reveal that enclosures, as they have been implemented, may actually compound grazing problems for most residents and the wider ecosystem. Household enclosures (often fenced) intensify hypercritical stocking rates on highly vulnerable rangeland, exacerbating wind and soil erosion processes. Good resource management requires attention to other factors, such as secure tenure, equitable access to community resources and meaningful institutional support in the form of rural credit, production services and legal protection (see Chapter 15).

References

Katoh, K., Takeuchi, K., Jiang, D., Nan, Y. and Kou, Z. (1998) Vegetation restoration by seasonal exclosure in the Kerqin Sandy Land, Inner Mongolia. *Plant Ecology* 139, 133–144.

Li, C. and Bai, O. (1984) Effects of amending wind-sand soil in fixed sand forest. *Journal of Desert Research* 4(1), 36–41.

Williams, D.M. (1996) Grassland enclosures: catalyst of land degradation in Inner Mongolia. *Human Organization* 55(3), 307–313.

Zhao, H., Li, S.G., Zhang, T.H., Okhuro, T. and Zhou, R.L. (1994) Sheep gain and species diversity in sandy grassland, Inner Mongolia. *Rangeland Ecology and Management* 57(2), 187–190.

Zhu, Z. (1994) Development trend of sandy desertification in interlaced agropastoral region of northern China in recent ten years. *Journal of Desert Research* 14(4), 1–7.

9 Case Study 3: Xilingol Grassland, Inner Mongolia

Jianhui Huang, Yongfei Bai and Ye Jiang
State Key Laboratory of Vegetation and Environmental Change, Institute of Botany, Chinese Academy of Sciences, Beijing, China

Synopsis

Xilingol has undergone major land degradation and parts have become desert within the past few decades. The dynamic pulse of stocking rate on the spatio-temporal patterns of grassland degradation was altered and this was a key contributing factor, along with large-scale rangeland conversion to cropland and rising populations of people and their livestock.

The respective roles of traditional nomadic and sedentary grazing in the land-use history of Xilingol grassland is examined in the light of drastic land-use changes and a dramatic increase of human population since 1950. We explore the major social economic factors driving grassland degradation during the past 50 years and put forward practical approaches for improving the current practices of grassland management based on the theoretical framework of adaptive ecosystem management and sustainability science.

Keywords: land-use change; adaptive management; nomadic culture; sedentarization; drought index; climate change; indigenous knowledge; carrying capacity; overgrazing

9.1 Preamble

Xilingol is one League (prefecture) of the Inner Mongolia Autonomous Region and is located in the east part of middle Inner Mongolia (Fig. 9.1). It covers an area of 203,000 km^2 and ranges from 41°35' to 46°46' in latitude and from 111°09' to 119°58' in longitude. Xilingol borders on Hebei Province in the south and Mongolia in the north and has an 1100 km boundary with Mongolia. The Luanhe, Wulagaihe and Baiyinhe are three river systems in this region. Xilin River is one important tributary of the Wulagaihe River system and also a major inland river which passes through this region from south-east to north-west of north-eastern Xilingol (Chen, 1988). The Xilin River is important, especially to Xilihaote City and its outskirts. Around 97.2% of the land in Xilingol is classified as rangeland, about 97.5% of which is used for animal husbandry, either directly for grazing or for hay harvesting. The husbandry production of Xilingol is based largely on rangelands.

9.2 Brief Statement of the Major Environmental Problems

The degradation of grassland in Xilingol started in the 1950s to the 1960s and speeded up in the 1980s. By the end of the 1980s, the area of degraded rangelands accounted for 48.6% of the total land in Xilingol, and this had risen to 70–80% by the end of the 1990s. The degradation of rangelands in Xilingol generally takes the following forms: (i) encroachment by shrubs over

Fig. 9.1. Location of Xilingol, China.

a large area; (ii) a remarkable decrease of the density, height and biomass of high-quality forage plants; (iii) replacement of perennial dominant species; (iv) change of soil texture by an increase of sand composition; and (v) secondary salinization. Long-term overgrazing and irrational use of rangeland resources are, however, the predominant causes for grassland degradation, although climate change in this area is also an important factor (Fig. 9.2 and Chapter 3). This can be divided into five categories.

1. The number of domestic animals has long exceeded the loading capacity of the natural rangelands. In the early 1950s, the total number of domestic animals in this region was less than 3 million sheep units; that means each sheep unit could occupy around 6 ha of rangeland. In the 1990s, however, the number of domestic animals increased sharply to 13 million sheep units and to 24 million in 1999. Thus, by 1999, each sheep unit had only 0.7 ha of rangeland (Fig. 9.3), far exceeding the loading capacity of the natural rangelands, resulting in the degradation of all grazing lands and 40% of hay-harvesting rangelands.

2. There is a long-term inconsistency between the number of overwintering domestic animals and the seasonal change of forage production

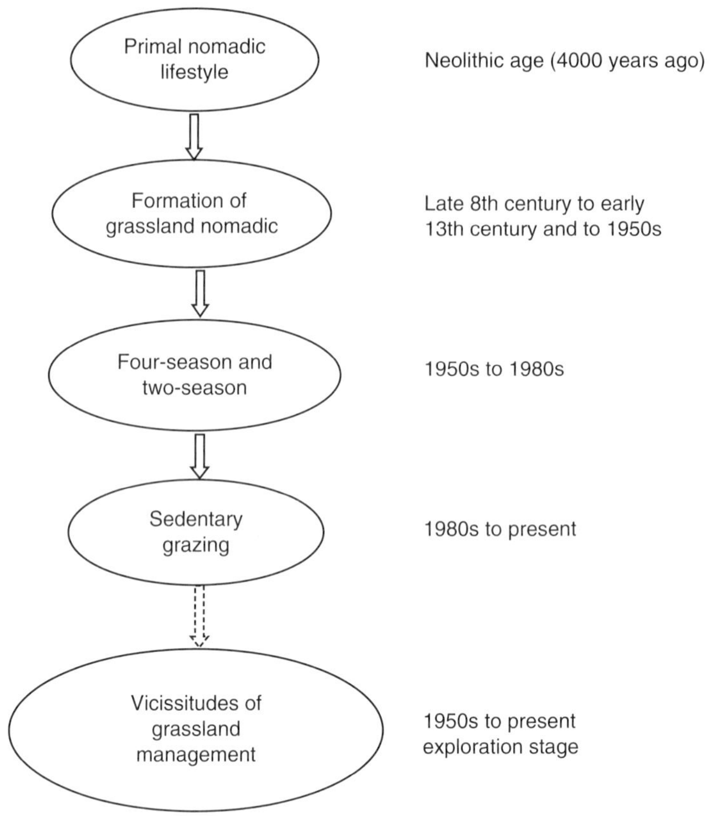

Fig. 9.2. Reconstruction of a typical nomadic pastoral progression on the Xilingol Grassland.

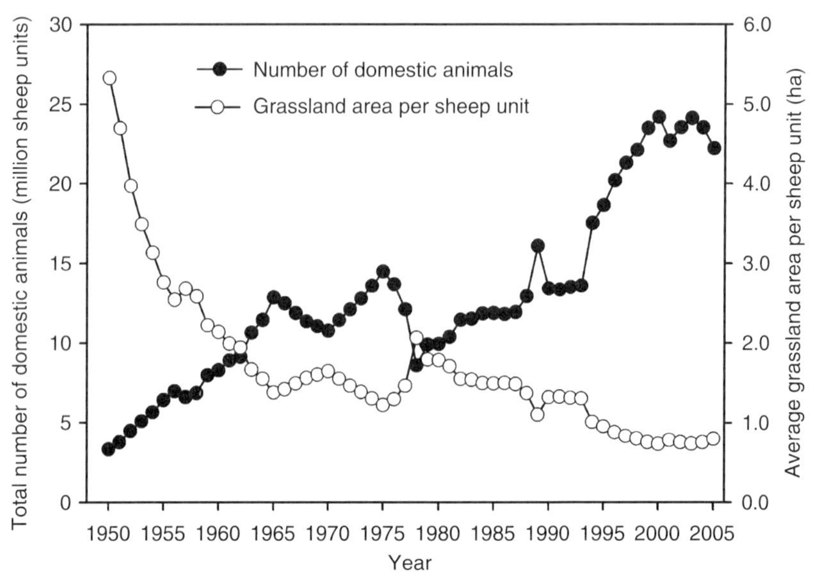

Fig. 9.3. Changes in the number of domestic animals and the grassland areas per sheep unit in the Xilingol steppe region from 1949 to 2005.

in the natural rangelands. The growth of plants in Xilingol lasts around 150 days only, in which sufficient forage can be provided for sheep for only 120 days and for cattle for only 100 days. Thus, for at least 240–260 days a year, the domestic cattle and sheep have to live without a fresh forage supply and supplementary hay forage is needed if the domestic animals need to live over winter (and spring).

3. As the hay field area was only 15% of the available rangelands, the hay-harvesting rangelands degraded substantially because hay-harvesting had been practised on those rangelands continuously, allowing no time for them to recover. Artificial (sown) pastures made up only 1% of the area in Xilingol and, additionally, appropriate techniques for forage processing and storage were lacking. The reserves for winter fodder were extremely limited, leading to severe malnutrition in those animals living through the winter.

4. The population growth of domestic animals also became a vicious circle, i.e. 'thriving in summer, getting fat in autumn, thinning in winter and dying in spring'. As a matter of fact, husbandry production in Xilingol was low, although the number of animals raised was high. More importantly, those animals living on rangelands through autumn, winter and spring would gnaw at the remaining aboveground biomass and even the roots in the surface soil, which is now thought of as being a major driving factor leading to rangeland degradation and even widespread desertification.

5. Long-term irrational use and grazing regimes of the rangelands in Xilingol were the main causes of land degradation. This may be concluded as follows: (i) unreasonable arrangement of human settlement and water points; (ii) inappropriate hay-harvesting regimes and an imbalance in the proportion of hay-harvesting rangelands to natural grazing rangelands; and (iii) irrational rangeland conversion and excessive digging of medicinal plants on a large scale in the rangelands.

9.3 Physical Environment

9.3.1 Climate

The climate in Xilingol has a typical temperate, semi-arid continental climate and is characterized by chilliness, strong winds and uneven distribution of precipitation. Overall, the climate in Xilingol has the following characteristics: long daily solar radiation, the concurrent occurrence of high radiation, temperature and precipitation, large daily temperature difference and high variation in precipitation. The mean annual temperature is around 0–3°C, the mean monthly temperature in January is around −17 to −21°C (Fig. 9.4). In general, the average wind speed ranges from 3.5 to 4.0 m/s and the highest winds can reach 24–28 m/s. There are around 50–80 days in a year in Xilingol with strong winds, which occur mostly in March–May.

The mean annual precipitation is around 200–350 mm but is distributed unevenly, both spatially and temporally. The precipitation in Xilingol varies greatly from year to year and there is a gradient from about 400 mm in the east to about 200 mm in the west. The rainfall is summer dominant, with 70% falling between June and August. The seasonal variations in precipitation result in a periodic change to the rangelands as appearing green in summer, yellow in autumn, senescing in winter and dry in spring. Concomitantly, livestock thrive in summer, are fat in autumn, lose about 30% of their body weight in winter and many die in spring. Natural climate disasters such as drought in summer, heavy snowfalls (snow disasters), or the opposite, an absence of much snow during the wintertime, may significantly affect both the primary and secondary productivity of the rangeland ecosystems in this area. Elevation ranges from 900 m in the north-east to 1300 m in the south-west and the average is 1000 m. The highest place is 1900 m.

9.3.2 Distribution of soils

As the precipitation increases from east to west Xilingol, the distribution of soil also shows corresponding change. Thus, the distribution of soils shows a zonal spatial pattern. But there are also some non-zonal soil types existing in the region, formed under microtopography and microclimate (Xilingol Grassland Management Station of Inner Mongolia Autonomous Region, 1988). Chestnut soil is predominant in this region and the corresponding vegetation (typical steppe) is also the most important rangeland type and the most widely distributed form. The use of chestnut soil in Xilingol can be divided into two types, i.e. for agriculture or for husbandry.

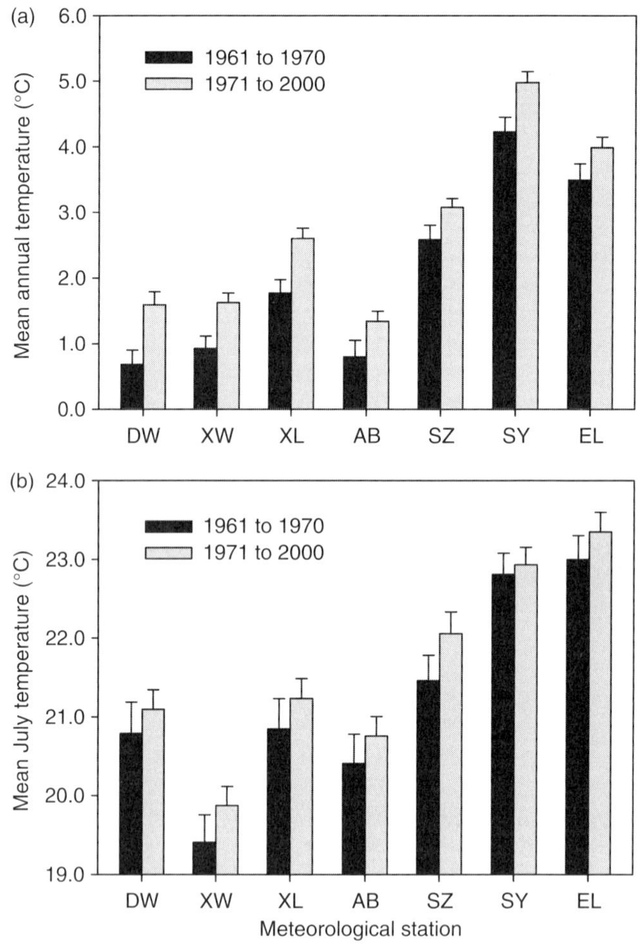

Fig. 9.4. Changes of mean annual temperature (MAT) and mean July temperature (MJT) in the Xilingol steppe region since the 1960s (error bars denote SEM): a, comparison of MAT between 1961 and 1970 and 1971 and 2000 at different locations of the region; b, comparison of MJT between 1961 and 1970 and 1971 and 2000. DW, Dongwuzhumuqin Qi; XW, Xiwuzhumuqin Qi; XL, Xilinhot; AB, Abaga Qi; SZ, Sunitezuo Qi; SY, Suniteyou Qi; EL, Erenhot.

Sandy soil is distributed primarily in the Hunshandake sandy area and other small sandy areas such as Gahaielesu and Dongwuzhumuqin Qi (East Ujimqin Banner). The vegetation that has developed on it is composed mainly of psammophytes. The ecosystems that have evolved are extremely fragile because of the unstable soil conditions and low vegetation coverage. The area with sandy soil distribution is important in the husbandry economy of Xilingol.

Meadow soil has developed on the meadow steppe in Xilingol and is distributed mainly along the rivers and on the flood plains. The soil water may come directly from the groundwater and the water table is only 1–2 m deep. Meadow soil has a very high productivity because of its high humidity and fertility and thus is ideal for establishing artificial rangelands (sown pastures).

9.3.3 Distribution of rangelands

Xilingol Grassland is part of the Eurasia Steppe, which extends over 8000 km from north-eastern China westward all the way to Hungary. As environmental conditions change greatly, resulting in different types of soil developing in the Xilingol area, the types, structures and functions of vegetation on the various soils may be significantly different. Temperature and precipitation are the two most important environmental factors in Xilingol, which may decide the distribution and growth of plant species and plant functional groups, and thus the distribution of communities. For example, the aboveground biomass and dominance of C_4 plant functional groups decreases with Xilingol's mean annual precipitation and increases with its mean July temperature (Fig. 9.5).

True grassland is a kind of vegetation type that is developed under temperate arid and semi-arid climate conditions, and is composed of drought-resistant plants, such as bunch grasses, rhizome grasses, forbs and maybe some dwarf

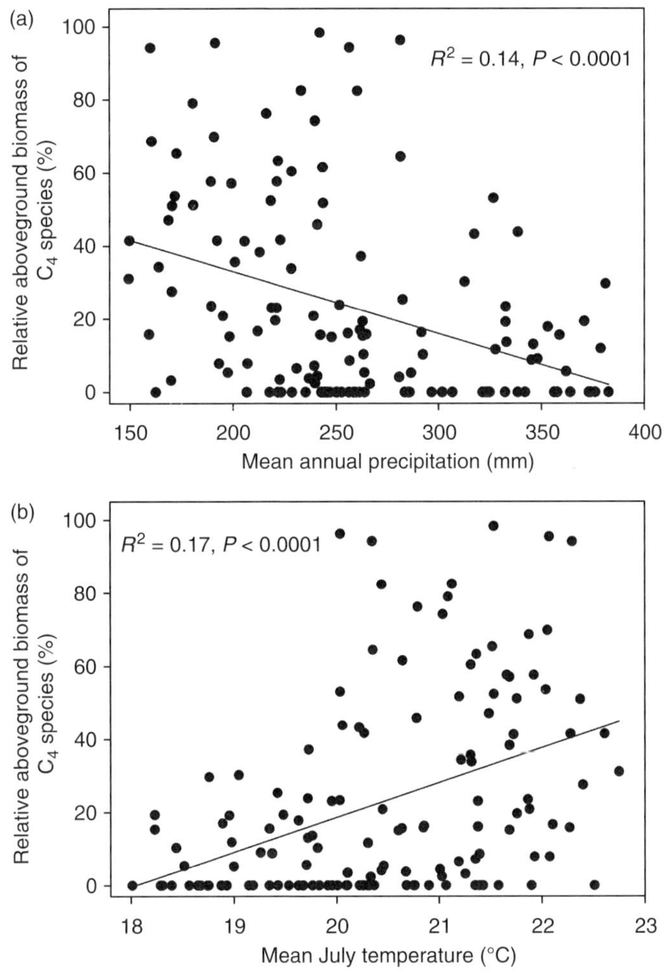

Fig. 9.5. Changes of the relative aboveground biomass of C_4 plants along precipitation and temperature gradients in the Xilingol steppe region. a, Changes of the relative biomass of C_4 plants along the mean annual precipitation (MAP) gradient. Regression model for the relative aboveground biomass of C_4 plants (RBC_4, %): $RBC_4 = -0.17 \times +66.96$ ($R^2 = 0.14$, $N = 133$, $P < 0.001$). b, Changes of the relative biomass of C_4 plants along the mean July temperature (MJT) gradient. Regression model for the relative aboveground biomass of C_4 plants (RBC_4, %): $RBC_4 = 9.40 \times -169.37$ ($R^2 = 0.17$, $N = 133$, $P < 0.001$). These results indicate that every 5.9 mm decrease in MAP or 0.1°C increase in MJT will lead to 1% increase in the relative biomass of C_4 plants.

shrubs. In Xilingol, the area of grassland is around 14 million ha (Mha), accounting for 71.88% of the total land area of the League. The growth of Xilingol grasslands may be significantly limited by three environmental factors, i.e. insufficient water supply, low temperature and the dramatic change in water and energy conditions. During the long-term development of rangeland ecosystems, only those drought- and cold-tolerant plant species and functional groups were left in the communities. These plants and functional groups can not only sustain long-term drought in summer and extreme chilly conditions in winter, but also endure dramatic change in water supply and energy balance. Well-developed xeric structures, such as low leaf area, leaf involution at the edge, sunken stomata and well-developed mechanical and protective tissues, are the marked traits of these plants. Besides, under dry conditions, rangeland plants have well-developed surface roots and can capture brief and/or limited precipitation quickly. Although many plants can blossom and fruit, they generally use asexual reproduction, which reflects their adaptation to the stressed environment.

There is a longitudinal gradient in the aridity (drought) index as the driest conditions appear in the west and relatively mesic conditions in the east. The distribution of different soil types and rangeland communities also shows a similar spatial distribution pattern. Based on the structure of communities, Xilingol grasslands can be divided into three groups, which are distributed in different parts of the region from east to west.

Meadow steppe

Meadow steppe is generally distributed in the forest–grassland ecotone area, i.e. in the most eastern part of the region. It is a relatively wet type of community in Xilingol. The constructive species of this kind of vegetation include middle xeric or wide spectral xeric perennial plants, mixing with many mesic forbs, rhizome grasses and bunch grasses. Typical xeric bunch grasses may also appear in this subvegetation type, but definitely not as dominant plants, and xeric dwarf semi-shrubs are almost eliminated. The total area of meadow steppe in Xilingol is around 1,710,000 ha, accounting for 12.1% of the total grassland area of the region. According to species composition, meadow steppe can be divided into the following formations: Form. *Stipa baicalensis*, Form. *Leymus chinensis* and Form. *Filifolium sibiricum*.

Typical steppe

Typical steppe is also called dry grassland, or authentic grassland, which is generally located in the middle of a temperate grassland zone. The climate in the location area shows typical semi-arid characteristics. Typical steppe is a most important vegetation type in Xilingol in its distribution area (around 6,420,000 ha), which makes up 45.4% of the total grassland area of the region. The dominant species are mainly typical xeric or eurytopic xeric plant species. This kind of sub-vegetation is dominated by xeric bunch grasses and some xeric forbs and xeric rhizome sedges are also common species. Small semi-shrubs such as *Artemisia* appear at local sites only. With the soil becoming sandier, *Caragana microphylla* may become dominant and gradually develop into a shrubland. Typical steppe has the following formations: Form. *S. grandis* (21.62%), Form. *S. krylovii* (18.60%), Form. *L. chinensis* (31.48%), Form. *Cleistogenes squarrosa* (10.93%), Form. *Agropyron cristatum* (6.17%), Form. *Artemisia frigida* (7.66%), Form. *Thymus serpyllum* var. *asiaticus* (0.75%) and Form. *Allium polyrrhizum*.

Desert steppe

Desert steppe is a kind of vegetation type which generally develops in very dry habitats (located in the most western and driest part of Xilingol). The constructive species in desert steppe are generally composed of some xeric bunch grasses, often mixed with many super xeric species such as semi-shrubs, alliums and some desert plant species. Compared with that of typical steppe, the habitat of desert steppe is characterized by a more continentally controlled climate and can be categorized into an arid habitat with extremely stressful environmental conditions. The area of desert steppe is around 2,040,000 ha and accounts for 14.4% of the total rangelands in Xilingol. The diversity of plant species, height, coverage and productivity of communities and aboveground biomass of desert steppe are remarkably lower than those of typical steppe. The desert subvegetation in Xilingol has the following formations: Form. *S. klemenzii* (72.12%), Form. *S. gobica* (21.74%) and Form. *Hippolytia trifida* (2.43%).

9.4 Grazing History and Environmental Problems

9.4.1 A history of nomadic grazing in the Xilingol rangelands

Nomadism, a primary lifestyle of Mongolians with the characteristics of regionality, self-sufficiency, autarky and self-consciousness, has already had a history of thousands of years (Chen, 2004). The use of rangelands by nomadic grazing is an important part of the Mongol culture. With the development of a nomadic economy, nomadic culture has been gradually formed, which is significantly different from the cropping culture of central China. Nomadic culture is the dominant culture of Mongolians and is not only an important foundation for lifestyle, traditional habits and religion, but also a window for outsiders to understand the history of the Mongols' development (Zhang *et al.*, 2007).

With the development of societies and the emergence of some environmental problems, people have now started to reconsider and to understand the value of nomadism. Scientists now agree that nomadism is a relatively rational way of using rangelands and it can not only sustain and ensure the future development of the Mongols, but also maintain the functioning of rangeland ecosystems and their sustainable utilization (Shichinohe, 1994; Humphrey and Sneath, 1999; Miller, 1999; Berkes *et al.*, 2000; Zhang *et al.*, 2007). Mobility lies at the heart of the traditional nomadic culture of the Mongols, who capitalize on the spatial and temporal heterogeneity of the rangelands by shifting their herds in response to changing seasonal conditions (Zhang, Q., 2006). This is a practical system to manage grassland ecosystems (Brunson and Steel, 1996; Bailey, 2005) which has evolved over a long time and has been handed down from generation to generation (Fig. 9.2). For them, this practical system is the right way to conduct a rangeland-based livestock husbandry. There are implications for modern husbandry. Planners who are fixated on sedentary systems should consider a return to shifting grazing to help overcome the problems inherent in the highly variable environment in Xilingol.

9.4.2 Nomadic use of rangelands

In ancient times, rangelands were the living places of the minority people only who lived in northern China and followed a nomadic grazing lifestyle. Their use of the generally abundant rangelands (see Chapter 2) at that time was generally without restriction. But this traditional nomadic life is only appropriate when population size is low, the number of domestic animals is limited and people are free to move (Swallow *et al.*, 1994; Fernandez-Gimenez, 2000, 2001). Minority people, especially the Mongols, have also been leading a simple and primal nomadic lifestyle in this region (Qi, 2002). This kind of lifestyle can be dated back thousands of years ago to the Neolithic age. In some remote areas, people followed a nomadic grazing lifestyle up until the 1950s, when China started to implement a collective economy (Chapter 2).

Nomadic people might choose a particular rangeland to graze at a certain time of year based primarily on geographical location, climate, water supply, amount of aboveground biomass, number of livestock and availability of specific fodder in different seasons. They would drive the domestic animals, sheep or cattle from one rangeland to another, depending on the seasons (Suttie and Reynolds, 2003). As the topography is extremely simple in Xilingol, the rangelands have been distributed at an almost similar elevation; thus, the microtopography, aspect and slope mainly decide water and energy conditions and their availability in the rangelands (Shichinohe, 1994). According to our survey, nomadic grazing in Xilingol follows mainly two patterns: large scale with long distance and small scale with short distance.

The traditional indigenous knowledge of these herdsmen was that when the number of livestock was higher, to obtain enough forage, they needed to go further; when the rangelands had lower productivity, they needed to move more frequently. In summer, herdsmen generally drove the animals to the upland, with a sufficient and high-quality water supply, to find a cool habitat and to avoid harassment by flies and mosquitoes. When autumn came, the herdsmen would move their livestock to a lower level to find sufficient forage in a warmer habitat. In winter, the herdsmen would drive the livestock to woodlands or shrublands, where the temperature was a little higher than on the open rangelands. Sand dune, *Achnatherum splendens* tussocks and high reeds were also good habitats for animals. When the following spring came, the herdsmen would still stay in a warmer place, such as south-east-facing sand dunes, and close to a water supply to await

Box 9.1. Profile of a herder.

Zhaxima, one of our interviewees, lives in Wulasitai gacha (village), Dabuxilete sumu (town). She became a herdsman when she was very young. Before the 1950s, her family consisted of two people, 20–30 sheep, 9–10 cattle and 2–3 horses. The family had no other income except to herd animals for others so as to get some pocket money. At that time, she needed to divide and drive the domestic animals to more than ten rangeland sites each year. Herds/flocks would stay at each site for 2–3 months and move quickly from one rangeland to another in summer (three to four moves) or winter (one move per month), but stayed in one place a relatively longer time in spring or autumn. These rangeland sites were within a 10 km radius (see Fig. 9.6). Nowadays, the household has 27 members and more than 200 sheep, 30 cattle and 4 horses and lives solely on husbandry. The household was allocated 1500 mu (1 ha = 15 mu) of grazing land under the semi-private property rights (Li *et al.*, 2007). Thus, each sheep unit can have 5 mu of rangeland. Now, they think they live a relatively easy life. The only remaining question, then, is how to make sure this lifestyle is sustainable since the stocking rate is so high.

the birth of the baby animals. In Box 9.1, we provide an example of early shifting grazing in Xilingol.

9.4.3 Half-sedentary grazing systems

Since the 1950s, with China implementing a collective economy (Chapter 2), most herdsmen have been following half-sedentary or sedentary nomadic grazing, but still show some characteristics of nomadism (Qi, 2002; Wang, 2006). Compared with a full nomadic lifestyle, sedentary nomadic grazing is definitely a more advanced land-use style, which has been developed based on the paradigms of traditional grazing, i.e. seasonally driving animals from one rangeland to another. Thus, the rangelands can be divided into spring, summer, autumn and winter rangelands (i.e. four-season rangelands), or winter–spring and summer–autumn rangelands (i.e. two-season rangelands), as winter–spring can be considered the non-growing season, while summer–autumn represents the growing season (Suttie and Reynolds, 2003).

Spring rangelands are generally used from mid-March to early June. The climate in spring might change dramatically. Animals consume a lot of their fat during the long stressful wintertime. At the same time, they are in a period of reproduction and high-nutrition forages are desperately needed. However, dry forage left in the spring rangelands is generally scarce and of low quality; thus, the mortality of these animals is generally high during the period from winter to spring. The spring rangelands are generally located near a water source, are south-east facing, downwind of sand dunes and always characterized by a dominance of early spring plants, which may green up quickly and provide relatively high nutrition for the famine-stricken animals. Although these ephemeral plants can live for a short period of time only, they are important to animals, especially pregnant and lactating ones, in the transition from winter to the breaking of dormancy by the perennials.

Summer rangelands are generally used from middle June to early September. Plant growth reaches its peak and there is ample forage with relatively high nutrition during the summer period. Summer rangelands are generally located in uplands with a low temperature and more winds, but with a sufficient water supply. The growing season of plants on summer rangelands is generally short, but this also favours a low population of flies and mosquitoes, which are such a nuisance to livestock on lowland areas at this time of year.

Autumn rangelands are generally used during the period of late August or early September to middle October or early November. They are often located in open riverside, bottomland or saline–alkali land with a high quality of forage and available minerals. It is cool in autumn and gradually turns to cold. The growth of plants ceases and they dry off. Most plants are in a state of fruiting and the aboveground parts begin senescing. The autumn rangelands are often dominated by *Artemisia* spp. and *Allium* spp. Autumn rangelands are the last place for livestock to find enough food to prepare for the chilly winter and long-lasting spring. This is an extremely important phase of the seasonal cycle in husbandry and it is vital in ensuring winter survival to allow profitability.

Winter rangelands are generally used from early November to mid-March. It is extremely cold during winter in the rangelands, the ground is often covered with snow and the remaining plants are dry with a low nutritional value. The fat that animals accumulated in autumn is often consumed to survive in the stressful conditions and most animals lose 30% of their body weight over the period November–March. Like spring rangelands, winter rangelands are also located in low, downwind, south-east-facing sand dunes or in shrublands. High *A. splendens* tussocks in bottomlands and tall reeds (*Phragmites* spp.) are good choices for winter rangelands.

9.4.4 The advantages and disadvantages of four-season shifting grazing

1. During the 1950s and 1960s, the use of rangelands in Xilingol could be categorized into nomadic, half-sedentary nomadic and full-sedentary nomadic grazing. The full-sedentary nomadic grazing preserved the advantages of nomadic grazing. However, with the application of shifting grazing among four-season rangelands, animals were no longer required to walk a long distance to find enough to eat and herdsmen also avoided some inconvenience to their daily life.

2. Local government could now try to organize herdsmen to graze their domestic animals at certain locations at certain time periods based on the growing status of the rangelands, thus greatly decreasing the risk of unsustainable use and uneven grazing (Bedunah and Harris, 2004). Besides, local government could also provide herdsmen with additional forage in drought or after severe snow disasters, a water supply, epidemic prevention, transportation to bring in supplies and move livestock to market or to other more favoured sites, thus increasing productivity and efficiency.

3. Under the collective ownership of rangelands, local government tried to increase their productivity by increasing human population and stocking rates. The utilization of rangeland resources was highly intensified, resulting in their later degradation. Besides, with the increase of population size, a certain amount of rangeland was converted to cropland to alleviate food shortage. The functioning of rangeland ecosystems had been changed permanently, resulting in soil and water loss and further malfunctioning of these ecosystems.

4. Lack of understanding of the function and processes of rangeland ecosystems inevitably led to ineffective grazing management, which depleted the rangelands further, especially as it affected spatial distribution. For example, there was underutilization in the upland rangelands of the Xilin River Basin, while stocking pressure was too high in the lower range area. In the areas around villages and other settlements, rangelands were generally severely degraded because of heavy stocking rates.

9.4.5 Sedentary grazing systems

Since reform of the economy in the 1980s, the management of grassland use has changed greatly. Each household can have a certain amount of land for their own management, called semi-private property rights (Li *et al.*, 2007). Herdsmen can decide the grazing stock rates and what kind of animals to raise. This was meant to avoid the 'tragedy of the commons' (Hardin, 1968). This change has greatly stimulated the enthusiasm of herdsmen for raising domestic animals. The self-management of grazing limited to a piece of land led to real sedentary grazing in rangelands. Herdsmen changed the traditional nomadic grazing, or sedentary nomadic grazing, to a system involving grazing for about half the year and feeding the livestock in pens for the other half of the year.

Sedentary grazing, however, has both advantages and disadvantages. Sedentary grazing is a fundamental advance for a society which has developed from a primal nomadic society to a half-nomadic and half-sedentary society during the past thousands of years. Herdsmen no longer drift from place to place. With the rapid development of the economy in Xilingol, a herdsman's living standard, education, medical facilities and social security have improved fundamentally, with a leap in living quality. Sedentary grazing can also be beneficial to administrative management of the area and to urban planning. However, as herdsmen concentrate grazing in a specific area, the rangelands have no time to rest and recover fully from grazing, especially in some overgrazed areas. Because herdsmen have self-management rights, sedentary grazing then lacks instruction and steering activities from the government. Patchiness in rangeland management can neither improve the ineffective grassland management status, nor reach the target of sustainable resource use. Under the stimulus of the market economy, many herdsmen have increased their stock number without considering the rangeland loading capacity, resulting in accelerated degradation (Williams, 2002; Zhang, Q., 2006). As a result of the practice of hiring other people to look after the livestock, there is no clear idea on the part of the local administration as to how many livestock a household owns and the balance between animal number and land area allocated under the grazing user rights (GUR) (Fernandez-Gimenez and Allen-Diaz, 1999). The lack of mobility is also an important factor as the spatio-temporal variation of the rangeland cannot be properly taken into account by individual herders who are restricted to their allocated sites (often fenced) and this has led to accelerated rangeland degradation (Williams, 2002).

9.5 Ideas Underlying the Mongols' Nomadic Culture

The primal nomadic economy was characterized by self-sufficiency and autarky and this reflected the traditional use of rangelands by Mongols (Zhang, Q., 2006; Zhang *et al.*, 2007). Under the traditional system, herdsmen could utilize the rangelands fully and still let them have time to rest by shifting grazing within a large area and by choosing rangelands depending on the characteristics of the region, the geographical and geomorphological conditions, the water supply, vegetation properties, the amount of biomass, the seasons and the forage edibility for different livestock. This nomadic lifestyle reflected the herdsmen's profound understanding of the rangelands and the strategy of scientific grassland management (Fernandez-Gimenez, 2000).

Herdsmen can recognize and know basic plant composition and functions of rangeland ecosystems and they use forage resources rationally and scientifically. This is reflected by the following aspects. First, herdsmen can recognize the characteristics of most rangelands and even the relationship between plants, animals and the environment, and can make classifications based on the potential use of the forage resources. Then, they can formulate a preliminary plan on how to use these rangelands rationally and make corresponding management decisions based on the above-mentioned knowledge. Second, herdsmen can identify and make full use of plant resources (Launchbaugh and Howery, 2005). Herdsmen can distinguish the properties of different plants, including their living conditions, phenology, root structure, response to grazing, geographical location, edibility for different kinds of livestock and edibility of their different organs in different seasons. Third, herdsmen may have some basic knowledge about the functioning and dynamics of rangeland ecosystems (Farnsworth *et al.*, 2002). They can understand that spatial heterogeneity is caused mainly by the corresponding spatial distribution of soils, temperature, precipitation volume and pattern and solar radiation resources, human land use and other activities (Carande *et al.*, 1995). These all prove that the traditional nomadic grazing culture reflects the Mongols' understanding of rangeland ecosystems, which gives important inspiration to the present modern ecosystem management concept.

From the point of view of ecology, the nomadic grazing culture is an important carrier of traditional ecological knowledge and is also a precious experience for understanding, protecting, managing and, more importantly, for restoring the rangelands. Ecologists now realize that the traditional ecological knowledge held by herdsmen will generate profound influences on biodiversity conservation and endangered species protection, building up the natural reserves and other bases for further use of rangeland resources. These have much in common with the adaptive management of ecosystems in modern ecology, which pays more attention to understanding fully and responding actively to the uncertainties and unpredictability of ecosystems (Berkes *et al.*, 2000). The adaptive rules of grassland ecosystem management contained in the nomadic grazing of the Mongols included keeping

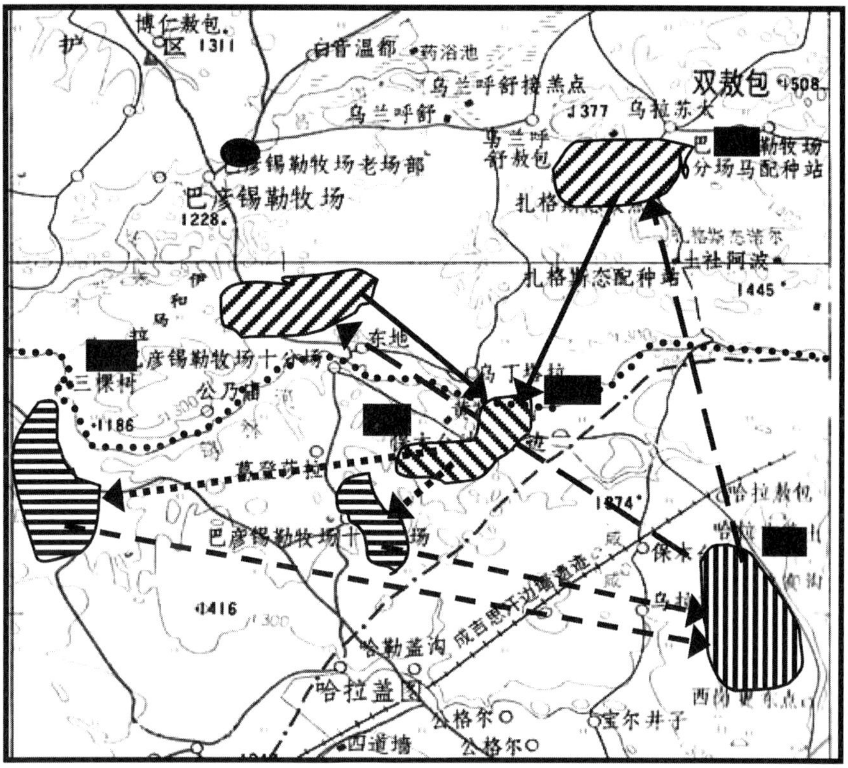

Fig. 9.6. The formation and evolution of the grassland nomadic culture and ecosystem management paradigm.

a high diversity of raised livestock, shifting use of resources and conserving as many backup rangelands as possible (Pamo, 2004).

9.6 Restoration and Adaptive Ecosystem Management: a Framework for Action

9.6.1 Modern ecosystem management theories

The science of modern ecosystem management studies how to manage ecosystems to ensure that they remain stable, healthy and sustainable (Christensen *et al.*, 1996). Modern ecosystem management seeks to meet human beings' increasing needs and, at the same time, still protect and maintain processes and recoverability of ecosystem functioning to ensure availability of resources for the next generation. Adaptive ecosystem management asks questions and tries to solve them in practice to test technological approaches and to validate and improve management results (Fig. 9.6) (Batabyal and Godfrey, 2002).

With the fast development of the economy, increase of population, dramatic change of land use, decrease of arable land area and extensive urbanization, the living conditions of human beings are facing stronger and stronger pressure (Pringle and Landsberg, 2004). This situation is even truer in arid and semi-arid areas because of the frangibility of the grassland ecosystems, the unpredictability of natural precipitation and the stressful environmental conditions (Eakin and Conley, 2002). Such questions as how to use and manage rangelands rationally to protect the diversity of life, how to maintain resource support for ecosystem functioning and how to sustain and harmonize local society, the economy and the eco-environment may be widely concerning and challenging.

9.6.2 Recovery: actions for sustainable rangeland management

To improve the ecological environment of Xilingol, the population size needs to be controlled and decreased further as it is directly dependent on agriculture and husbandry (Turner *et al.*, 2003). Some of the people currently engaged in both agriculture and animal husbandry can be educated and trained and then transferred to industry and transportation, thus reducing dependence on the natural resources of Xilingol's rangelands. Local government also needs to advise herdsmen to cut down their stocking load and to increase the slaughter rate at the end of each year. Thus, the grazing rate can be reduced and adjusted to a degree that the ecological environment can sustain.

In developed countries, such as the USA, the UK and Australia, the modernization of husbandry has been closely related to the application of science and technology. Thus, industrial development may help support the development of agriculture and husbandry in Xilingol to maintain prosperity. The development of the social economy in Xilingol has been relatively slow and the management of husbandry has long lagged behind. Thus, there is huge potential to change this inappropriate status by adjusting the structure of the economy rationally, strengthening industrial development, increasing the quota of industrial production and, at the same time, pushing the development of hotels and restaurants. With the development of industry and tourism, agriculture and husbandry can be compensated to alleviate the pressure on the rangelands (Zhang, J.T., 2006).

The traditional nomadic style of grazing, which is fully dependent on the natural environmental conditions, can be replaced as much as possible with modern and intensified feedlot raising (Zhang, J.T., 2006). For example, the profit produced by five milch cows can be compared with 300 sheep. The impacts of 300 sheep on grassland ecosystems, however, are much larger than those of five cows because cows can be raised in feedlots, while it is difficult for this to be done for sheep or other free-grazing animals.

By using advanced science and technology, advanced production and processing facilities and with corresponding personnel training and scientific management, the high labour load-producing processes can be mostly replaced by instrument manipulation, which needs only mechanical automatic production. Thus, producing efficiency can be increased greatly and, in addition, yields can be augmented with extra profit from products through value adding.

Construction of artificial pastures and their rational use can increase the productivity of grassland ecosystems at some local sites and thus provide sufficient forage for those domestic animals such as cows and beef cattle raised in feedlots and further may facilitate intensified secondary productivity. For this purpose, herdsmen can also be encouraged to plant small areas of highly productive pasture around their area of residence to compensate for forage shortage (Miller, 1999).

Short-term fencing of rangelands to allow short-term grazing bans can help slightly degraded rangelands recover quickly. Rotational grazing is also possible in fenced areas and can bring great benefits. Rotational grazing can not only enhance animal productivity, but also alleviate the degradation of the rangelands and, in the end, return the primary productivity of the rangelands to their peak levels (Savory and Parsons, 1980).

Compared with the traditional nomadic grazing lifestyle, the resource use of the present rotational grazing is more scientific and more sustainable under the current pressure of an increasing population around the world (Niamir-Fuller and Turner, 1999). Nomadic grazing is more like a type of blind shifting of grazing areas. In some extremely fragile areas where vegetation has been destroyed or its growth is strictly limited by environmental factors such as low precipitation or highly sandy soil, people should move out of the area to ensure that those degraded or highly fragile ecosystems have a chance to recover, thus avoiding a complete collapse of the ecosystem.

9.6.3 Successional pathways in recovering rangelands

The extent and speed of the degradation of grazing grasslands are highly dependent on stocking rates and the duration of grazing time, especially when in an overgrazed condition. Light degradation may not change species composition and species number much, but can decrease community density, height and aboveground biomass significantly. For example, light grazing in an original *S. grandis* steppe did not change the dominance of *S. grandis* and the species number, but did decrease the coverage (around 50–30%), density (from 26 to 18 cm) and aboveground biomass (from 125 to 87 g/m^2) along a grazing gradient from low to medium, indicated by a decreasing distance from residential areas and water source (Li, 1988). When the grazing pressure is increased further, especially near residential areas, replacement of dominant plant species might occur. In Xilingol, some *Artemisia* species (such as *A. frigida* or *A. pectinata*) might take the place of *S. grandis*. At the same time, community height, density and aboveground biomass reduced even further. Similarly, *L. chinensis* steppe would have the same change pattern along the grazing gradient and, under high grazing conditions, the change of the original community might follow the same route and, in the end, be replaced by an *Artemisia*-dominated community in which spatial heterogeneity and community productivity were extremely low (Wang *et al.*, 1996a). Fuzzy cluster analysis showed that the divergence of the similar communities under different grazing intensities and the convergence of the different communities under heavy grazing pressure, i.e. both *L. chinensis* steppe and *S. grandis* steppe, would change into *A. frigida* steppe through regressive succession if long-term high grazing pressure were imposed (Li, 1988). The degraded *A. frigida* community was generally stable if a certain grazing intensity was maintained (Wang *et al.*, 1996a).

According to an earlier study by Li *et al.* (1994) on an *L. chinensis* steppe which was enclosed for recovery from 1983 to 1991, and some other relevant studies (Wang *et al.*, 1996a,b), the restoration succession of the degraded steppe included some unstable communities in the middle of the progressive succession process. During the recovering process, the species composition in the community was not changed obviously, but the dominance or the roles of some specific species, such as those zonal species, might change significantly. For example, the dominancy of *L. chinensis*, *S. grandis* and *A. cristatum* increased while *A. frigida*, *Potentilla acaulis* and *C. microphylla* declined annually with the increase of enclosed years. Besides, the productivity of degraded grassland communities seemed to recover most easily compared with other ecosystem functions and, of course, this was also controlled by other environmental factors such as precipitation and adjustment of grazing intensity (Wang *et al.*, 1996b; Li *et al.*, 2008).

9.7 Lessons Learnt

Based on the above analysis, we conclude that the rapid increase in population and number of livestock and the partial change of rangeland use (principally, land conversion for cropping) are critical reasons for the severe degradation of large areas in Xilingol. The unintended consequences of past policies have led to the current situation. Inward migration, uncontrolled conversion of rangelands for cropping and flaws in the GUR (land allocation, lack of security of tenure and restrictions on mobility) have exacerbated the situation (see Chapter 15). Further consideration needs to be given to the GUR question to ensure that the emerging problems (Williams, 2002, 2006) are dealt with (Chapter 15).

By analysing the natural environment and social economy in Xilingol, and using the knowledge of adaptive ecosystem management and theories of sustainability sciences, we propose here some suggestions for ecosystem conservation and sustainable husbandry development of Xilingol.

1. The number of people employed in agriculture and husbandry needs to be decreased and strictly controlled by providing farmers and herdsmen with more occupations in manufacturing, hotel and restaurant services and tourism.
2. The local government of Xilingol should adjust the industrial structure by placing greater emphasis on manufacturing, followed by hotel and restaurant services and ecotourism, thus decreasing the pressure on crop production and livestock husbandry.
3. Traditional nomadic grazing, which depends fully on natural grassland productivity, should be replaced with intensive feedlot animal raising. Besides, the productivity of domestic animals can be increased by breed improvement and scientific feeding through nutritional rations.

4. Forage productivity can be increased significantly by building up artificial pastures, especially those highly productive fodder species suited to ensiling. Rational use of rangelands is also highly recommended through fencing in some severely degraded areas and by rotational grazing.

5. Large-scale land-use change, especially rangeland conversion, must be strictly prohibited through taking measures of administration, economy and legislation. Abandoned cropland and severely degraded rangeland should also be restored naturally by enclosure or re-seeded.

6. A campaign to raise the awareness of herdsmen and farmers about the high value of rangeland ecosystems, such as important social, economic and cultural values, can help increase their consciousness of environmental protection.

References

Bailey, D.W. (2005) Identification and creation of optimum habitat conditions for livestock. *Rangeland Ecology and Management* 58, 109–118.

Batabyal, A.A. and Godfrey, E.B. (2002) Rangeland management under uncertainty: a conceptual approach. *Journal of Range Management* 55, 12–15.

Bedunah, D.J. and Harris, R. (2004) Past, present and future: rangelands in China. Here we examine rangelands and changes in pastoral use in an ethnic-minority region, Gansu, China. *Rangelands* 24, 17–22.

Berkes, F., Colding, J. and Folke, C. (2000) Rediscovery of traditional ecological knowledge as adaptive management. *Ecological Applications* 10, 1251–1262.

Brunson, M.W. and Steel, B.S. (1996) Sources of variation in attitudes and beliefs about federal rangeland management. *Journal of Range Management* 49, 69–75.

Carande, V.G., Bartlett, E.T. and Gutierrez, P.H. (1995) Optimization of rangeland management strategies under rainfall and price risks. *Journal of Range Management* 48, 68–72.

Chen, B. (2004) A discussion on the formation, changes and characteristics of traditional Mongolian culture. *Journal of Inner Mongolia University (Humanities and Social Sciences)* 36(3), 66–69.

Chen, Z. (1988) Topography and climate of Xilin River Basin. *Research on Grassland Ecosystem* 3, 3–22 (in Chinese).

Christensen, N.L., Bartuska, A.M., Brown, J.H., Carpenter, S., Dantonio, C., Francis, R., Franklin, J.F., MacMahon, J.A., Noss, R.F., Parsons, D.J., Peterson, C.H., Turner, M.G. and Woodmansee, R.G. (1996) The report of the Ecological Society of America Committee on the scientific basis for ecosystem management. *Ecological Applications* 6, 665–691.

Eakin, H. and Conley, J. (2002) Climate variability and the vulnerability of ranching in southeastern Arizona: a pilot study. *Climate Research* 21, 271–281.

Farnsworth, K.D., Focardi, S. and Beecham, J.A. (2002) Grassland–herbivore interactions: how do grazers coexist? *American Naturalist* 159, 24–39.

Fernandez-Gimenez, M.E. (2000) The role of Mongolian nomadic pastoralists' ecological knowledge in rangeland management. *Ecological Applications* 10(5), 1318–1326.

Fernandez-Gimenez, M.E. (2001) The effects of livestock privatization on pastoral land use and land tenure in post-socialist Mongolia. *Nomadic Peoples* 5(2), 49–66.

Fernandez-Gimenez, M.E. and Allen-Diaz, B. (1999) Testing a non-equilibrium model of rangeland vegetation dynamics in Mongolia. *Journal of Applied Ecology* 36, 871–885.

Hardin, G. (1968) The tragedy of the commons. *Science* 162, 1243–1248.

Humphrey, C. and Sneath, D. (1999) *End of Nomadism? Society, State and the Environment in Inner Asia.* Duke University Press, Durham, North Carolina.

Launchbaugh, K.L. and Howery, L.D. (2005) Understanding landscape use patterns of livestock as a consequence of foraging behavior. *Rangeland Ecology and Management* 58, 99–108.

Li, W.J., Ali, S.H. and Zhang, Q. (2007) Property rights and grassland degradation: a study of the Xilingol pasture, Inner Mongolia, China. *Journal of Environmental Management* 85, 461–470.

Li, Y. (1988) The divergence and convergence of a *Eurolepidium chinense* steppe and *Stipa grandis* steppe under the grazing influence in Xilin River valley, Inner Mongolia. *Acta Phytoecologica et Geobotanica Sinica* 12, 189–196.

Li, Y., Wang, W., Liu, Z. and Jiang, S. (2008) Grazing gradient versus restoration succession of *Leymus chinensis* (Trin.) Tzvel. grassland in Inner Mongolia. *Restoration Ecology* 16(4), 572–583.

Li, Z., Pei, H., Liu, Z. and He, T. (1994) The research on the dynamics of community characters of degenerate *Leymus chinensis* steppe during recovering processes. *Acta Scientiarum Naturalium Universitatis Neimenggu* 25, 88–98.

Miller, J.B. (1999) The feasibility of agroforestry interventions for traditionally nomadic pastoral people. *Agriculture and Human Values* 16, 11–27.

Niamir-Fuller, M. and Turner, M.D. (1999) A review of the recent literature on pastoralism and transhumance in Africa. In: Niamir-Fuller, M. and Turner, M. (eds) *Managing Mobility in African Rangelands: The Legitimization of Transhumance*. Intermediate Technology Publications, London, pp. 18–46.

Pamo, E.T. (2004) Water development strategy as a driving force for sustained rangeland management by local communities in sub-Saharan Africa. *Environmental Monitoring and Assessment* 99, 211–221.

Pringle, H.J.R. and Landsberg, J. (2004) Predicting the distribution of livestock grazing pressure in rangelands. *Austral Ecology* 29, 31–39.

Qi, B. (2002) *The Annals of Xilingol League*. Inner Mongolia People's Publishing House, Huhhot, China.

Savory, A. and Parsons, L.D. (1980) The Savory grazing method. *Rangelands* 2, 234–238.

Shichinohe, C. (1994) *The Nomadism in China*. Hokkaido University Press, Sapporo, China.

Suttie, J.M. and Reynolds, S.G. (2003) *Transhumant Grazing Systems in Temperate Asia*. Food and Agriculture Organization of the United Nations, Rome.

Swallow, B.M. and International Institute for Environment and Development, Sustainable Agriculture Programme (eds) (1994) *The Role of Mobility within Risk Management. Strategies of Pastoralists and Agro-pastoralists*. Sustainable Agriculture Programme of the International Institute for Environment and Development, London.

Turner, B.L., Kasperson, R.E., Matson, P.A., McCarthy, J.J., Corell, R.W., Christensen, L., Eckley, N., Kasperson, J.X., Luers, A., Martello, M.L., Polsky, C., Pulsipher, A. and Schiller, A. (2003) A framework for vulnerability analysis in sustainability science. *Proceedings of the National Academy of Sciences of the United States of America* 100, 8074–8079.

Wang, J. (2006) Inhabited pasturage, pasture landscape and the social political construction of eastern Mongolia (1950–1980). *Nankai Journal (Philosophy, Literature and Social Science Edition)* 2006(5), 71–80.

Wang, W., Liu, Z.-L., Hao, D.-Y. and Liang, C.-Z. (1996a) Research on the restoring succession of the degenerated grassland in Inner Mongolia. I. Basic characteristics and driving force for restoration of the degenerated grassland. *Acta Phytoecologica Sinica* 20, 449–459.

Wang, W., Liu, Z.-L., Hao, D.-Y. and Liang, C.-Z. (1996b) Research on the restoring succession of the degenerated grassland in Inner Mongolia. II. Analysis of the restoring processes. *Acta Phytoecologica Sinica* 20, 460–471.

Williams, A. (2006) Improving rangeland management in Alxa League, Inner Mongolia. *Journal of Arid Land Studies* 15(4), 199–202.

Williams, D.M. (2002) *Beyond Great Walls: Environment, Identity, and Development on the Chinese Rangelands of Inner Mongolia*. Stanford University Press, Stanford, California, xii + 251 pp.

Xilingol Grassland Management Station of Inner Mongolia Autonomous Region (1988) *Grassland Resources of Xilingol*. Inner Mongolia People's Publishing House, Huhhot, China.

Zhang, J.T. (2006) Grassland degradation and our strategies: a case from Shanxi Province, China. *Rangelands* 28, 37–43.

Zhang, M.A., Borjigin, E. and Zhang, H. (2007) Mongolian nomadic culture and ecological culture: on the ecological reconstruction in the agro-pastoral mosaic zone in Northern China. *Ecological Economics* 62, 19–26.

Zhang, Q. (2006) May they live with herds: transformation of pastoralism in Inner Mongolia, China. MSc thesis, University of Tromso, Norway.

10 Case Study 4: Ordos Plateau, Inner Mongolia

Yuanrun Zheng and Qiushuang Li
Institute of Botany, Chinese Academy of Sciences, Beijing, China

Synopsis

The Ordos Plateau in Inner Mongolia is an important region north of the Yellow River. It has undergone major transformation over the past 70 years. Here, we examine the changes and underlying causes of land degradation, especially the rapid and serious increase in the areas covered by sand (sandification). Efforts to effect recovery are reviewed and finally we outline some suggested measures that combine socio-economic and biophysical approaches to arrest and reverse land degradation.

Keywords: sandification; dunes; land conversion; coal mining; firewood; climate dynamics; socio-economic dynamics; vegetation restoration

10.1 Statement of the Degradation Problem

Sandification is a process by which usable land is converted into, or covered by, sand. The process may involve both sand erosion and accumulation. Seriously sandified areas occur mainly in the Kubuqi Desert and the north-west part of the Mu Us Sandy Land. The rate of sandification has varied over time from the late 1950s to the 1970s but, generally, it has been rapid. It decreased after the 1980s due to the efforts of local government. However, the general situation was still severe. For example, in 2000 fixed dunes, semi-fixed dunes and drifting dunes occupied 7.2%, 21.4% and 44.5%, respectively, in the Mu Us Sandy Land. Seventy per cent of the total usable rangeland (over 40,000 km^2) was degraded, of which over 30,000 km^2 was seriously degraded. Most degradation occurred after 1949, especially from the 1960s to the end of the 1970s. From 1960 to 1979, biomass on the Ordos Plateau declined by 26% and the area of usable rangeland shrank by 17%.

In the 1950s, 1960s and 1970s, there was large-scale conversion of rangeland. There was a rapid increase in the area of cropping. Because of unsuitable management and low yields, large areas of cropland were abandoned shortly after cultivation, but the farmers shifted to new rangeland areas and this was cultivated. Statistically, the total area of cropland was not significantly different, but in reality, the figures masked the loss of rangeland. Land conversion, some of it illegal, was a major reason why vegetation on the Ordos Plateau was damaged so quickly. Populations of both humans and their animals also rose rapidly. A certain portion of the rural population in the Ordos region is still partly dependent on firewood for cooking and heating due to traditional customs and lack of money. The plants used as firewood are mainly shrubs and semi-shrubs. It was estimated that an average of 2–3 ha of rangeland shrubs were cleared per household each year in western Ordos in the 1970s. Loss of shrubs has affected stability and exposed the land surface to erosion by wind and water. The Ordos Plateau contains the largest coal resources in China and open-cut coal mining aggravates the desertification problem.

10.2 Basic Introduction to the Ordos Plateau

The Ordos Plateau (about 120,000 km^2) lies in the south-west of Inner Mongolia. It is characterized as an ecotone in aspects of atmospheric circulation, climate, geology and geography, vegetation, industry and social culture. The environment here is very fragile and sensitive to outside disturbance (Zhang, 1994) and the land becomes desertified quite easily (Zeng, 1985). The Ordos Plateau had fine pasturage with enough water, plenty of grass and huge potential productivity. However, in the past 100 years, especially in the recent 50–60 years, climate change, irrational cultivation of grasslands and overgrazing have resulted in serious desertification. It has been the main centre of desertification in China and the main source of raw sands and sediments to the Yellow River. Further, the district around Ordos City, which occupies most of the Ordos Plateau, contains the largest coal resources in China and open-cut coal mining aggravates the desertification problem. Therefore, it is urgent to combat desertification and restore the degraded ecosystem. In the past 20 years, there has been great progress in combating desertification but success has only been partial and, overall, desertification is worsening. Here, we focus mainly on the Ordos District as a representative of the Ordos Plateau and analyse the factors that contribute to both degradation and recovery.

The Ordos District is bordered by the Yellow River in the west, north and east (Fig. 10.1). The total administrative area, which includes Dongsheng District, Jungar, Ejin Horo, Uxin, Dalad, Hangjin and Otog Banner (similar to the administrative level of county), is 87,400 km^2, with a population of 1.17 million in 1998. The region lies between 850 and 1600 m above sea level. The geology of the area is primarily of Cretaceous and Jurassic origin (Walter and Box, 1993) and is an ancient depositional basin. Most of the soil on the

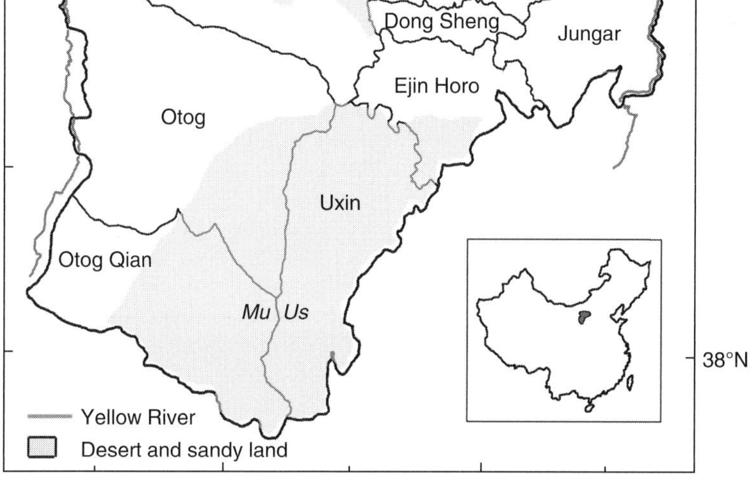

Fig. 10.1. The Ordos District of Inner Mongolia, the major part of the Ordos Plateau.

plateau is sandy. The Ordos District has five major geomorphological subdivisions: (i) in the east are the Jungar loess hills; (ii) in the west are the slopes of the Zhuozi Mountains; (iii) in the north is the Kubuqi Desert; (iv) in the south is the Mu Us sandy area interlaced with deserts, lowland and lakes; and (v) the centre of the area is an undulating plateau.

10.2.1 Climate

The climate is typically continental with extreme seasonal and diurnal temperature variation, low rainfall, cold dry winter and hot moist summer. The continentality of the climate decreases markedly from west to east. Atmospheric circulation changes from continental, temperate high-pressure ridge (Mongolia–Siberia) in the north-west to the influence of the monsoon through to the south-east, and covers the arid, semi-arid and moist climate types on the Ordos Plateau.

Annual precipitation varies from 300–400 mm in the east to only 100–150 mm in the west, distributed mainly from June to September. The fluctuation of annual rainfall is a main feature in the region. In 28.7% of years, annual rainfall drops to 100–150 mm, or lower. The potential evapotranspiration is about 2300 mm. The hours of sunlight range from 2900 to 3200 h/year and there are between 130 and 165 frost-free days/year.

10.2.2 Topography and soil

The topography changes from gobi and sand dune in the west, encroaching on the loess plateau in the east. A combination of wind and water erosion has resulted in a mix of low valley floors with soft sand dunes and harder dissected residuals. Soils on the Ordos Plateau change from brown pedocals, castanozems to dark loessial soils from west to east. Because of severe desertification and dune movement, the soil is generally sandy.

10.2.3 Vegetation

Vegetation varies from western continental desert, central desert grassland, to eastern grassland, with flora and fauna from the ancient Mediterranean central-Asian zone to the eastern forest zone. According to Chen (1964), vegetation on the Ordos Plateau can be divided into three types: steppe (east part), desert steppe (middle part) and steppe desert (west part). In the eastern steppe zone, the plant community coverage can reach 40–50% and include 20–30 plant species. The dominant grass species are *Stipa bungeana*, *Aster altaicus* and *S. breviflora*. The dominant semi-shrub species are *Lespedeza dahurica* and *Artemisia frigida*. In the central desert steppe zone, the plant community coverage is lower than that of the steppe zone and may reach only 30–40% and include 20–30 species. The dominant species include semi-shrub species *A. ordosica* and *S. gobica*. Compared with the steppe zone, the desert steppe zone is characterized by ecologically important shrub and semi-shrub species. In the west, the steppe desert, the plant community coverage can reach 30–40% and coverage of the dominant species, *Caragana tibetica*, can reach 30–35%. This dwarf community is usually lower than 20 cm in height. In this zone, the plant species are characterized by water storage in the leaves, leaves with a lower bioplasm potential, leaf carnification or very small leaves. There are some endangered species protected by the Chinese government and these are remnants of the Tertiary period, including *Tetraena mongolica*, *Ammopiptanthus mongolicus*, *Potaninia mongolica*, *Reaumuria trigyna*, *Holianthemun soongoricum* and *Zygophyllum xanthoxylon*.

10.2.4 Socio-economy

The population was about 1.3 million in 1998 but has risen since then, and is mainly Han and Mongolians. Traditionally, local people depended on livestock husbandry, but agriculture has developed gradually over the past 100 years. The main crops include maize, potato, buckwheat and millet. Generally, agriculture has developed in the eastern part and livestock husbandry in the west. Coal mining boosted the local economy considerably during the late 1990s to early 2000s.

Besides being used as grazing land for cattle, sheep and goats, the Ordos Plateau is also a source of many useful wood and non-wood products. During the rainy season, the rangeland produces up to 50 species of edible herbs (such as

Scorzonera divaricata and *Allium mongolicum*) and mushrooms, although herb and mushroom collection has little effect on vegetation (CCICCD, 1999). The collection of firewood (mainly *A. ordosica* and *C. intermedia*) and medicinal plants contributes greatly to rangeland degradation.

Although the living standard of the local people has been greatly improved and coal and electricity have been widely available for household energy for the last two decades, a certain portion of the rural population in the Ordos region is still partly dependent on firewood for cooking and heating, due to traditional customs and lack of money. The plants used as firewood are mainly shrubs and semi-shrubs. Shrubs are usually cut with axes at about 10 cm above ground by either clear or selective falling, but in many other cases roots are also excavated. It was estimated that an average of 2–3 ha of rangeland shrubs were cleared per household each year in western Ordos in the 1970s. For example, in 1979, 4500 t of *A. ordosica* were burned in the region, with considerable effects on rangeland structure (CCICCD, 1999).

The Ordos rangeland produces many medicinal plants, which are valued in both traditional Chinese medicine and modern medical science. *Ephedra sinica*, *Glycyrrhiza uralensis*, *Bupleurum chinensis* and more than 40 other species of medicinal plants occur widely in the plant community. The roots of medicinal plants are usually the most valuable parts. Medicine gathering often means damage to both the plants and the topsoil (CCICCD, 1999).

10.3 Situation of Vegetation Degradation

10.3.1 Sandification

Sandification is a process by which usable land is converted into, or covered by, sand. The process may involve both sand erosion and accumulation. Most of the Ordos Plateau (over 70% of the total area) suffers from sandification. Seriously sandified areas occur mainly in the Kubuqi Desert and the north-western part of the Mu Us Sandy Land (Uxin, Otog Qian, Ejin Horo, Otog and Hangjin Banner) (CCICCD, 1999). The rate of sandification has varied over time, but from the late 1950s to the 1970s it was generally rapid. It decreased after the 1980s due to the efforts of local government. However, the general situation was still severe. For example, in 2000 fixed dunes, semi-fixed dunes and drifting dunes occupied 7.2%, 21.4% and 44.5%, respectively, in the Mu Us Sandy Land. The high percentage of drifting/shifting dunes is a major contributor to sandification.

10.3.2 Vegetation degradation

Vegetation changes were characterized by a loss of forage biomass. The coverage, height and biodiversity of vegetation decreased. A decrease in high-quality plants and an increase in poor-quality species were also observed. Seventy per cent of the total usable pasture area is degraded to a greater or lesser extent. Over 40,000 km^2 pasture is classed as 'degraded' and over 30,000 km^2 as 'seriously degraded'. Most pasture was degraded after 1949, especially from the 1960s to the end of the 1970s. From 1960 to 1979, forage biomass on the Ordos Plateau declined by 26% and the area of usable rangeland shrank by 17% (CCICCD, 1999). Examples of pasture degradation were found all over the Ordos. On sand dunes and the sand-covered western high plateau, when *A. ordosica* was degraded, *A. sphaerocephala* became the dominant species and seriously degraded land was covered by sparse (less than 10%) *Agriophyllum squarrosum*. The typical plant community succession from steppe to moving dune is *S. bungeana* → *A. ordosica* → *A. ordosica* + *A. sphaerocephala* → *A. sphaerocephala* → *A. squarrosum*.

Typical steppe vegetation became seriously degraded. The original steppe was dominated by *S. bungeana* and had a dry biomass of more than 1300 kg/ha. Most areas are now dominated by *A. ordosica*. Degradation of sandy steppe and meadows is also apparent in the Mu Us Sandy Land. Biomass was estimated to have decreased by 30–40% from 1949 to the end of the 1970s (CCICCD, 1999). Compared with sandy steppe, valley meadow has been more degraded by overuse. Historically, this meadow was high and dense enough to hide cattle and horses, but, by the end of the 1970s, it had become so short and sparse that even a running rabbit was visible. As degradation increased, high-quality species disappeared,

Table 10.1. Monthly mean temperature changes in eight sites on the Ordos Plateau.

Sites	Dongsheng	Jungar	Ejin Horo	Uxin	Dalad	Hangjin	Otog
February	↗*	↗*	↗*	↗*	↗*	↗*	↗*
September	↗*		↗*	↗*			↗*
December	↗**		↗*	↗**	↗**	↗**	↗**

The arrow ↗ represents the ascending trend of the changes of monthly mean temperature. Significance level: $^{*}P < 0.05$; $^{**}P < 0.01$.

while biennial, inferior and even poisonous species took their place.

10.3.3 Soil erosion

Water and wind are the two main agents of soil erosion. Water erosion occurs mainly in the eastern river catchment. Other parts of the Ordos are affected by wind erosion. The transitional area in between is affected by both water and wind. The degree and extent of wind erosion are obviously relevant to sandification. Severe wind erosion occurs on the unprotected cropland, where erosion depth can reach 5–7 cm/year. Sand dunes move 3–10 m every year, pushed by the prevailing winds (CCICCD, 1999).

Water erosion has occurred mainly in the east and middle of the Ordos Plateau. For example, western and south-western Jungar Banner is severely eroded, with an erosion amount ranging between 15,000 and 20,000 t/km^2/year.

10.4 Causes of Desertification

Climatic change and human disturbance together cause desertification all over the world. It is also true on the Ordos Plateau. Here, about 30 years of climatic data (1971–1998) and about 50 years of socio-economic data (1947–1998) were used to clarify the reasons for desertification (see also Chapter 3).

10.4.1 Long-term climate dynamics

Change of temperature

The mean monthly temperature in February, September and December in most banners (eight sites) increased significantly, but in other months there were no significant changes (Table 10.1). Since September is the late growing period on the Ordos Plateau and February and December are in winter, the fall in temperature during these 3 months may aggravate the dryness of the soil and air in early spring and be disadvantageous for vegetation restoration in spring. It may be one of the reasons why desertification has been accelerating in recent years.

The mean temperature of the growing season (May–September) in the eight sites showed an insignificantly increased trend. The mean temperature of the non-growing season in contrast to the growing season (October–April) had increased significantly (Fig. 10.2). The mean annual temperature in the eight sites increased significantly from 1971 to 1998.

Changes of precipitation

Precipitation decreased significantly in September in most sites, while it decreased significantly in January in four sites (Table 10.2). Generally, precipitation increased in the growing season but decreased in the non-growing season.

The annual precipitation of all sites showed no significant changes. Although the effect of precipitation on specific ecosystems is important, the distribution of precipitation in different periods on structures and functions of ecosystems is more important. Distribution of precipitation may change the effect of ecosystems on the environment and affect the rhythm of change in ecosystems. On the Ordos Plateau, the ratio of precipitation in September to the annual total in all sites had a significant downward trend. The ratio of February to the 1-year total at Hangjin Banner, Dongsheng, Ejin Horo Banner and Uxin Banner also showed a significant downward trend. Precipitation decreased in late autumn and winter.

The change of potential evapotranspiration (PET), potential evapotranspiration rate (PER)

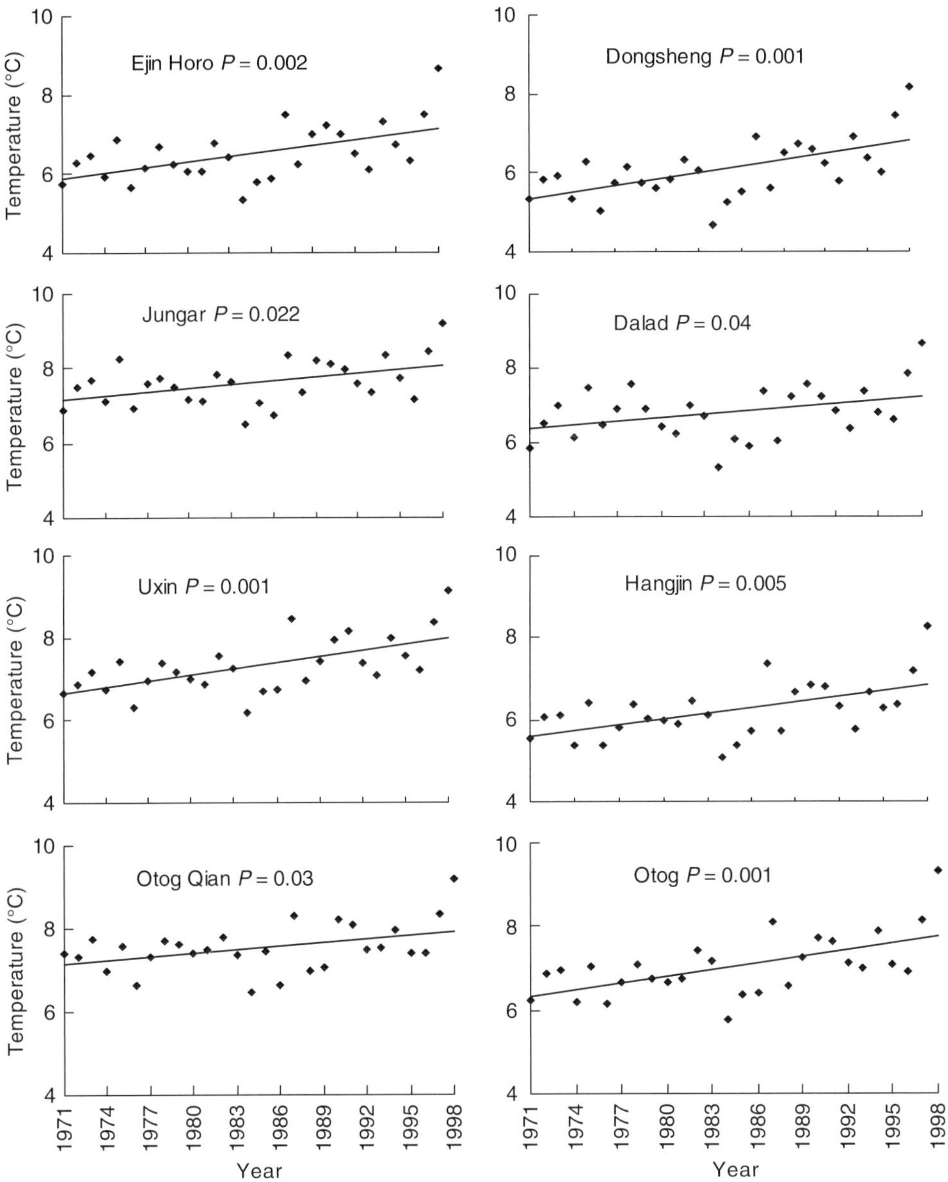

Fig. 10.2. The mean annual temperature for eight sites on the Ordos Plateau.

Table 10.2. Monthly mean precipitation changes in eight sites on the Ordos Plateau.

Station	Dongsheng	Jungar	Ejin Horo	Dalad	Uxin	Hangjin	Otog	Otog Qian
January	↘*		↘*			↘**		
September	↘*	↘*	↘**		↘*	↘	↘**	↘*

The arrow ↘ represents the descending trend of the changes of monthly mean precipitation. Significance level: *$P < 0.05$; **$P < 0.01$.

and radiative dryness (RDI) are also important. PET, PER and RDI are the synthetic reflection of total environmental factors within the ecosystem; these indices are used to clarify climatic change on the Ordos Plateau. Many indices are needed to calculate these three indices and, as some of them could not be obtained from the eight sites, empirical equations were used to calculate them.

$$\mathrm{RDI} = (0.629 + 0.237\ \mathrm{PER} - 0.00313\ \mathrm{PER}^2)^2$$

where RDI is the radiative dryness and PER is the annual potential evapotranspiration rate.

$$\mathrm{PER} = \mathrm{PET}/r = \mathrm{BT} \cdot 58.93/r$$

where PET is the potential evapotranspiration (mm), BT is the mean annual biological temperature (°C) and r is the annual precipitation (mm).

$$\mathrm{BT} = \sum t/365 = \sum T/12$$

where t is the mean daily temperature ($0°\mathrm{C} < t < 30°\mathrm{C}$) and T is the mean monthly temperature ($0°\mathrm{C} < T < 30°\mathrm{C}$).

PET increased significantly in all sites except Dalad and Jungar Banner (Table 10.3), which indicated that the climate had become drier. Although the change was not statistically significant, PER and RDI in the eight sites showed an increasing trend.

10.5 Long-term Socio-economic Dynamics

On the face of it, desertification and degradation of ecosystems are natural phenomena, but they are also socio-economic problems. The conflict between the sharp decline of land productivity induced by desertification and increased population and improvement of living standards through overexploitation of the resource base is a vital cause of accelerating land degradation.

10.5.1 Population

Dalad Banner had the biggest population, Otog Qian had the smallest population and the population increased by about 300% from 1947 to 1998 in some banners. Significant population increment will inevitably disturb the environment, and to a much greater extent in areas where local people rely on livestock and agriculture (Fig. 10.3).

Table 10.3. Significance level of increase of potential evapotranspiration in eight sites on the Ordos Plateau.

	Dongsheng	Jungar	Ejin Horo	Uxin	Dalad	Hangjin	Otog	Otog Qian
P	0.002	0.122	0.006	0.01	0.065	0.007	0.009	0.046

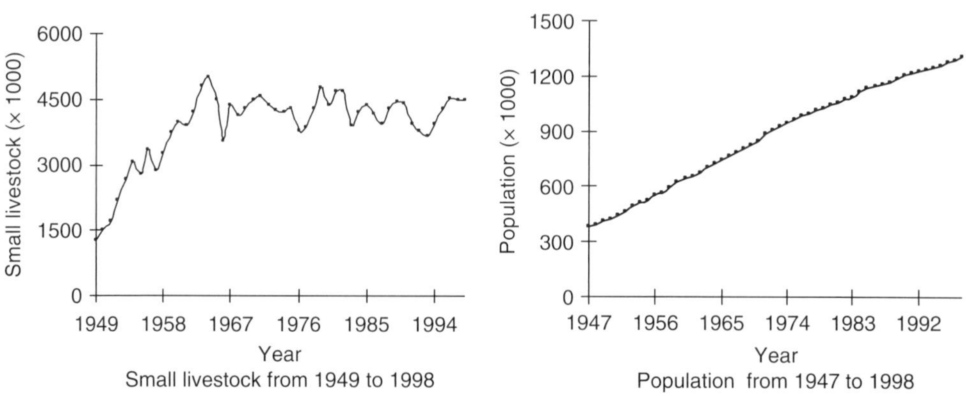

Fig. 10.3. Populations of both people and livestock increased rapidly in the period 1947–1998 and both continued their upward trend in the early 21st century.

10.5.2 Total area of cropland

The largest area of cropland, including grain and beans, is in Jungar Banner and Dalad Banner and the smallest is in Uxin Banner and Otog Qian Banner. The cropland of all banners has shown a significant downward trend in the past 38 years but it peaked in the 1950s, 1960s and 1970s, which coincided with large-scale conversion of rangeland during this period. Because of unsuitable management and poor harvests, large areas of cropland were abandoned a short time after the first cultivation, but new grassland was then cultivated to replace it. Statistically, the total area of cropland did not show significant change because it was masked by the conversion of rangeland. This was a major reason why vegetation on the Ordos Plateau was damaged so quickly.

10.5.3 Output of grain and beans

Although cropland had decreased, the output of grain and beans had increased significantly, especially in Dalad Banner, whose output increased fastest. Water is the main limiting factor for agricultural production on the Ordos Plateau. Dalad Banner lies near the Yellow River and is affected by its flooding, which results in fertile soil and plenty of groundwater and makes it suitable to develop irrigation agriculture. Output has increased significantly with the advancement of technology, including the introduction of new crop varieties, wise management, water-saving irrigation, higher efficiency in fertilizer usage, etc. This could be useful for mitigating desertification.

10.5.4 Number of small livestock

The Ordos Plateau is located in the transitional zone between cropland and rangeland and is used mainly for livestock husbandry. In the past, livestock husbandry was the most important land use. Otog Banner has the most livestock and also had the greatest rate of increase in livestock numbers over the period 1949–1980. Dongsheng District had the least and others are in between. The annual growth in numbers of small livestock in Dongsheng District kept stable, but Otog Banner's increased significantly and fell sharply in 1980, which was because Otog Qian Banner was established on separating from Otog Banner and thus their data were collected and reported separately. At that time, but to a lesser extent, the numbers in other banners all showed an increasing trend. The rapid increase in livestock numbers, causing a rise in overgrazing, is another important reason for rangeland desertification on the Ordos Plateau.

Although the total area of cropland decreased, productivity increased significantly, which is attributed to a slowing down of the rate of desertification. The significantly increased trend in the number of small livestock indicated that the trend of overuse of all sorts of rangeland was still severe. Overgrazing and loss of prime rangeland to land conversion were the major reasons that caused the expansion of desertification in this region. More recently, improvement in industry, commerce, science and culture has reduced the direct dependence on land for human needs and has led to more rational land use. Breaking the link between humans and the land through the creation of secondary and tertiary industry is a long-term solution to the overexploitation of rangelands (see below).

10.5.6 Holistic reasons for desertification on the Ordos Plateau

Significant temperature increment and precipitation decrement in late autumn and winter may induce much more drought and make desertification more serious. This could be the climatic reason for desertification on the Ordos Plateau.

Socio-economic development has been rapid due to inward migration and the conversion of rangeland to crop production to support the burgeoning population. Unwise land and natural resource use has put great pressure on the fragile environment. Further, not only population growth but also improved living standards place greater demands on the resource. This could be the socio-economic reason for desertification on the Ordos Plateau.

10.6 Socio-economic Development Pattern on the Ordos Plateau

Analysis indicated that each separate economic indicator of the Ordos District developed differently at each site. However, the socio-economic development had certain commonalities that related to specific environmental conditions there and to natural resource exploitation. Insufficient attention was given to this aspect. Each banner developed in its own way, with little cooperation between them. Now, it is realized that it is necessary to carry through the regional economic orientation and division in accordance with the characteristics of the natural environment and the resources in each banner so that the advantages of resources and environment can be used wisely on a regional scale. Because there were many indices of economic development, it was difficult to reveal their complicated internal relations and reduce them to a single index. We analysed the orientation and division of the regional economy considering the climatic character and socio-economic variables in the Ordos District using cluster analysis and proposed the optimized socio-economic development pattern considering the total Ordos Plateau as an integral unit in region planning.

In the ordination, seven banners and one major city could be classified into four types at a point where the distance coefficient was over 20,000 (Fig. 10.4), and we classified the pattern of economy development into four types.

10.6.1 Industrial, commercial, political and cultural types

Dongsheng District is the metropolis of Ordos District and is suitable for developing industry and commerce and enhancing the cultural, information and political functions. It is not fit to develop agriculture because of its high topography and shortage of water resources. The key points in developing its economy were: reinforcing the construction of basic establishments and strengthening the function of the information and cultural centre of the total area; developing industry without pollution and with low power consumption and social services; and developing protected agriculture to provide special vegetable and fruit crops. Generally, Dongsheng City should be the centre of information, culture, industry, services, commerce and specially protected agriculture for the total area.

10.6.2 Agricultural, commercial and industrial types

The economic characteristics of Dalad Banner are very significant. Dalad Banner has a particular advantage in developing agriculture because of its location near the Yellow River. At the same time, Dalad Banner connects the Ordos Plateau with Baotou City, a famous industrial city in China (second largest city in Inner Mongolia) and is the vital communication line to Hohhot City, Inner Mongolia. On the basis that its agriculture, commerce, transportation and industry progress well, its development could be as follows: first, it could actively develop highly efficient agriculture, making use of its advantage of being near the Yellow River (with enough water for irrigation),

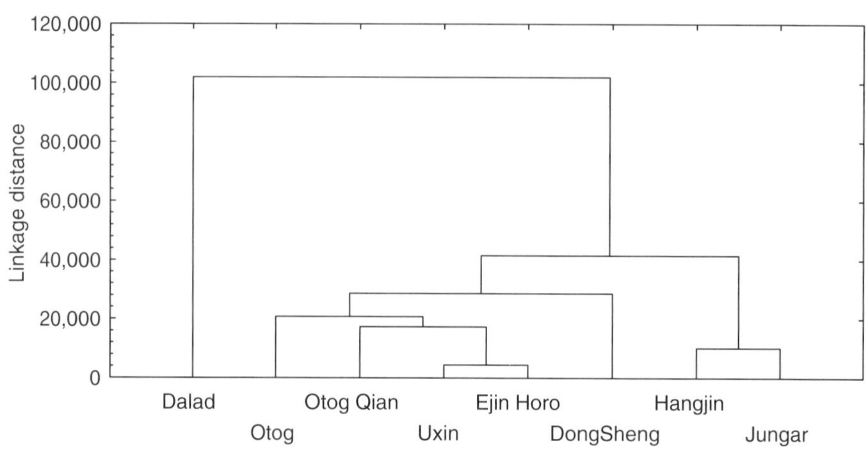

Fig. 10.4. Tree diagram of regional economic functions.

to establish an agriculture base for high-quality products. Second, the function of being a traffic hub could be considered fully to develop commerce actively. Third, a typical secondary industry like manufacturing should be developed. In general, Dalad Banner should become a base for agriculture and commerce.

10.6.3 Agriculture and livestock husbandry in the loess hills

This type includes two banners. Jungar Banner lies to the far east of the Ordos Plateau and is located in the transition belt of loess plateau and sandy land. Its terrain is dominated by loess hills with severe soil erosion and water loss (through runoff). Although there is a large cropping area, it is not appropriate to develop agriculture further on a large scale because of its fragile ecosystem. First, considering the management of erosion by water and soil, it could develop rain-fed farming and runoff agriculture to collect enough runoff to provide irrigation in the dry season, with the support of the coal industry. Second, special fruit orchards adapted to the loess plateau should be developed and further plantings of fruit shrubs such as *Hippophae rhamnoides* should be encouraged so that it brings economic benefit on the condition that soil and water erosion might be mitigated. Third, livestock husbandry should be developed actively on natural pastures, supplemented by artificial grassland and crop products, so that livestock husbandry could develop in a sustainable way.

The elevation in Hangjin Banner is higher and its soil type in the east is similar to that in Jungar Banner, so their economic development is similar too. Although some of the towns in Hangjin Banner had a basis in agriculture, livestock husbandry on sandy land and grassland was an important industry. The Kubuqi Desert lies to the west of Hangjin Banner, so we should adjust its economic regime appropriately, according to its limitations. Hangjin Banner should take full advantage of the sandy land and grassland resource to establish a base for livestock production to support crop agriculture. Generally, Jungar Banner and Hangjin Banner should be an agriculture and animal husbandry base.

10.6.4 Sandy land livestock husbandry

Uxin, Otog, Otog Qian and Ejin Horo Banners lie in the Mu Us Sandy Land and their soil is of a coarse texture. Underground water in the Mu Us Sandy Land is plentiful and fit for developing irrigation agriculture on a small scale, but livestock husbandry should be its major industry. Its socio-economic development should focus on: first, highly efficient irrigation agriculture on the condition of making full use of the underground water resource, but without complete depletion; second, making use of crop production, natural pasture and artificial grassland to develop livestock husbandry as its major industry. At the same time, Ejin Horo Banner could take advantage of its coal resources, strengthen the agricultural and livestock husbandry base and use the Ghinggis Khan Mausoleum to develop tourism. In general, these banners should become livestock husbandry centres in sandy land supplemented by small-scale agriculture.

The previous analysis took into account natural and economic conditions, rules of economic development and the resource advantage of each banner on the Ordos Plateau and proposed regional economic functions and orientation in the Ordos District. Only after the advantages of regional economic functions are exerted and each economic pattern is developed harmoniously will the sustainable and efficient development of the economic system on the Ordos Plateau be achieved. Dongsheng District's function as the industrial, commercial, information and cultural centre must be strengthened so that the economic development of each banner can be promoted, and support in terms of funds, technology and information should be provided. The agriculture and livestock husbandry production of other banners will provide the material guarantee for Dongsheng District so that the regional resource advantage of each banner can be realized. The only highly efficient economic link has been achieved in Ordos City. Later, sustainable development of the whole society and economy can be obtained. The conflict between economic development and conservation of the eco-environment must be resolved. Therefore, a sustainable ecological economic pattern on the Ordos Plateau must be formed in order to ensure that combating the region-wide desertification can proceed and high efficiency can be achieved in the restoration of degraded ecosystems.

10.7 Vegetation Restoration

Over their long history, Chinese people have developed local knowledge and appropriate technologies for combating desertification and, although they are simple, they are effective (Chapter 5). We should take full account of this traditional knowledge when we develop new technologies.

10.7.1 Some useful technologies

Mulching sand surface with straw chequerboards to fix moving dunes

Straw chequerboards are used widely to fix moving dunes; this technology has been used successfully in arid and semi-arid areas in China, such as the Taklimakan Desert, the Tengger Desert and the Ordos Plateau (see Chapter 5). The chequerboard can be made using different materials found easily in the local area. Traditionally, straw material was used, but now other material has been used, for example *Salix* on the Ordos Plateau. Mechanical barriers were erected to prevent sand moving from mobile dune areas by placing tree branches as vertical barriers 1–1.5 m high and using the straw chequerboard barriers on a 1 m × 1 m grid to a height of 15–25 cm (CCICCD, 1999). The size of the straw chequerboard could be changed according to the local environment and climate, and especially the direction and strength of the prevailing wind. Usually, the straw chequerboard can last for 5 or more years, after which plants may colonize naturally and replace the function performed by the chequerboard. In practice, when straw chequerboards were established, a native species was often transplanted into each chequerboard square to promote vegetation recovery.

Stabilizing mobile dunes

The straw chequerboard is both time- and cost-effective. Instead, artificial plantations of trees, shrubs and grass are effective measures to stabilize mobile dunes in semi-arid regions where the annual precipitation varies from 200 to 400 mm. Both seedling and seeding methods are used. The methods used in a plantation include: plantations of trees, shrubs and grasses in the interdune and lowland and plantations of shrubs that have a high resistance to wind erosion on the front or middle part of the windward slopes of dunes.

Collection of runoff for irrigation in the east of the Ordos Plateau

Water erosion is a major environmental issue in the east of the Ordos Plateau. In order to re-vegetate this area, runoff was collected and stored to irrigate trees and crops and to reduce water erosion. Small terraces and furrows were built or dug to harvest runoff, then trees and shrubs were be planted on the terraces and in the furrows.

Optimum rotation grazing

Overgrazing is the main reason for rangeland degradation. The rangeland was fenced and divided into several parts and a grazing plan was proposed based on the total area of the rangeland and animal development. Grazing took place in different parts of the rangeland at different times; therefore, forage plants had a chance to recover during the rest period. This practice greatly helps rangeland on the Ordos Plateau to recover.

Establishment of fuel energy

The measures for establishing fuel energy are suitable to the conditions in the land areas affected by desertification where the fuel source is limited and insufficient. The main species used for fuel energy are fast-growing trees and shrubs. Firewood energy plantations are adopted widely on the Ordos Plateau, thereby reducing disturbance of natural shrubland and helping degraded shrubland to recover.

Aerial sowing for recovering degraded ecosystems

Aerial sowing is one of the effective methods for re-vegetating degraded ecosystems. It is used mainly in regions where precipitation exceeds 400 mm/year. This technology has been used successfully on the Ordos Plateau during the past 30 years. The key points for aerial sowing include: the species used for aerial sowing must be native ones for rapid growth and psammophytes resistant to wind erosion. The appropriate species used on the Ordos Plateau included: *Hedysarum laeve*, *A. sphaerocephala*, *A. ordosica*, *C. korshinskii*, *C. intermedia*, etc.

Most of the species used for aerial sowing are small in size, light in weight and, as a consequence, they can be blown away easily by the wind. Seed coating is common and combines nutrition, to accelerate seedling growth, and a chemical to protect seeds from being eaten by insects like ants and by birds and animals.

10.8 Comprehensive and Sustainable Regimes for Vegetation Restoration

Overuse of land resources induced by heavy population pressure is the main reason for desertification in semi-arid regions. If human disturbance has decreased or stopped, desertification will be mitigated and desertified lands may recover to their original vegetation, provided that soil loss has not occurred to the extent that nutrients and seed stores are lost. Therefore, the most effective measure to combat desertification is to increase land productivity, to feed as many as possible on a limited land area. In this way, people will be able to maintain high living standards on minimum land, the remaining large land areas will be relieved of overuse and degraded land could recover naturally or with appropriate assistance from humans. The only reliable way to combat desertification is to regulate land use holistically and carry out the regionalization of land use, regarding each district as an integral whole.

Although the regional eco-environment has become better through the great efforts of the local people since the 1970s, this change was restricted to some areas only. Unless the living standards of the local people are improved, the recovered land will be degraded again because the local people still rely on natural resources for their survival. Therefore, it is urgent to apply the principles of restoration ecology to arrange economic development and land use on the Ordos Plateau on a regional scale; then the objectives, which include combating desertification, socio-economic development and sustainable resource use, will be achieved.

The landscape of the Ordos Plateau can be classified into three types, namely hard hill, soft hill and lowland. Soil nutrition and water condition are best in the lowland and worst in the hard hill. Therefore, land use in the three parts should be stratified according to their characteristics. Further, the optimized pattern for desertification should take account of the following principles: biodiversity, shrub dominance and priority, water balance, landscape dependence and sand dune semi-fixation. Based on the above idea, a 'three-circle' pattern for restoration and reconstruction for a degraded ecosystem was proposed and demonstrated on the Ordos Plateau. This pattern means: to develop 'high efficient agroforestry (high input and high output)' in lowland (the first circle), 'runoff garden' in soft hill (the second circle) and 'shrub-protected garden or vegetation recovery zone without human disturbance' in hard hill (the third circle). The proportion of land area in the first, second and third circles is 1:3:6 (Fig. 10.5).

10.8.1 The first circle

This encompasses the oasis in lowland characterized by high output. The soil water could be supplemented by rainfall redistribution in this zone and mainly irrigated agriculture and artificial grassland are located in this area. Land productivity in this area could be several hundreds of times that of the surrounding zones because of underground water, energy inputs, machinery, fertilizer and pesticide, etc. A high input and high output base for agriculture, forest, fruit, artificial grassland and Chinese medicinal plants could be established on this land. This base could have a good shelter belt, fruit trees intercropped with crops, forage crops and other speciality crops in order to achieve high productivity. Therefore, the local people should concentrate on this system in the lowland to make a living.

10.8.2 The second circle

This is the site for the runoff economic garden. Semi-artificial grassland in soft hill and fruit trees in low sand dune and sandy land could be established in this zone. The soft hill and low sand

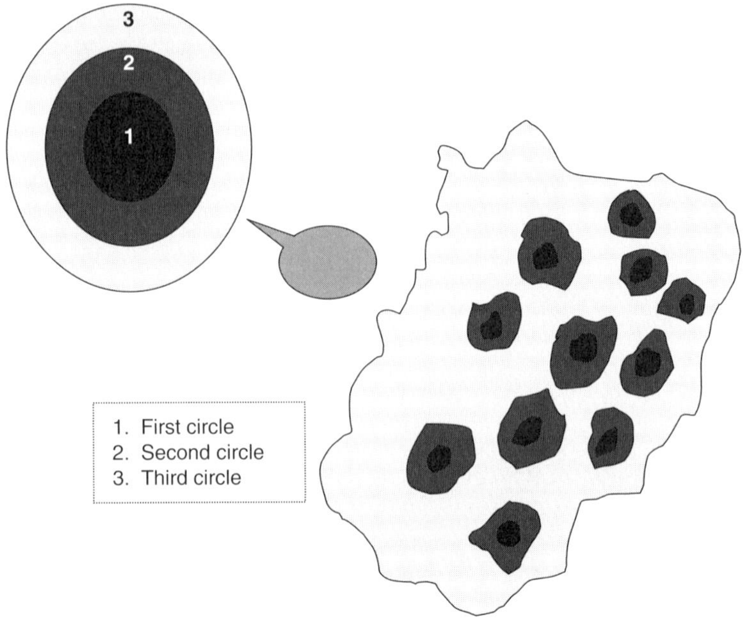

Fig. 10.5. 'Three-circle' pattern for combating desertification on the Ordos Plateau.

dune in the Mu Us Sandy Land is the transitional zone between oasis in the lowland and hard hill (the third circle). The environment is better than hard hill but worse than lowland. This zone functions as a buffer and barrier for the oasis in the lowland and supplies some fruit products with high economic value. Based on the micro-environment, runoff could be collected and used for water-efficient drip irrigation, combined with plantation in belts to establish gardens for growing economic crops using runoff water.

10.8.3 The third circle

This is the shrub garden on the Ordos Plateau. Shrub- and rangeland could be recovered on hard hills, high sand dunes, semi-fixed dunes and drifting dunes. The soil and water conditions are the worst in this part of the Mu Us Sandy Land. Overgrazing and firewood collection have resulted in serious environmental degradation in this part; therefore, it has almost no value for human use. The vegetation should be recovered naturally or promoted by human input, then limited gazing could be available, but it serves mainly as an eco-environment reserve zone. Further, the Ordos Plateau is famous for its rich shrub flora of over 100 species; these native shrub species and other shrub species from the surrounding zone could be introduced to establish a shrub protection area for the first and second circles. Also, the shrub garden could serve as a shrub reservation and shrub gene zone and a research base for shrubs.

Through the 'three-circle' pattern for combating desertification, the local people could live in a good environment (Zheng, 1998). This pattern has been demonstrated and extended in the local area with support from the United Nations Development Programme (UNDP). Because of a shortage of funds, not every family was able to establish a complete 'three-circle' pattern, but they began to arrange their agriculture and livestock husbandry activity similarly to this pattern, but in a simpler way.

10.8.4 Effect of vegetation restoration

Through great effort to combat desertification on the Ordos Plateau, especially after 2001, when the 'Return Cropland to Forest and Grassland' project proposed by China's central government was launched, the degraded ecosystem has been significantly restored. Environmental degradation in Ordos has been largely mitigated in some areas and the improvement in specific locations is spectacular. Both the area of moving sand and the total area of the desertified zone have decreased on the Ordos Plateau. For example, the rangeland has recovered well in east Ordos, especially in the Jungar Banner.

Efforts in rangeland protection have decreased degradation and productivity has begun to increase, plant species composition has changed to typical steppe and plant coverage has increased to 40%.

Soil erosion resulting from wind erosion and water erosion has decreased because vegetation cover has increased and because of the establishment of effective shelter belts. Native shrub species were selected when constructing shelter belts to reduce wind erosion. *C. intermedia* and *C. korshinskii* were widely used on the Ordos Plateau because they can not only reduce soil erosion, but also improve soil nutrition by fixing nitrogen.

The living standards of the local people have improved significantly because of economic development. Some younger people moved to the city for work and therefore decreased pressure on the rural environment. Coal mining has helped the local people to change their energy structure and firewood collection has reduced considerably. Free grazing was greatly reduced by grazing bans and this has contributed much to the recovery of degraded vegetation.

10.9 What Lessons Have We Learnt from the Past?

10.9.1 Socio-economy should be emphasized before environment recovery

The reason that the ecological restoration effort has been limited for a long time on the Ordos Plateau is the failure to consider combining environment recovery and improvement of the socio-economic environment. Good natural vegetation can degrade because of overuse and the recovered vegetation is generally worse than the original vegetation. If the reason that caused the vegetation degradation is not eliminated, the recovered vegetation will soon face the situation of overuse and will degrade again. In general, if the socio-economy develops well and the local people can reduce their dependence on agriculture and livestock husbandry, the degraded vegetation could be released from overuse and soon recover. In other words, one specific zone is able to feed a specific number of people to a high standard. If there are too many people (as now), a high standard of living cannot be maintained; some people would have to rely on industrial development. For example, Japan's resources are limited, but only a few people rely on agriculture and livestock husbandry; most people rely on developed industry. Therefore, the environment in Japan has not become degraded.

An effective measure to combat desertification would be to transfer some people from agriculture and livestock husbandry to industry. Under the current situation of a less-developed socio-economy, it is impossible to transfer too many people to industry, but this should be a target in future. It should be emphasized that, when an environment recovery project is going to be launched, the socio-economy development should also be considered.

10.9.2 Raising public awareness to combat desertification

Combating desertification is a long-term process concerning every person in the desertified area. The local government and local people should be made aware of the possible long-term consequences of desertification. However, because of limited knowledge, local people are not usually this well informed and, therefore, public awareness to combat desertification is extremely necessary. This will help the officials and people of the desertified regions not only to understand the importance and urgency of combating desertification, but also to call for nationwide concern about combating desertification, and to increase people's awareness to combat and rehabilitate desertification.

10.9.3 Protection first, reconstruction second

It is important and necessary to ensure that the ecosystem is used sustainably to avoid degradation.

Later, we could begin to recover other degraded land. Thirty years have already passed since the project to combat desertification on the Ordos Plateau was launched in the 1970s. However, the situation is still severe, so we can understand how important it is to prevent a healthy ecosystem from degrading, rather than trying to rehabilitate it later.

10.9.4 Zonal versus azonal vegetation

The distribution pattern of zonal vegetation should be fully taken into account. If we focus on the ecological effect only, native vegetation is ideal. The species selected should be chosen on the basis of the specific stages of the different desertification problems, such as moving dune, semi-fixed dune or fixed dune. Because precipitation is limited, zonal vegetation is grassland and shrubland; the species composition for restored vegetation should be based on the principle of 'grass first, shrub second and tree last in appropriate and limited sites' on the Ordos Plateau.

10.9.5 Appropriate vegetation coverage is necessary

Dense vegetation usually needs much more water, which may result in depletion of underground water. In the early stage of vegetation restoration, vegetation growth could be maintained, but after several decades, owing to a shortage of soil water, the vegetation will begin to decline. We should decide on the appropriate coverage for restored vegetation based on the long-term observation of natural vegetation. For example, the local government in Zhanggutai, Liaoning Province, introduced the Scotch pine, *Pinus sylvestris* var. *mongolica*, with high density over large areas in the 1970s and it was very successful at first, but it later became degraded as the water table dropped by 10–20 m.

10.9.6 Increasing biodiversity to maintain vegetation stability

In the past, forestation to combat desertification was overemphasized. To ensure the fast growth of vegetation, the poplar species was the main species for vegetation restoration and it was grown in monospecific stands. The protection function of restored vegetation is low in springtime, when there are no leaves, and low biodiversity decreases the stability of vegetation, rendering it more susceptible to pests and diseases. For example, a shelter belt of poplar species was destroyed almost totally by insect pests in Ningxia and elsewhere. Therefore, it is necessary to keep a high biodiversity with native species for future plantations.

References

CCICCD (1999) *Traditional Knowledge and Practical Techniques for Combating Desertification in China.* China Environment Science Press, Beijing.

Chen, C.D. (1964) Where is the boundary in middle part of typical steppe subzone and desert steppe subzone (Ordos plateau) in China? *Acta Phytoecologia Sinica* 2(1), 143–150 (in Chinese with English abstract).

Walter, H. and Box, E.O. (1993) Deserts of Central Asia. In: West, N.E. (ed.) *Ecosystems of the World*, Vol. 5: *Temperate Deserts and Semi-deserts.* Elsevier, Amsterdam, pp. 193–236.

Zeng, Z.X. (1985) *Chinese Terrain.* Guangdong Sci-technology Press, Guangzhou, China (in Chinese).

Zhang, X.S. (1994) Principles and optimal models for development of Mu Us sandy grassland. *Acta Phytoecologia Sinica* 18, 1–16 (in Chinese with English abstract).

Zheng, Y. (1998) New sustainable highly efficient pattern for desertification combating: theory and practice of 'three circle' pattern in Mu Us sandy land. *Forest Science and Technology Management* 1998(2), 20–23 (in Chinese).

11 Case Study 5: Hexi Corridor, Gansu

Yuhong Li[1] and Victor R. Squires[2]
[1]*Bureau of Water Resources, Gansu Province, China;* [2]*University of Adelaide, Australia*

Synopsis
The Hexi Corridor in Gansu Province is a major production base for many agricultural products. To achieve this status required large-scale conversion of rangelands to irrigated artificial oases. The impacts of this conversion were far-reaching. This chapter details the history of development, catalogues the changes and reviews the current status of natural resource use. Particular emphasis is given to a case study of the Shiyang Basin, one of the key inland river systems in north-west China.

Keywords: water resources; desertification; oasis development; desert encroachment; dust storms; livestock impact; over-abstraction of groundwater; ecological flows; plant succession

11.1 Brief Statement of the Problem

The extent of environmental degradation in the arid Hexi Corridor has increased rapidly in the past three decades as a result of increasing population pressure, massive conversion of marginal land into farmland, unsustainable use of water resources, overgrazing, inappropriate farming systems and the destruction of natural vegetation and loss of biodiversity due to desert encroachment. Desertification has been leading to soil loss, ecological degradation and silting up of reservoirs. Monitoring results show that the deserts are advancing towards the oases at a rate of 8–10 m/year and the frequency of sand and dust storms is increasing.

The main cause of land desertification is the excessive economic activities of humans, caused by rapid population growth and low productivity levels. Thus, in order to increase agricultural productivity, there was large-scale conversion of rangelands to croplands, which destroyed natural vegetation; then the newly cultivated land was quickly abandoned and desertification occurred owing to the lack of protective systems and mismanagement of water resources. Animal husbandry and agricultural areas were developed unsuitably because of the unbalanced condition between forage production and livestock carrying capacity, with the result that desert vegetation was destroyed. Population growth caused great demand on living resources and these were plundered. The ecological benefit of the surrounding rangelands was not appreciated and their role in oasis protection was ignored completely. The process of desertification resulting from natural conditions (mainly climatic factors) also accelerated. These human and natural factors are the underlying causes of the accelerated land degradation in the Hexi Corridor.

Overuse of water has become a serious chronic problem. There was no water allocated for environmental protection. Unsustainable use of water has desiccated wetlands and disrupted the lower reaches of river systems, including the drying up of terminal lakes. Natural vegetation in downstream areas has been devastated completely, with widespread death of riparian vegetation, including *Populus euphratica* and *Tamarix* spp.

The artificial oases established in the middle and lower reaches of inland rivers such as the Hei River and the Shiyang River are under constant threat of encroachment by sand from the adjacent deserts. Sand encroachment not only affects croplands and village housing, but also is a problem for irrigation canals, roads and other infrastructure.

High rates of the exploitation of groundwater downstream has led to a rapid drop of the regional water table, and increasing abstractions of surface water from the river systems in the upper and middle reaches have severely depleted water resources available to the communities and ecosystems downstream. Groundwater mining is happening, particularly in some downstream areas. Water shortages, increasing salinity of groundwater and environmental degradation on the edges of the encroaching deserts have led to numerous families becoming ecological refugees.

The Hexi Corridor is faced with some tough issues, including water-use conflict between upstream and downstream users, large population of people living in poverty, deteriorating environment, degradation of the downstream ecosystems and non-sustainability of a high water-use economy in the oases encroached by deserts. These complex issues, which are interrelated with social equity, poverty and environmental deterioration, competing resource use and resource management policies, become intractable problems in this region. Poverty is often amplified since the chronic lack of water restricts agricultural productivity and leads to low household incomes in rural areas, especially serious in Minqin County. Along Minqin's northern frontier, villages like Xiqu, Zhongqu, Shoucheng and Hongshaliang have been totally or partly abandoned. Sand dunes smother empty homes. Poplar, sand jujube, plum and date trees are stacked for firewood.

11.2 The Setting

11.2.1 Natural resources, geography

The Hexi Corridor lies along the inclined plain north of the Qilian Mountains. It is bordered by the Qinghai–Tibet Plateau to the south and by Inner Mongolia to the north (Fig. 11.1). There is a series of low-lying areas of low mountain and knoll with denudation and dryness, located between N37°01' and 42°46', E92°12' and 104°30', which include five prefectures, 20 counties and 171 townships in the Gansu part of the corridor. The corridor consists of three large inland river basins, namely Shiyang River Basin, Hei River Basin and Shule River Basin. The area of rangelands in the Hexi Corridor exceeds 50% of the total area of Gansu; about 5.2% is occupied by artificial oases created in the past 70 years.

11.2.2 Climate overview

Located in the interior part of the continent in the middle temperate zone, winters are very cold and summers hot. The temperature rises quickly in spring and drops quickly in autumn. The temperature varies greatly between day and night. The annual average temperature is 4.9–13.0°C. The temperature drops gradually from the south and south-east to the north and north-west.

Solar radiation in the Hexi Corridor is abundant. Sunlight exposure time is more than 2500 h annually and total radiation is more than 5500 MJ/m^2. Annual precipitation is as much as 500–700 mm in the high mountain zone of the northern mountain slopes, 300–500 mm in the middle altitude area and 100–200 mm in the low basin area. The average wind speed in these three areas is 3–4 m/s, but the maximum wind velocity in the lowlands may exceed 30 m/s and can cause severe sand- and dust storms.

The middle part of the Hexi Corridor, with the Qilian Mountains in the south and Badan Jilin Desert in the north, is characterized by a typical continental climate. Mean annual precipitation is 110–120 mm. The mean annual evaporation is 2337.6 mm. The mean annual temperature is 7.6°C and the mean annual accumulated temperature ≥10°C is up to 3000°C. The frost-free period is 160 days. The groundwater is abundant at a depth of 1–3 m, but water tables are falling. The waters from the Hei River and the Liyuan River originating from the Qilian Mountains are available for cropping, farming and forestry plantations.

Yearly precipitation in the western sector of the Hexi Corridor is only 61.3–85.3 mm and varies dramatically, with 25–41.5 mm in some

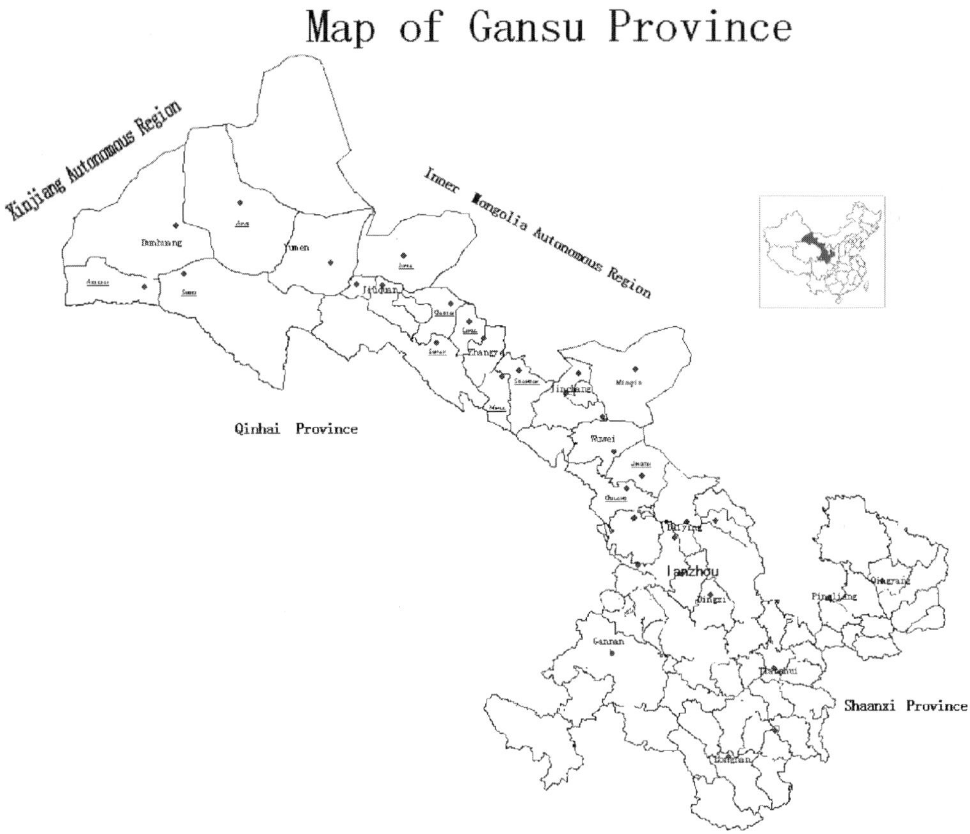

Fig. 11.1. Map of Gansu showing the principal locations mentioned in the text. The Hexi Corridor is the narrow section west of Lanzhou and stretching westward to the border with Xinjiang.

years, in contrast to a high water year of about 165 mm. The amount of September rainfall comprises a quarter of the whole year's precipitation, in contrast to the 140 days of no rainfall in spring and winter. In the gobi area, winds are very strong, with average wind velocity about 3.3–4.7 m/s in spring and winter. Gale winds average about 42 days a year. All of this is disadvantageous to the growth of plants. The climatic characteristics of different parts of the Hexi Corridor are set out below (Tables 11.1 and 11.2).

11.2.3 The people, ethnicity, occupations and relative level of income

Humans began livestock development and agricultural production in the Hexi Corridor region about 3000 years ago. At present, the total population in the region is 4.4655 million people, including 42,200 people in Qilian County of Qinhai Province and 15,700 people in Ejinaqi of Inner Mongolia Autonomous Region. Besides the Han nationality (the majority of the population), there are more than ten other ethnic groups, including Mongol, Hui, Kazakh, Tibetan, Yugu and Dongxiang. The features of population composition and distribution in this region can be summarized as follows. (i) Vast lands are sparsely populated; the majority of the population live in the oases and large industrial/mining districts (Table 11.3). The vast Gobi Desert and cold highlands are inhabited by only a few people. Population density is 16 people/km^2, far below the provincial average of 51 people/km^2. (ii) The population growth rate is 1.1%, which is lower than the provincial average level (1.5%) and national average level (1.2%). (iii) The majority

Table 11.1. Climatic characteristics in the Shiyang River Basin (eastern end of the Hexi Corridor).

Location		Qilian Mountains			Yongchang–Wuwei Basin		Minqin–Changning Basin
Subregion		Eastern part	Middle part	Western part	Wuwei	Jinchang	Minqin
Temperature (°C)	Annual average	0.9	3.6	−3.1	8	8	7.8
	Maximum	30.5	32.4	28.4	38.6	38.4	44
	Minimum	−31.1	−36.6	−39.6	−31.7	−32.3	−34.6
≥10°C cumulative temperature		785	1631	233.3	2896.6	2954.4	2950.4
Precipitation (mm)	Long-term average	340.8	386.9	238.8	162.3	133.5	110
	June–September	253.6	257.8	186.1	136.5	53.7	50.7
Annual evaporation (mm)		867.1	980.3	1017.1	1895.3	1994.8	2248.8
Dryness index		2.5	2.5	4.3	6.8	7.1	8.2
Frost-free days		111	123	118	153	161	136
Annual average wind velocity (m/s)		2	2.5	2.1	2.2	2.6	4.2
≥8[a] degree windy days		29.9	12	54.4	24.5	29	43

[a] On the Beaufort scale, force 8 is gale force.

Table 11.2. Climatic characteristics in the Hei River Basin (middle part of the Hexi Corridor).

Location		Qilian Mountains			High plains in the corridor		Plateau
Subregion		Eastern part	Middle part	Western part	Zhangye	Jiuquan	Ejinaqi
Temperature (°C)	Annual average	0.7	3.8	−3.2	7	7.3	8.2
	Maximum	31.2	33.1	31.5	38.9	39.1	43.1
	Minimum	−27.1	−27.6	−38.6	−29.5	−33.6	−37.8
≥10°C cumulative temperature		825	1632	231.5	2897.2	2957.4	2978.5
Precipitation (mm)	Long-term average	342.8	389.2	242.8	193.5	129.8	46.3
	June–September	265.6	258.8	194.1	142.5	66.2	30.7
Annual evaporation (mm)		887.1	990.3	1067.9	2128.3	2203.2	2258.8
Dryness index		4.3	4.5	5.7	6.8	7.2	8.4
Frost-free days		120	122	119	159	157	130
Annual average wind velocity (m/s)		2.1	2.5	2.8	2.9	3.5	4.4
≥8[a] degree windy days		24.6	17	46.4	16.9	28	83

[a] On the Beaufort scale, force 8 is gale force.

of the population (77%) depends on farming. (iv) The variety of ethnic groups is large, but the population of the ethnic groups is small, accounting for only 2.94% of the total population.

11.3 Brief History of the Hexi Corridor and the Impact of Changes Since the 1950s

The Hexi Corridor is an important agricultural region, even though only 5% of the land is irrigated. Its development timeline since 1949 can be divided into two main periods.

1. *From 1949 onwards.* Comprehensive development stage. This is the new era for socio-economic development in the Hexi Corridor. The timeline is presented as follows:

(i) After the Anti-Japanese War (from 1937 to 1945) and Chinese Civil War (from 1945 to 1949), the new Chinese government launched the 'Land Transformation Reform' and the 'Cooperative Campaign' to tap

Table 11.3. Major social economic statistics of the Hexi Corridor.

Municipality name	Population (ten thousand)			Industrial product value (100 million yuan)	Irrigated area (10,000 mu)[a]		
	Urban	Rural	Total		Irrigated farmland	Vegetable growing area	Total
Jiuquan Municipality	33.56	61.55	95.11	91.21	232.71	14.65	247.36
Jiayuguang Municipality	13.55	2.42	15.97	40.98	4.53	0.84	5.37
Zhangye Municipality	21.44	104.32	125.76	40.87	280.85	12.41	293.26
Jinchang Municipality	21.00	24.25	45.25	63.55	64.44	2.28	66.72
Wuwei Municipality	28.21	162.89	191.10	55.57	250.17	11.41	261.58

[a] 15 mu = 1 ha.

underused natural resources to encourage people to build water resource infrastructures, to expand farmland and to improve living and production conditions steadily.

(ii) From 1957, the whole country entered into the stage of 'Great Leap Forward' and 'People's Commune'; arbitrary and blind policies prevailed, which led to the abandonment of farming and to economic recession. China suffered severe starvation in the early 1960s. The legacy of this disaster resulted in the new policy of 'Grain Production is the Core for the Whole Nation', which led to large-scale land conversion from rangelands to cropland, deforestation and massive soil erosion that buried lakes and wetlands. Ecosystems in the region were severely damaged at that time.

(iii) During the Cultural Revolution (from 1966 to 1976), because of large-scale conversion of rangeland to cropland, deforestation, overgrazing and other human activities, an area of artificial oases was increased and terminal lakes and wetlands began to disappear. The eco-environment was deteriorating rapidly.

2. *From 1980 to present.* The central government listed the Hexi Corridor as one of the national key commercial grain production bases and supplied tremendous financial and equipment support. In the process of China's rural reform, the household contracting responsibility system was introduced and most collectively owned land was redistributed to individual households under contractual arrangements (Chapters 2 and 15). The Hexi Corridor entered a new phase of development. For instance, a large fine-wool sheep production base with a capacity of one million sheep and large lean-meat pig farms were constructed in Zhangye, Wuwei and Anxi. Large dairy farms and chicken farms were constructed in Jiuquan and Jinchang. Consequently, rangeland conversion and use of water resources were accelerated. The area of artificial oases was increased, while the area of natural oases was decreased significantly. No doubt, further serious eco-environmental problems were caused.

Industrial development and urbanization also imposed considerable pressure on the natural resource base and the environment in this region. For instance, construction of the Lanzhou–Xinjiang railway line in 1962, construction of medium-sized thermal power plants in Yongchang County and Jiayuguang in the 1960s, as well as development of iron and steel, nickel and copper mines in Jiuquan and Jinchang, have caused a huge migration of the labour force into the Hexi Corridor, which in turn has created modern cities in the vast land and caused a population boom. These cities have become raw material production bases for the country and severe environmental pollution has been caused in all industrial districts. In this interaction process between artificial oases and natural oases, the eco-environment was damaged further. With intensified human activities, some parts of the natural oases were inevitably replaced by artificial oases. The carrying capacity of natural oases is significantly lower than that of artificial oases (Li and Chen, 2002). However, both types of oases are very vulnerable environmentally. When there is surplus water, there is more salinity and swamps occur. When water is insufficient, the buffer zone starts shrinking, then vegetation degradation, wind erosion and sandification are triggered, which leads to accelerated land degradation in the whole region. In the Hexi Corridor, the process of artificial oases development consists of these primary activities: to divert water for irrigation, to plant shelter forestry, to create conditions for farming, to develop industrial enterprises and to construct cities. The immediate causes of desertification include deforestation, massive land conversion, overgrazing, loss of biomass from other human activities, over-extraction of water, degradation of sandy soil and soil salinization.

11.4 Climate Change and Use of Water Resources in the Hexi Corridor

There are two primary triggers that have resulted in eco-environmental change in the Hexi Corridor, namely human activities and climate change. In the past 50 years, because of global warming, the air and water circulation patterns have been changing. For example, the total area of ice glaciers in the Hexi Corridor region is 1657.2 km^2, with a storage capacity of 80.13 billion m^3, accounting for 84% of the total glacier area of the Qilian Mountains. The glaciers are considered as an important 'long-term regulatory reservoir' for the Hexi Corridor. Runoff from melting glaciers accounts for more than 10% of the annual replenishment of rivers in the Hexi Corridor (Li and Chen, 2002). However, in recent years, snowlines have been moving up, glaciers have been retreating and wetlands and lakes in the mountainous region have been disappearing. It is estimated that, in the case of a temperature rise of 3°C in the Qilian Mountains, the snowlines will retreat 500 m further and most of the glaciers will have melted in a few years (see Chapter 3). Nevertheless, further study is needed to investigate the impact magnitude of climate change.

Water resources and agricultural development in arid inland river basins are dependent on the isolated oases surrounded by gobi desert in the Hexi Corridor. Climate has certain impacts on the survival of isolated oases. Because of water availability, these isolated oases can survive and unique ecosystems are created. However, there is now less precipitation and its distribution in the arid region is spatially and temporally uneven. Natural rainfall cannot support the steady growth of oases; it can support growth of sparse desert vegetation only. Survival and growth of natural oases are completely dependent on nearby rivers and lakes, as well as on the maintenance of a high groundwater table. There are transition zones between the fringe of oasis vegetation and desert vegetation, and vegetation growing in the transit zones is drought-resistant; relying on limited rainfall; it can survive in depressed areas or flood trenches. Small sand dunes can be formed adjacent to these small plants and these sand dunes are favourable for collecting moisture from the air.

As a result, drought-resistant vegetation such as *Tamarix*, *Elaeagnus angustifolia*, *Hippophae rhamnoides* and *Hedysarum scoparium* can grow well, particularly in wet seasons, and vegetation coverage can be 30–40%. It becomes a natural buffer zone for oases to prevent the encroachment of desert. For example, there are vast ecosystems with an area of 66,668 ha downstream of the Shiyang River Basin that protect the Minqin oasis. Unfortunately, for the immediate economic interest, some people have made large-scale illegal land conversion for growing watermelons by extracting groundwater excessively. The rapid drop of the water table has accelerated degradation of vegetation and the disappearance of the oasis. The prerequisite for the stability of the oasis is the effective prevention of desert encroachment and vegetation degradation, as well as sufficient water supply for the eco-environment, which makes water available for the root zone of the natural vegetation. Human interventions change the proportion of water allocated for domestic use, production use and ecosystem consumption; as a result, human activities often change the water consumption proportion between the natural ecosystem and the artificial ecosystem. They can be used as indicators to judge the development status and stability of oases. The situation in the inland river basins is presented in Table 11.4.

11.5 Change of Land Use and Impact of Desertification in the Hexi Corridor

The proportion of land for agriculture, forestry and livestock purposes has been changed over the

Table 11.4. Ecosystem water consumption and allocation in the inland river basins of the Hexi Corridor.

River basin name	Water resources (10^8 m^3)	Water consumption (10^8 m^3): Agric. use	Water consumption (10^8 m^3): Industry and domestic use	Ecosystem water consumption (10^8 m^3): Natural systems	Ecosystem water consumption (10^8 m^3): Semi-artificial ecosystem	Ecosystem water consumption (10^8 m^3): Artificial ecosystem	Allocation of ecosystem water consumption (%): Natural ecosystem	Allocation of ecosystem water consumption (%): Semi-artificial ecosystem	Allocation of ecosystem water consumption (%): Artificial ecosystem	Water resources utilization rate (%)
Shule River	12.83	3.61	0.75	8.36	0.96	0.33	86.6	9.9	2.4	37
Hei River	38.56	17.34	1.0	11.76	6.8	1.66	58.1	33.6	8.2	52
Shiyang River	17.73	13.93	1.93	1.13	4.12	0.89	18.4	67.1	14.5	89

past 70 years. The proportion of land for forestry and livestock development has been increased in the three river basins over recent decades. Land for livestock development is concentrated mainly in rangeland with low carrying capacity. In terms of annual gross product value in the Hexi Corridor, cropping accounts for 70.35%, forestry accounts for 1.69% and livestock accounts for 27.83%. However, because of irrational use of land and water resources, land desertification has become the primary environmental problem in the Hexi Corridor. Survey results showed that the land desertification area increased by 17.49%, with a total area of 0.7903 million ha (Mha) in the 46-year period from 1949 to 1995. The annual average increase rate is 0.40% in the Hexi Corridor. Both human and livestock populations (Fig. 11.2) rose rapidly in the period at the beginning of the late 1950s, for the reasons outlined above. There were profound effects on the eco-environment, the legacy of which we are seeing today.

At present, the area of oases accounts for about 5% of the total land area of the Hexi Corridor. An artificial ecosystem was created in the core areas of the oases. Salinity, and even the formation of sand dunes within the core area of the oasis, has occurred in some counties (see Table 11.10). Tables 11.5 and 11.6 show the impact and damage caused by desertification.

The major problems in land use and their causes in the Hexi Corridor include the following. (i) There is no master plan for land use in the region for unified planning and macro-level regulation. Abusive land use and blind development policy have caused land desertification. More specifically, deforestation for land conversion, destroying rangeland for cropping and overuse of water resources are the triggers for land salinization and desertification. (ii) Irrational land use and

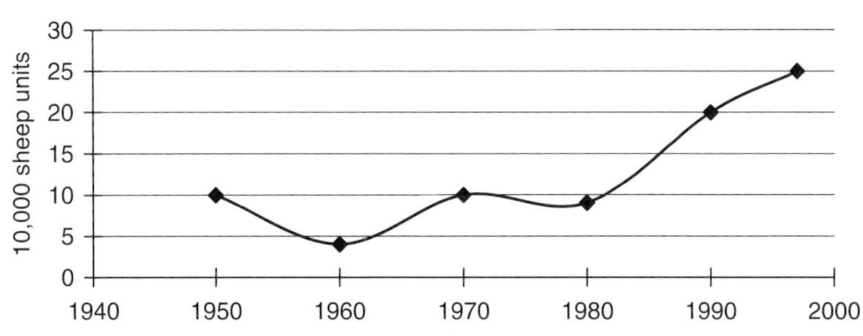

Fig. 11.2. Time-series data for growth in livestock populations in Minqin County, Hexi Corridor.

Table 11.5. Impacts and damage caused by desertification in the Hexi Corridor.

Region	Jiuquan	Jiayuguang	Zhangye	Jinchang	Wuwei	Shule Basin	Hei River Basin	Shiyang Basin	Whole Hexi Corridor
Affected farmland (10,000 Mha)	14.69	0.29	6.37	0.91	4.55	2.27	9.07	5.47	16.81
Affected rangeland (10,000 Mha)	1,353	0.53	17	1.73	22	340	30	24	393
Villages	1,442	17	515	23,171	2,170	1,160	814	25,341	27,315
Affected villages	537	14	254	1,286	2,078	374	431	3,364	4,169
Railway lines (km)	712	133	163	165	94	444	563	259	1,267
Affected railway lines (km)	384	35	140	50	76	249	304	126	679
Highways (km)	3,869.2	118.9	672.6	348	2,106	2,542	2,118.5	2,454	7,114.7
Affected highways (km)	1,258.7	36	398.2	114	727	1,030.5	662.4	841	2,533.9
Canals (km)	8,938.6	87.6	2,340.1	9,055	2,537	5,441.5	5,924.8	11,592	22,958.3
Affected canals	4,980.1	29.8	1,025.9	1,553	946	2,567.9	3,467.9	2,499	8,534.8
Annual losses (10,000 yuan)	20,412.1	589	1,670.9	650.6	3,231	1,846.5	20,236.5	3,881.6	25,964.6

Table 11.6. Ecological and social impacts of land degradation in the Hexi Corridor, Gansu.

Ecological impacts	Indicators	Social impacts	Indicators
Loss of land resources on farms and rangelands due to erosion and desertification	Areas of desertified and eroded land (and trend)	Rural poverty and outmigration to urban and other areas	Average income trend Per cent of households in poverty Number of people outmigrating
Declining ecological productivity and less vegetation cover	Area of land with less than 5% cover (and trend) (2 figures – mountains and plains)	Reduced food security at local and other levels	Nutrition: days of secure food for average family
Increased runoff and decreased groundwater; streams, lakes, oases, wells dry up; increased salinity and flooding	Peak flood rates in small upper catchments; falling groundwater levels, number of wells drying, trends in groundwater salinity	Poor farm household nutrition, health, waterborne diseases	Percentage of malnourished children Rate of respiratory disease Rate of eye disease Rate of gastrointestinal disease Fluoridosis
Poor air quality and increased sand- and dust storms; human loss of life and livestock	Trend in severity and frequency of dust storms; loss of life; loss of livestock	Feminization of agriculture (temporary male migration for off-farm employment)	Number of female-headed households
Increased chemical pollution and toxicity from over-application of fertilizers/agrochemicals to compensate for lost soil/plant nutrients	Soil acidification, water quality (surface and groundwater)	Loss of livelihood opportunities and traditional practices and skills, especially for minorities	Per cent of minority households Decrease of family income Dropout of schoolchildren
Loss of biodiversity due to habitat destruction	Trend in number of species (plants and animals)	Increased household risk, stress and reduced quality of life	Incidence of mental illness, suicide, family break-up
Reduced opportunity for carbon capture	Carbon sequestration potential reduced; bare ground, soil surface loss, loss of annual production	Reduced infrastructure and links to markets	Days of transport and communication failure

fragmented management practice for urban construction, mining development and road construction works have caused huge waste of farmland. (iii) Neglecting land protection and inconsistent policies have exacerbated land degradation. Before the 1980s, all farmlands, rangeland and wasteland were used collectively by villages or communes under the 'Cooperative Communes System', the ownership of land belonged to the state, users were interested only in using them for maximum benefits, but stewardships were not in place. There is no doubt that the 'tragedy of the commons' happened in many of the public lands.

11.6 Underlying Causes of Land Degradation

11.6.1 Desert grasslands: the key to desertification control in the Hexi Corridor

The rangelands in the Hexi Corridor are found mainly in the transitional zones of farmlands and sandy deserts, or gobi areas. In the bottom area between the mountains along the two sides of the corridor, the rangelands are scattered among the peripheral areas of the oases. In the montane areas on both sides of the corridor, the rangelands stretch to the middle or upper areas of the hills and mountains. The sandy deserts with sparse vegetation in remote areas are sometimes also used as grazing lands at the foot of the mountains. The desert grasslands of the Hexi Corridor are a marginal resource, naturally low in productivity and diverse in character in terms of both precipitation and availability of forage. They also represent a diverse cultural landscape, shaped concurrently by physical forces and human use. In this context, it is important to view desert grasslands as something more than just a resource to sustain livestock. They should be viewed as a complex environment with a diverse array of amenities and possibilities and a rich cultural milieu. In fact, desert grassland covers 950,000 ha in the Hexi Corridor and is an important natural resource providing critical environmental and economic functions. It is important to remember that for each piece of land there is often more than one user (Fig. 11.3).

In most areas of the Hexi Corridor, human population pressure and environmental mismanagement have given rise to a situation in which significant areas (90% in most counties) are seriously degraded and desertification is spreading (Table 11.5 is a locality-by-locality listing). This calls into question their long-term sustainability under current use. Rangeland degradation is often manifested in a decrease of plant species diversity, reducing vegetation height and cover; an increase of invasive weeds and then desertification may take place and, finally, conversion into sandy deserts. Therefore, rangeland degradation is not only a precursor of desertification, but also a main source of sand-dust, which damages other cultivated lands and environments.

The traditional herders who occupied and used the desert grasslands prior to the massive relocation of people on to the newly expanded oases employed a semi-nomadic lifestyle that involved altitudinal migration as the seasons changed. In summer, they grazed the mountain meadows, in spring and autumn, the foothills and, in winter, they took the animals into the desert fringes. Diversity and mobility characterized the pastoral production system of these herders. It is an ecological reality that livestock must be mobile to maintain grassland health, the basis of extensive grazing systems (Chapter 4). The reality is that, as conditions become harsher, the further herders must move to acquire forage for their livestock. If one can identify the factors that led to changes in mobility, one can often address the causes of grassland degradation. Factors that have led to restriction of mobility include:

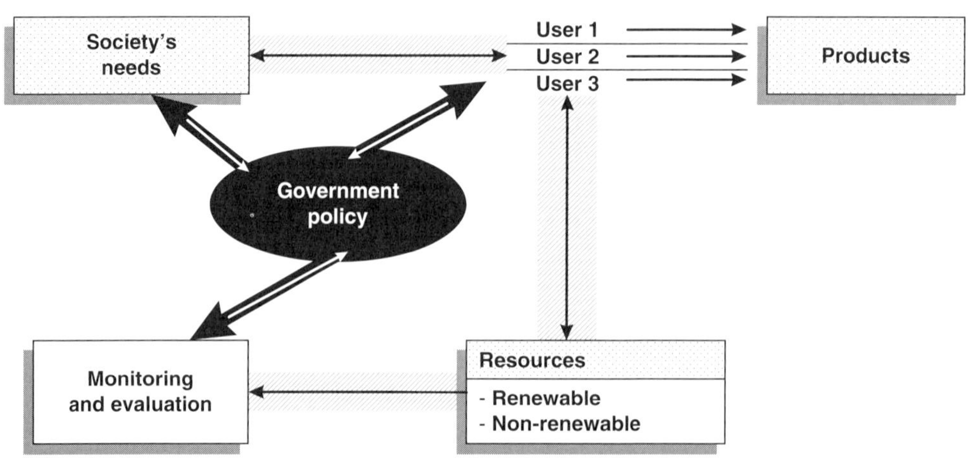

Fig. 11.3. For each piece of land there is more than one user (farmer, herder, miner, urban dweller, etc). Each draws from the resource base in an attempt to cater for society's needs. Each is affected by government policy. Monitoring and evaluation have a significant role to play in shaping policy and in protecting the resource base.

- growing populations of people and livestock;
- expanding oases into the best-quality grassland area;
- forestry and protected area initiatives that restrict grazing rights;
- government policies that promote settlement; and
- changing aspirations of the herders themselves.

The sedentarization of the herders, the expansion of the oases into the refuge areas along the river courses, the sheer pressure of human population and the incentives to produce more meat for a rapidly expanding market by more affluent city dwellers led to problems. Herd size increased well beyond the carrying capacity of the desert grasslands.

As a result of these changes to the way the desert grasslands were used, including increasing disturbance through the collection of firewood and the harvesting of medicinal plants such as *Glycyrrhiza guralensia*, *Cynomorium songaricum*, *Cistanche deserticola*, *Ephedra intermedia* and *Notopterygium incisum*, the surface layers were grossly disturbed. Under the action of wind and water, there was a sorting of soil particles. Fine material, chopped up by the sharp hooves of livestock, blew away to add to the dust load in the atmosphere. Coarser particles (sand) became mobile and so the process that feeds the formation of sand dunes began.

Efforts to stop the sand dunes closer to the oasis are mostly futile, unless the source of sand is treated and feedstocks of sand are cut off. The desert grasslands are the key to desertification control in the Hexi Corridor. Natural regeneration will occur if the areas are closed to livestock. But it may take years before the favourable co-occurrence of events that will allow the germination and establishment of shrubs and grasses. Species such as *Nitraria*, *Salsola*, *Alhargi* and *Tamarix* will establish and survive without watering, without planting. The main management intervention is to prevent access by livestock.

While the focus in this section so far has been on the rangelands, it is clear that the Hexi Corridor is dominated by a series of glacier-fed streams that originate in the Qilian Mountains to the south and flow northward, forming 'green necklaces' as they pass through the desert to either wetlands or terminal lakes (Chapter 12).

The impact of water management (or lack of it) has had severe repercussions on the entire inland river system. This impact is multifaceted and includes the obvious effect on the surface water flows, their periodicity and onflows but also involves the impact on the riparian areas and on the adjacent rangelands and deserts. Of course, it also impinges on the urban centres, on the intensive agricultural industry, on hydropower generation, on secondary industry and on groundwater recharge and exploitation potential.

11.6.2 Degradation impacts on vegetation and soil

The rangelands in the Hexi Corridor once maintained the production of animal husbandry in the area under the control of herder families and the rangelands functioned as a protector of agricultural production in the neighbouring oases. Rangelands are now managed as a common property resource by many thousands of farmers and herders. Because no single bureau or agency accepts responsibility for managing the desert grasslands and because the customary use and access rights have broken down, they suffer from the fate of many common resources. That is, no one owns them, everybody uses them, but no one takes responsibility. There is no tradition in Chinese bureaucracy for rangeland management (the integrated approach to livestock and the resource on which they depend).

However, the rangelands have degraded rapidly during the past 50 years. The driving forces of the accelerated rangeland degradation in the corridor can be divided into two major categories: one is the change in natural conditions and the other is human activities. The former category includes climate changes, long drought spells and unpredictable spatial and temporal distributions of rainfall, etc. Human activities include poor management, abuse of water resources, overgrazing, farmland expansion and invasion, firewood collection and the collection of traditional Chinese medicines from the fragile semi-stabilized or stabilized sandy rangelands. The combination of the two types of factors accelerated and resulted directly in the degradation of the rangelands in the area. Rangeland degradation is a visible phenomenon of vegetation succession of the rangelands. The main driving factors for plant succession change with location and time.

The general rangeland succession in the Hexi Corridor in the past decades is characterized by the following outcomes:

1. Emergence of sandy lands or other degradation forms: sands emerged on the spot or invaded from nearby areas into the rangelands and formed plains of shifting sandy lands or piled up into sand dunes. Salinization sometimes occurred instead of sandy landforms in the areas where the underground water table was shallow.
2. Decrease of biomass: the total biomass of rangeland vegetation decreased dramatically due to the decrease in height, coverage, number, stem diameter and density of plants on the rangelands. The decrease in yield of plants on the rangelands led to a more serious consequence of livestock pressure.
3. Decrease of palatability: palatable and valuable species disappeared and creeping plants such as *Potentilla* spp. increased. Cushion plants such as *Taraxacum mongolicum* and *Plantago* spp. also increased, especially in the residential areas or near water points in the rangelands. Poisonous plants such as *Stellera chamaejasme* and other detrimental plants dominate the rangelands where overgrazing is serious. *Achnatherum splendens* and *Peganum harmala*, which have less palatability due to coarse fibre or strong smell, also increased in dry areas.
4. Changes of dominant plant species: shrubs or semi-shrubs such as *Nitraria tongutorum*, *N. sibirica*, *Tamarix* spp., *Calligonum* spp., *Alhagi sparsifolia* and *Caragana* spp. dominated the rangelands. The original high-quality grass species such as *Stipa* spp. and *Leymus* spp. disappeared.
5. Biodiversity degradation: number of species, plant density and vegetation coverage decreased and some areas became bare lands or sand dunes. The distribution of plant species in the mountain areas moved upward to the top of hills or upper parts of the mountains along the two sides of the corridor. The fragile balance of the rangeland ecosystems was broken and the ecosystem service function was decreased considerably.
6. Direction of succession: according to the 2007 investigation by the Gansu Desert Control Research Institute in the eastern part of the corridor, the rangeland dominant plant species in the oasis–desert transitional zones changed from mesophytes to xerophytes, and finally to super-xerophytes. The succession direction was closely related to the disappearance of surface water resources, the rapid decrease of the groundwater table and the reduction of soil moisture.

A study on the community succession of rangelands was conducted in the eastern part of the Hexi Corridor in 2007 by Wang Jihe and his colleagues from the Gansu Desert Research Institute in Wuwei (personal communication). The study used the historical records, aerial photographs and satellite images to investigate the changes and dynamics of the plants in the oasis–desert transitional zones (Tables 11.7 and 11.8). The results are summarized as follows:

1. During the past 50 years, the dominant plant species of the rangelands changed in the following direction: *T. ramosissima* + *Karelinia caspia* or *T. ramosissima* + *E. angustifolia*, *P. euphratica* or *P. simonii* alone 50 years ago, which were replaced by *T. ramosissima* + *Reaumuria songarica*, and then were replaced by *T. ramosissima* + *N. tangutorum* or *T. ramosissima* + *N. tangutorum* + *Haloxylon ammodendron* (planted) and, finally, the transitional zone was occupied by *N. tangutorum* or *N. tangutorum* + *H. ammodendron* (planted).
2. The area of the rangelands decreased rapidly due to the expanding farmland on the oasis side and the encroachment of sandy deserts from the other side.
3. The main driving factors for the rangeland succession are reduction of water resources, including a lowering of the underground water table, increased spells of drought, unpredictable distribution of rainfall and decrease of surface runoff due to the increased consumption of water resources in the upper reaches of the rivers. The encroachment of sandy desert and the occurrence of salinization accelerated the succession process.

11.7 The Chronology of Degradation in the Hei River Basin (an Example)

The Hei River Basin is the second largest (143,000 km^2) inland river watershed in the arid zone of north-west China. Its main stream, with a length of 821 km, rises in the Qilian Mountains, flows through the Hexi Corridor and enters into the western part of Inner Mongolia (Chapter 12).

Table 11.7. Vegetation changes in the oasis periphery of the eastern part of the Hexi Corridor during five decades 1949–2002. (Unpublished data of Ma Quanlin and Wang Jihe, 2008 (personal communication).)

	Mean height (cm, mean ± SD)			Species coverage (%, mean ± SD)		
Species	1959	1992	2002	1959	1992	2002
Tamarix ramosissima	200 ± 13.0[a]	168 ± 11.3[b]	152.5 ± 16.8[b]	25 ± 4.03[a]	17.3 ± 2.59[b]	7 ± 1.47[c]
Nitraria tangutorum	30 ± 1.8[a]	18 ± 1.7[a]	22 ± 2.1[a]	0.5 ± 0.22[c]	3 ± 0.72[b]	7.9 ± 1.58[a]
Kalidium foliatum	30 ± 3.4			20 ± 3.91		
Glycyrrhiza uralensis	30 ± 2.1			10 ± 1.52		
Karelinia caspia	30 ± 4.6			2 ± 0.43		
Phragmites australis	20 ± 3.6			0.3 ± 0.08		
Sonchus oleraceus	10 ± 1.2			0.1 ± 0.02		
Limonium aureum	30 ± 3.3[a]		20 ± 3.4[a]	0.1 ± 0.01[a]		0.1 ± 0.02[a]
Reaumuria songorica		21 ± 2.5			0.5 ± 0.11	
Agriophyllum squarrosum			12 ± 1.3			0.3 ± 0.04
Bassia dasyphylla			8 ± 2.1			0.4 ± 0.06
Halogeton arachnoideus		2 ± 0.7[b]	8 ± 1.9[a]		0.1 ± 0.02[a]	0.3 ± 0.03[a]
Haloxylon ammodendron		265 ± 75.9[a]	175 ± 53.2[b]		11 ± 1.6[a]	3 ± 1.1[b]
Overall coverage (%, mean ± SD)				58 ± 10.27[a]	31.9 ± 7.13[b]	19.0 ± 3.42[c]
IV of *Tamarix* (mean ± SD)	0.957 ± 0.13[a]	0.897 ± 0.11[a]	0.752 ± 0.18[b]			
Simpson index (mean ± SD)	0.702 ± 0.16[a]	0.589 ± 0.13[b]	0.712 ± 0.21[a]			

[a,b,c]Values followed by different superscript letters are significantly different among the selected years ($P < 0.05$).

Comprehensive development of the Hei River began in 1944 and, from that time onwards, especially after 1949, large-scale development of water resources in the Hei River Basin increased significantly with the construction of artificial oases and industrial facilities. Between 1949 and 1978, 93 reservoirs were built on the plains of the Hei River Basin. By 1985, total water storage was almost 20 times as much as in 1949. Since the 1970s, the use of groundwater has also been increasing, to the extent that the water table has been falling (by about 5 m in some areas).

It is the large-scale diversion of water and upstream reservoir development for irrigation that have been depriving downstream riparian areas of the quantities of water historically available. Important changes of hydrological conditions and ecological environment in the Hei River Basin have occurred in the past 50 years. The discharge in the lower reaches of the river has decreased significantly and more than 30 tributaries, as well as terminal lakes, have dried up. Such hydrological changes have resulted in a marked degradation of the environment, secondary salinization and desertification of land in the entire river basin. Vegetation degeneration and increasing blown sand activity have accelerated land degradation in a short time. Desertification is of special concern in the middle and lower reaches of the river.

Since the 1960s, water shortages have become a serious problem in the Hei River Basin due to the increasing urban, industry and domestic consumption of water, especially irrigated

Table 11.8. Vegetation changes in the mountainous areas in the eastern part of the Hexi Corridor during the past 50 years. (Unpublished data of Ma Quanlin and Wang Jihe (personal communication).)

	Mean height (cm, mean ± SD)			Species coverage (%, mean ± SD)		
Species	1959	1992	2002	1959	1992	2002
Tamarix ramosissima	200 ± 13.0[a]	168 ± 11.3[b]	152.5 ± 16.8[b]	25 ± 4.03[a]	17.3 ± 2.59[b]	7 ± 1.47[c]
Nitraria tangutorum	30 ± 1.8[a]	18 ± 1.7[a]	22 ± 2.1[a]	0.5 ± 0.22[c]	3 ± 0.72[b]	7.9 ± 1.58[a]
Kalidium foliatum	30 ± 3.4			20 ± 3.91		
Glycyrrhiza uralensis	30 ± 2.1			10 ± 1.52		
Karelinia caspia	30 ± 4.6			2 ± 0.43		
Phragmites australis	20 ± 3.6			0.3 ± 0.08		
Sonchus oleraceus	10 ± 1.2			0.1 ± 0.02		
Limonium aureum	30 ± 3.3[a]		20 ± 3.4[a]	0.1 ± 0.01[a]		0.1 ± 0.02[a]
Reaumuria songorica		21 ± 2.5			0.5 ± 0.11	
Agriophyllum squarrosum			12 ± 1.3			0.3 ± 0.04
Bassia dasyphylla			8 ± 2.1			0.4 ± 0.06
Halogeton arachnoideus		2 ± 0.7[b]	8 ± 1.9[a]		0.1 ± 0.02[a]	0.3 ± 0.03[a]
Haloxylon ammodendron		265 ± 75.9[a]	175 ± 53.2[b]		11 ± 1.6[a]	3 ± 1.1[b]
Overall coverage (%, mean ± SD)				58 ±10.27[a]	31.9 ± 7.13[b]	19.0 ± 3.42[c]
IV of *Tamarix* (mean ± SD)	0.957 ± 0.13[a]	0.897 ± 0.11[a]	0.752 ± 0.18[b]			
Simpson index (mean ± SD)	0.702 ± 0.16[a]	0.589 ± 0.13[b]	0.712 ± 0.21[a]			

[a,b,c]Values followed by different superscript letters are significantly different among the selected years ($P < 0.05$).

agriculture in the middle reaches. Because of the dramatic growth of population and over-abstraction of water resources, as well as massive land conversion driven by food security policy, serious degradation of the ecosystem and chronic water conflicts between users in the middle and lower reaches have become pressing problems in the basin. Since the 1960s, the Inner Mongolia Autonomous Region in the downstream area has appealed to central government to solve the water allocation problem. Some agreements were reached, but these agreements were not binding. Over time, water allocation problems have become more acute in the form of frequent water disputes and conflicts. Some farmlands have become salinized, some river channels and lakes have dried up, oases have shrunk and rangelands deteriorated. New deltas have formed in the lower reaches as a result of the decrease in water discharge and these are a source of sand- and dust storms.

The renewable water resources in the basin are not only for agriculture and livestock development, but also for maintaining the natural ecological balance and protecting the environment in this arid zone. Important changes of hydrological condition and ecological environment have taken place in the past 50 years. Since the 1960s, water shortage has been growing due to increasing urban, industry and domestic water use, especially irrigated agriculture in the middle reaches, causing severe water conflicts and ecological degradation. The Hei River in the downstream reaches often ceases to flow. The dry period has increased from 100 days each year in the 1950s to more than 200 days in the 1990s.

The two terminal lakes of the Hei River, the Western Juyan Lake and the Eastern Juyan Lake, had a water surface area of 267 km^2 and 35 km^2, respectively, in the 1950s, but dried up in 1961 and 1992, respectively. Many springs and wetlands have also disappeared and the groundwater table in the downstream Ejina oasis has dropped sharply (Li and Chen, 2002).

The development and use of water resources has affected the size of the artificial oasis established in the middle basin and the natural oases in the downstream reaches. The problem of water shortage was caused mainly by the rapid population growth and the massive development of irrigated agriculture in the middle basin. Agricultural water consumption accounts for 80% of the total water used. The irrigated area has increased to 220,000 ha, which is three times the area in the 1950s, and the downstream discharge has reduced from 1160 million m^3 (MCM) to 7 MCM in the past 50 years (CWGR, 2002). The Ejina oasis in the desert receives just 3–5 MCM each year, less than half of that received in the 1950s.

Present land use in the Hei River Basin can be classified into six types, namely forest (natural forest, artificial forest), farmland (irrigated and non-irrigated), rangeland (steppe and meadow), open water, desert and salinized land. The forest and mountainous meadow are generally in the upper reaches. Most of the farmland is distributed in the middle reaches, but some of it has become saline land due to the unsustainable irrigation system in the region. The rangeland and the desert are located dominantly in the lower reaches. The land classification is shown in Table 11.9.

Table 11.9. Land use in the Hei River Basin.

Land type	Area (km^2)	Percentage of the total area
Farmland	4,663	4.02
Forest	6,879	5.93
Rangeland	58,139	50.12
Salt pond	23	0.02
Open water	557	0.48
No-use land	45,739	39.43
Total	116,000	100.00

Land degradation is one of the primary problems in this region due to waterlogging, salinization, water scarcity and desertification. Excessive rangeland conversion caused soil degradation. Today, the total area of cultivated land in this basin is more than four times that in the 1950s. Agricultural irrigation has deprived the ecosystem of the water supply for maintenance. Some cropping fields became salinized land or were even abandoned due to over-irrigation and poor drainage. Land degradation is very severe in the middle and lower reaches. For example, the area of desertified land exceeds about half of the total area of cultivated land in Gaotai and Linze Counties. The area of desertified land in each county in the middle reaches is shown in Table 11.10.

Because of severe water scarcity in the lower reaches, land desertification is more serious. The total area of desertified land is 6.386 million ha (Mha), accounting for 79% of the total area of the lower reaches. There are 0.953 Mha of shifting sand dunes. The shifting sand dunes account for 14.9% of the total area of desertified land. Compared with the situation in the upper and middle reaches, the ecosystem in the lower reaches is extremely vulnerable.

Loss of biodiversity has become a serious problem in the whole river basin. For example, herbaceous plants decreased from 200 species in

Table 11.10. Distribution of desertified land in the middle reaches of the Hei River Basin.

County name	Minle	Ganzhou	Linze	Gaotai
Desertification area (ha)	15,650	24,130	12,290	107,080
% of desertified land in total county area	22.05	6.92	40.37	24.32
Shifting dunes in desertified land area (%)	35.07	16.37	10.44	31.82
Shifting dunes and semi-stationary dune area in desertifiied area (%)	60.30	20.47	54.24	39.06
Shifting dune area in cultivated land area (%)	0.53	0.23	0.69	2.51
Desertified land area in cultivated land area (%)	1.51	1.40	6.64	7.90

the 1950s to 80 species in the 1990s and forage species decreased from 130 species to 20 species. Vegetation degradation also resulted in the decrease of the quantity and species of wild animals due to loss of habitat. Originally, there were 26 species of rare wild animals in the Hei River Basin; today nine species have disappeared and more than ten species have migrated to surrounding areas. Before 1949, there was about 0.07 Mha of forest in the Qilian Mountains in the upper reaches but, in the 1990s, this was reduced by 16.5%. The forest line retreated from 1900 m to 2300 m above sea level. For example, the forest line in Shandan County retreated by about 3 km on average.

More than 84% of the total discharge of water has been diverted for agricultural irrigation in the middle reaches, which has caused a rise of groundwater in cultivated oases, but it has also exacerbated secondary salinity. It is quite common to see many dry riverbeds with degraded vegetation (CWGR, 2002). The hydrophytes and swamp vegetation that once grew over large areas have declined or died out. However, under a national tree-planting programme, artificial forests have been developed in recent years and have efficiently prevented desert intrusion and land desertification. But, in some parts of the region, the desertification rate considerably exceeds the afforestation rate. In the 1960s, there were over 120,000 ha of hydrophytes or swamp vegetation in Ganzhou, but today the area has been reduced to 90,000 ha. The area of seriously degraded rangelands has reached 0.43 Mha. Because of improper irrigation practice, soil salinization and water pollution in the region increased dramatically. For instance, Ganzhou, Linze and Gaotai Counties have an area of saline cultivated land of about 15,000 ha.

The problems of ecosystem degradation become more and more serious downstream. The majority of the river discharge has been reduced dramatically and some parts of the river have dried up. As a result, it is obvious that vegetation has degraded, with the formation and encroachment of sand dunes. For example, the terminal lakes and springs, as well as swamps, disappeared in the period of the 1960s and 1990s. The delta of Ejina was covered with thick stands of *P. diversifolia* and some meadow vegetation, such as *Phragmites australis*, *A. splendida* and *L. secalinus*, before the 1960s. After massive land conversion driven by the food security policy, the area of the *P. diversifolia* and other riparian forest was reduced by 57,600 ha from 1958 to 1980 (CWGR, 2002). Bush areas with 70% canopy coverage and 30–70% coverage were reduced by 192,300 ha and 79,500 ha, respectively. Wind erosion has damaged the structure and composition of the soil and has led to a rapid decline of biomass production and soil productivity. According to aerial photographs and field surveys, the area of desertified land in the Ejina oases increased by 462 km^2 during the 1960s to 1980s, which means it increased by 23.1 km^2 every year (Li and Chen, 2002). It is reported that most of the sand in sandstorms originates from Ejina, at the terminus of the Hei River.

11.7.1 Specific measures to combat land desertification in the Hei River Basin

It is clear that sustainable management requires: (i) an effective integrated water resources allocation plan and natural resource management system to combat land degradation by means of controlled irrigation and grazing. Irrigation quotas for different crops and water permits should be raised. The low-productivity cropping land with low yield and high water consumption should be transformed into rangeland and/on reafforested, particularly those lands located on the fringes of oases in the middle reaches. More shelter forests and windbreaks should be planted. (ii) A ban on cropping activities on the lower reaches and the remaining rangeland should be protected. Farmland closure for natural regeneration, aerial seeding for tree and grass planting can be trialled. A land care campaign and environment protection advocacy should be carried out to raise people's awareness of environment protection and to change human behaviour.

In order to solve these problems, to maintain sustainability of social and economic development and the ecosystem, the first Hei River Water Allocation Scheme was approved by the State Council in 1997. The objectives of the plan include: to release 950 MCM of water to the downstream area through Zhengyixia; to maintain sufficient water supplies for agriculture; to convert 20,000 ha of irrigated farmland into rangeland and forest to protect the ecosystem; and to encourage farmers to generate more income by growing high-value crops with lower water consumption, such as cotton, fruits and vegetables.

Simultaneously, some significant institutional and legal arrangements were made. The State Council authorized the setting up of the Hei River Basin Management Bureau (HRBMB) in January 2000. A high-level Hei River Basin Management Steering Committee was set up, which was composed of representatives from different concerned ministries with the following responsibilities and authorities: to be responsible for integrated water resources management in the basin; to advise the steering committee on the formulation of regional water policy and standards on water resources planning, management, development, water quality and pollution control, hydrology, environmental protection; to propose an equitable, efficient and sustainable water allocation plan; to monitor the key hydrological stations; to carry out water resources assessment; to develop a database and web site for information sharing; to implement the water-use permit system; to inspect and supervise the implementation of the water allocation plan in the basin; to build, operate and manage the key water infrastructures in the basin; and to reconcile and settle water conflicts among the three provinces. The Hei River Basin Management Regulation was promulgated and enacted, which has created a legal foundation for integrated water resources management, a water abstraction permit system and enforcement of financial sanctions against non-compliance. Other approaches and measures have been taken, such as setting up a pollution permit system to implement the principle of 'polluter pays', reforming the current water pricing system, defining water rights, setting up water markets, establishing long-term stable investment mechanism for ecosystem conservation, etc.

In conclusion, some of the activities and measures have produced results, and some results are very positive while others are not very effective; further improvement is needed. It is believed that sustainability of social economy development and the ecosystem in the Hei River Basin can be maintained through effective implementation of the basin rehabilitation plan and institutional measures.

11.8 Water Management and its Ramifications for the Shiyang Basin

In this section, we focus on water management and its ramifications for the Shiyang River Basin, called by many a system in crisis. Scarcity of water resources and land degradation have become major constraints for regional socio-economic development and environmental sustainability. This crisis of water scarcity and land degradation is acute in the Shiyang River Basin.

Poor people are suffering disproportionately from increasingly severe water shortages. The growing trend of inadequate and inequitable access to water has been diminishing the regional capacity for poverty reduction and sustainable development. It is recognized increasingly that the focus on large-scale infrastructure construction and the conventional top-down and supply-based approaches in the water sector have, in many cases, failed to satisfy those people in the greatest need and do not produce sustainable outcomes.

An analysis of the pressures on the basin shows a wide range of issues that impinge on the available water resources and the capacity to meet demand:

- climate change – local and global impacts and trends;
- desertification – on-site and off-site affects economic growth – constraints and opportunities for industry (mining) and urbanization; and
- poverty – ecological refugees, population change – shift to urban areas, outward migration.

The Shiyang River Basin is situated near the eastern end of the Hexi Corridor, with an area of 41,600 km^2. The total population is 2.2689 million, of which the rural population is 1.7457 million. There are about 100,000 people in the basin with an income below the national poverty threshold of 1300 yuan/year (less than US$190/year). And more than 75% of these poor people live in the arid downstream areas of the basin (mostly in Minqin County). The basin is characterized as an area of spatially variable and acute water resource shortage with conflicting demands.

People rely on irrigation water from the Shiyang River and the alluvial aquifers for their agricultural incomes. There are more than 16,400 tube-wells in the basin, out of which 10,100 tube-wells are in Minqin County. High rates of exploitation of groundwater in the downstream reaches have led to a rapid drop of the regional water table, and increasing abstractions of surface water from the river systems in the upper and middle reaches have severely depleted the water resources available to the communities and ecosystem downstream. Water shortages, increasing salinity of groundwater and environmental degradation on the edges of the Badan Jilin Desert and the Tenggli Desert have led to numerous families migrating involuntarily out of Minqin County in recent years. The water-scarce Minqin County within Wuwei Municipality is known as an oasis separating the Tenggli and Badan Jilin Deserts to act as a vital buffer zone for the whole region. Without this buffer, the whole region will be threatened and encroached by the two converging deserts.

The basin is faced with some tough issues, including water-use conflict between upstream and downstream users, large poverty-stricken population, deteriorated environment, degradation of downstream ecosystem, poor basin management and non-sustainability of high water-use economy in the oases encroached by deserts. These complex issues, which are interrelated with social equity, poverty and environmental deterioration, competing resource use and resource management policies, become intractable problems in this region. Poverty is often amplified since the chronic lack of water restricts agricultural productivity and leads to low household incomes in rural areas. Over-abstraction of water resources and increased deterioration of the eco-environment are the major problems in the basin. In the 1960s, the annual flow of the Shiyang River into the centre of the basin was sufficient to support all the requirements of the existing human population at that time. A significant area of wetland in the Tenggli Desert was also maintained by the release of environmental flow in the river systems. Today, this flow has decreased by more than four-fifths and the considerable river course has been drying up for a long time, due principally to over-abstraction for domestic and industrial water supplies.

Water is used primarily by the upstream urban and rural areas of Wuwei and Jinchang Municipalities. It is used principally for irrigation of extensive farmlands in the foothills and middle reaches of the basin. The total water resource available for the basin is 1.660 MCM. Over 20 reservoirs with a total storage capacity of 450 MCM were constructed in the headwaters in the basin. The designed water supply capacity of these reservoirs is 1089 MCM, accounting for 37.85% of the total supply capacity. More than 16,400 tube-wells have been constructed in the basin in the past decades. Current actual groundwater extraction is 1447 MCM, accounting for 50.3% of the total supply capacity (GPDWR, 2007). Monitoring results show that the average water table drops at the rate of 0.3 m/year in the Wuwei Basin and at the rate of 0.57 m/year in the Minqin Basin. Current over-abstraction of groundwater is 432 MCM/year, out of which the annual over-abstraction is 296 MCM in Minqin County. It is indicated that steadily falling phreatic water levels confirm that the current rates of groundwater abstraction are unsustainable.

The river basin environment and ecosystem downstream are increasingly deteriorated. Domestic waste water and most of the industrial effluent from the urban areas are discharged directly into the river system, together with irrigation drainage waters, which are often highly salinized and contaminated with chemical fertilizers and pesticides. Mineral content in the water of most rivers is in the range of 3–10 g/l. About 30 MCM of waste water from Lianzhou District flows into the polluted Hongyashan Reservoir downstream of the city, which supplies all surface water for irrigation for Minqin County, while wastes from the national nickel mine and other major industries in Jinchang Municipality have contaminated alluvial aquifers in both urban and rural areas (GPDWR, 2006). As one of the richest municipalities in the province, Jinchang has significant financial independence and political influence. It is the world's second largest producer of nickel. Jinchang Municipality is one of the 108 nationally designated water-short cities in China. It uses up almost all the flow from the headwaters to the western half of the Shiyang River Basin. Minqin County, which is geographically downstream and ecologically vulnerable, has not received river flow for many years.

11.9 Rehabilitation and Recovery in the Hexi Corridor

11.9.1 Rehabilitation measures – technological interventions

The main measures that should be adopted to restore the degraded rangelands or to change the succession direction in the Hexi Corridor are restoration of underground water recharge, reduction of livestock load, promotion of the 'Grain for Green Project' (in which marginal land is retired and farmers receive a grain subsidy to compensate for production foregone) and integrated ecosystem management of the rangelands, oases and riparian areas. Technical measures such as controlling sand by aerial sowing for afforestation and pasture management have been carried out successfully in the sand areas. Aerial sown plants in drifting sand should possess the characteristics of covering sand easily, wind resistance, bearing sand cover, drought resistance, quick germination and high economic value. Aerial sowing reforestation and rangeland management have also succeeded in the Hexi Corridor but, because of cost and the sheer size of the problem, there has not been widespread scaling up.

11.9.2 Effectiveness of ecological approaches

The principal method for large-scale and cost-effective rehabilitation of degraded rangelands is through protecting them from the impact of grazing, browsing and trampling by livestock. Accordingly, government policy has been implemented to impose grazing bans that rest the areas for: (i) the whole year; (ii) part of the year; and (iii) specific seasons. There is a fuller discussion of these measures in Chapter 12.

Gaotai and Linze Counties, for example, have set aside and fenced large areas under total grazing bans. Regrowth and recovery have been spectacular, to the point where the 80 cm high *Artemisia* and *Stipa* represent a fire hazard. Lightning strikes can cause wildfire and there is a risk that a fire will engulf these areas. Some work on fire effects indicates that some rangeland species are unaffected by it because their growing points (meristems) are belowground, while others are quite vulnerable. A burning season is also an important factor affecting species response. Grazing bans are being used throughout the Hexi Corridor as part of a national programme of 'returning grazing lands to grasslands'. Many issues arise from this effort and one of them is the level of compensation that land users can be given to offset the loss of forage and income. The alternative animal husbandry system depends on pen feeding and this raises production costs and requires an ample and assured supply of fodder and grain – most of it from water-scarce irrigated farmland. A fuller discussion of these issues is in Chapter 12.

11.9.3 Successional pathways in the recovery phase

The initial stage of progressive succession was: *Typha orientalis* community → *P. communis* community → *Pennisetum alopecuroides* community or *P. alopecuroides* and *Agrostis matsumurae* community, and then the succession generally developed in two directions; one was that the *Salix microstachya* community from interdune swales developed into a *Betula populifolia* community and the other was that the succession from meadow developed into a *Hemarthria compressa* community → mixed *P. alopecuroides* community → *Arundinella anomala* community → *L. chinensis* community → *Cleistogenes chinensis* community. Succession path and rate were defined by climate and soil condition.

11.10 What Can We Learn from the Analysis of What Went Wrong?

11.10.1 What key points will help us prevent a recurrence of degradation episodes in future?

1. In the Hexi Corridor, the relationship between urbanization and total water utilization is close. As urbanization increased, more water was required for every unit increase in urbanization. The utilization ratio of water resources has exceeded 99%, surpassing the ecological safe limit of exploitation. If this condition persists, rapid urbanization in the Hexi Corridor would lead to higher water stress and create serious socio-economic and eco-environmental problems (Chao and Chuanglin, 2007).

2. Taking a broad view of experience over the past 50 years, it is considered that government policy has had a major adverse effect on the state of the rangeland resource. Uncontrolled harvesting of biological products through overgrazing, harvesting of medicinal plants and firewood is causing severe but localized problems in some places. The adverse effects of poor government policy may have lessened in recent years as policy (at least at the national level) has adjusted to reflect a somewhat more holistic view of how to manage rangeland ecosystems, although there is still potential for them to re-emerge if policy makers fail to remember the lessons of the past 50 years, one of which is that the environmental capacity of ecosystems should be taken into account when making development policy decisions.
3. Many of these developments in the pastoral areas have often been in response to political and economic development factors that sometimes conflict with the goals of maintaining rangeland ecosystem stability. Large areas of rangeland have also been converted to marginal cropland, which is one of the primary causes of rangeland degradation. By 1996, China had divided and contracted almost 40% of all usable rangeland to individual herders and farmers. In Gansu, the percentage of settled herders was 76%. The whole question of land tenure arrangements and the consequences of the pastoral land user rights systems are examined in Chapter 15.
4. Development of the large-scale artificial oases had a major impact on the surrounding rangelands, including the important riparian areas. Access by herders to their traditional refuge near the rivers and lakes and the excision of large areas of the more productive land for conversion to cropland had a profound effect.
5. Under the centrally planned economy system in past decades, national policies that had a severe impact on land degradation include:

- relocation of people from densely populated urban areas to the frontier counties;
- a national food self-sufficiency strategy;
- enforcement of grain quotas and the restriction on crops (other than food grains); and
- expansion of irrigated agriculture to produce crops with high water requirements.

6. Water allocation plans did not consider any allocation of water to the environment. Local irrigators claimed that water was too scarce to be allocated to the environment and the downstream riparian areas suffered and terminal lakes in Minqin and Erjina qi (in Alashan) dried up.
7. Most surface water resources in the upper part of the catchment have been regulated by dams and reservoirs, giving the ability to control the distribution of water resources. For example, nearly 90% of the available water in both the Hei and the Shiyang River Basins is used in agriculture and 80% of this available water is used in the urban areas and in the areas upstream.
8. The only viable strategy to save arid land is to move people out, reduce production, form conservation parks and let nature heal itself. The lower reaches of inland rivers, for example Minqin County, will not receive more water and can only support fewer people. In fact, a 200-year trend of migration into northern Gansu from overcrowded lands in south and central China has shifted sharply into reverse, with tens of thousands of farmers being relocated, some as far away as Heilongjiang Province in the north-east.

References

Chao, B. and Chuanglin, F. (2007) Quantitative relationship between urbanization and water resource utilisation in the Hexi Corridor. In: El-Beltagy, A., Saxena, M.C. and Wang, T. (eds) *Human and Nature Working Together for Sustainable Development of Drylands.* ICARDA, Alleppo, Syria, pp. 639–647.

CWGR (2002) *Integrated Water Resources Management for the Hei River Basin in China.* Chinese Work Group Report (CWGR) on China's Water Scarcity, Lanzhou, China.

GPDWR (2006) *Shiyang River Basin Water Resources Assessment Summary.* Gansu Province Department of Water Resources, Lanzhou, China.

GPDWR (2007) *Gansu Province Water Resources Bulletin 2007.* Gansu Province Department of Water Resources (GDWR), Lanzhou, China.

Li, S. and Chen, G. (2002) *Integrated Water Resources Utilization and Eco-environment Protection in the Hexi Corridor.* The Yellow River Water Conservancy Publishing House, Zhengzhou, China.

12 Case Study 6: Alashan Plateau, Inner Mongolia

Li Qingfeng
Inner Mongolia Agricultural University, Huhhot, China

Synopsis

This is an examination and analysis of animal husbandry development and the role of animal husbandry in the local economy. The interactions between climate, land degradation and changes in population density of humans and their livestock are based on a review of data for the past 30–50 years The extent to which accelerated land degradation has occurred and the relative contribution of anthropogenic factors and climate change are considered. The impact of social and economic development within the region on the exploitation of land and water resources has been a major contributor to land degradation. The role of the responsibility system and the allocation of grazing user rights is examined and some proposals about future actions are presented.

Keywords: primary productivity; carrying capacity; stocking rate; grazing ban; ecological migration; land tenure; grazing user rights; land conversion; policy issues; socio-economics; feed balance; re-seeding; aerial sowing; grazing system; pen feeding; fodder crops

12.1 Brief Statement of the Degradation Problem

One-third of the rangeland in the Alashan League is severely degraded. Vegetation coverage in the rangeland has decreased by 30–80%. The number of plant species has reduced from 130 to 30 in the rangelands and most of the palatable plants have disappeared. A large part of the rangeland in Alashan (73.91%) has very low productivity (less than 150 kg/ha DW of forage) and is considered as 'unusable' rangeland, which means of little value for animal grazing. Unusable rangeland accounts for more than half of the rangeland area (about one-third of the total land area in the Alashan League). For the rangeland, ever-increasing grazing pressure and excessive trampling, in particular during the spring season, causes large areas of rangeland degradation. Plant growth and vegetation recovery were checked at the critical time for plants that were beginning spring growth. This stimulated a vicious circle of grassland deterioration, i.e. grazing pressure was heavy, herbage growth became less and herbage regrowth was slow, so grazing pressure became even heavier.

Soil and water erosion has become obvious and desertification processes have accelerated. The area of shifting sand dunes has expanded and the oasis area has shrunk. For example, the Ejina Oasis area has reduced from 650,000 ha in the 1980s to 340,000 ha at present. Dust from the Alashan League and winds from Siberia combine to create major dust and sandstorms that harass northern China every spring. Under increased human disturbance in the form of animal production, depletive use of rangeland resources, land conversion for monoculture and dryland cultivation operations and the influence of climate change, Alashan rangeland is undergoing a fast degradation process.

12.2 Geographical Features of Alashan League Rangelands

12.2.1 Geographical location

Situated at the westernmost end of the Inner Mongolian Plateau, Alashan is the largest league (a prefecture-level administrative unit) in area in the Inner Mongolian Autonomous Region of China. The league borders Bayannur Municipality to the north-east, Wuhai City and Ordos Municipality to the east, Ningxia Hui Autonomous Region to the south-east and Gansu Province to the south and west, sharing a 733.48 km boundary with the Mongolian Republic in the north (Fig. 12.1). Bayanhot is its administrative centre.

The Alashan Plateau extends from the Tibetan Plateau northward into Mongolia's Gobi Desert. It is a region of low mountains separated by intermountain basins. The Alashan Plateau ridges attain elevations of 2000–2500 m, while the basins tend to lie at 1000–1500 m. The whole ecoregion of Alashan is enclosed by the Helan Mountains to the east and the Qilian Mountains to the south and the north-eastern part of the Tibetan Plateau to the south-west. Northward, the plateau extends into southern Mongolia, where it constitutes a large portion of the cold dry Gobi Desert. It is bounded on the north by the Altai Mountains, which separate the desert from grassland and forest ecosystems that are transitional to the Siberian taiga. Because the region is enclosed by mountains and lies a great distance from the sea, conditions here are arid. However, numerous oases, some of them natural, which are fed by mountain snowmelt, occur along the southern flank of the Qilian Range.

12.2.2 Climate

Alashan has a typical continental climate for its location in the hinterland of Asia, remote from the moisture of the sea. It is marked by

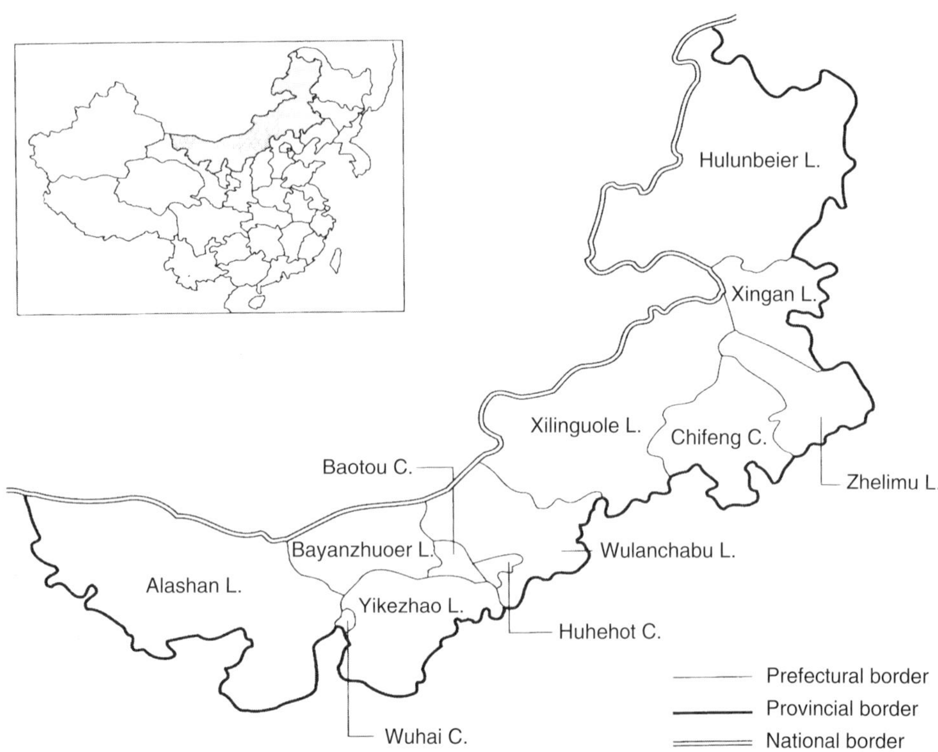

Fig. 12.1. Location of the Alashan League in Inner Mongolia, China.

the continental features of a long cold winter and a short hot summer. Spring is windy and dry. Over 60% of precipitation occurs in the 3 months of July, August and September. Wind-blown sand and dust are common, rain is rare and the air is dry (average RH is 30–40%). The winter is chilly and the summer is parched and there are marked seasonal fluctuations. Annual precipitation rarely exceeds 150 mm/year and it is both spatially and temporally variable. In the drier areas of the Gobi Desert, several years may pass with no measurable precipitation, or enough rain may fall in one summer to create a green flush of vegetation and fill hundreds of ephemeral ponds with fresh water. The Alashan Plateau has only 20–50 days of precipitation per year, with a coefficient of variation of 25–40%. Frequently, 100–200 days will pass with no precipitation and potential evaporation is 3000–4000 mm annually. The league has an extremely xeric climate. Dust from the Alashan League and winds from Siberia combine to create ideal conditions for entrainment and transport of the dust and sandstorms that harass northern China every spring.

12.2.3 Hydrology

Rivers rise in central Gansu Province and flow into the western Alashan Plateau. The rivers are formed by a series of small glacier-fed rivers flowing north from the Nan and Qilian Mountain ranges in Gansu, between Zhangye and Jiuquan. The Hei He flows across the Hexi Corridor (Chapter 11) and then flows northward across the Alashan Desert into terminal lakes near Erjina County.

The southern margin of the plateau lies at the foot of the Qilian Mountains. Streams that flow from the mountains seep underground when they reach the unconsolidated alluvial soils at the foot of the mountains, then emerge again near the valley floor in a 'green necklace' of desert oases. Surface water resources in the plateau are scarce, with only 900 million m^3 available annually.

A few rivers, such as the Dong He and Xi He, traverse the Alashan Plateau, and the Elbow Plains of the Huang He (Yellow River) lie near its eastern edge. Typical vegetation here consists of riparian forests dominated by poplar (*Populus diversifolia*, also known as *P. euphratica*), where the water is fresh, and *Tamarix* spp., where the water is brackish. Low-lying depressions in the western part support meadows and flooded reed beds of *Phragmites communis*.

12.2.4 Topography and vegetation

The Alashan Plateau exhibits flora typical of the deserts of Central Asia. Desert steppe species are found primarily at higher elevations and saxaul (*Haloxylon*) forests occur on mountain slopes. Moving south, the climate becomes increasingly arid and lower in elevation. Southern regions of the Southern Altai Gobi on the plateau are characterized by a special zone of stone-covered, super-arid desert where higher plants are largely absent, except in dry washes and depressions.

The Alashan Basin and range topography creates an arid climate. Great annual fluctuation of precipitation is another important factor affecting the vegetation distribution and agricultural production activities in the rangeland area. Yet increased rainfall in the mountain areas turns the desert green for a short time in the summer and shrub vegetation is found over a large area. The driest parts of the Alashan Plateau consist of shifting sand in the south and denuded, stony landscapes in the north. For example, more than 80% of the Badain Jaran Desert in the western part of the Alashan Plateau, an area of 33,000 km^2, consists of shifting sand blown into dunes 200–300 m high. Arid locations that are more stable acquire communities of the salt-tolerant, xerophytic shrub species, saxaul (*H. ammodendron*) and *Reaumuria songarica*. Once sand dunes become stabilized with sufficient cover of shrubs like these, they cease to shift and soil development can occur, albeit slowly, enabling a more diverse assemblage of plant species to colonize the site. Other places that are slightly less arid support semi-desert shrub communities comprised of wormwoods (*Artemisia salsoloides*, *A. ordosica*), beancaper (*Zygophyllum xanthoxylum*) and *Calligonum mongolicum*.

12.3 Social and Economic Features of the Alashan League

12.3.1 Population and distribution

The Alashan League is the least populated region in Inner Mongolia, and even in China. Han and

Mongolian are the dominant ethnic groups, contributing almost 95% of its total population. The population rose rapidly after 1949 with considerable inward migration (Fig. 12.2) The League is divided into three administrative units or banners (equivalent to a county in the Han populated region), i.e. Alashan Left Banner (LB), Alashan Right Banner (RB) and Ejina Banner (EB). Demographic features of the leagues are summarized in Table 12.1.

12.3.2 Infrastructure of social economy

The economy in the Alashan League relied heavily on the industrial sector (Table 12.2). Agriculture contributed less than 15% of the total net domestic production (NDP) value in 2005, although in the past it had a higher proportion of more than 30%. In the agricultural sector, animal husbandry and cropping have had an almost equal contribution in recent years.

12.3.3 Landscape and land distribution

The Alashan League has an average altitude of 1000–1400 m, with higher elevations in the north and lower in the south. It covers a total land area of 270,244 km². The Gobi Desert and desert rangeland comprise the main landscape of this area (Table 12.3).

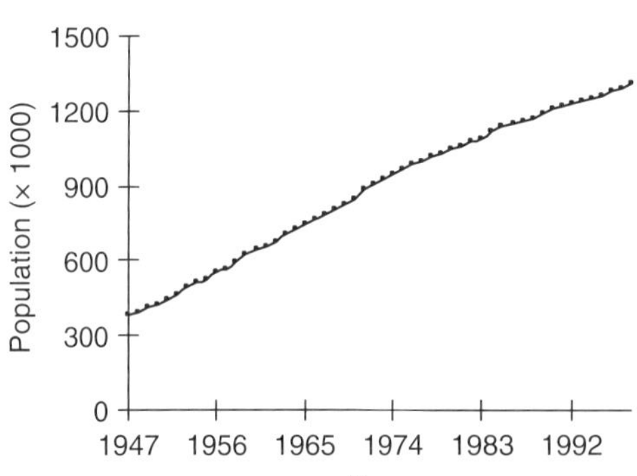

Fig. 12.2. Population from 1947 to 1998. The human population in the Alashan area rose rapidly from 1949 as inward migration became more common.

Table 12.1. Demographic features of the Alashan League.

	Total population	Land area (km²)	Population density (per km²)
Alashan Left Banner	123,000	80,412	60.2
Alashan Right Banner	20,300	75,226	30.2
Ejina Banner	15,000	114,606	70.1
Total in the Alashan League	158,300	270,244	110.3

Table 12.2. Social economic features in the Alashan League.

Indicators	1994	1999	2005
NDP (billion yuan)	0.652	1.906	5.032
Industrial output	0.306	1.269	4.295
Agricultural output	0.170	0.523	0.643
Cropping	0.044	0.248	0.275
Animal husbandry	0.093	0.2.61	0.292

Table 12.3. Land distribution in the Alashan League (ha).

	Land area	Natural rangeland	Usable rangeland
Alashan Left Banner	8,041,200	5,335,905	3,952,532
Alashan Right Banner	7,522,600	4,689,970	323,556
Ejina Banner	11,460,600	7,508,983	2,597,614
Total in Alashan League	27,024,400	17,534,358	6,873,702

12.4 Land Resources and Agricultural Use

12.4.1 Cropping land development

Although the Alashan League is a vast area, agriculturally usable land is sparse due to the scarcity of water. Cropping lands are distributed sporadically in depressions of the plateau along streams. Despite several efforts at developing irrigated land projects using water from the Yellow River, cropping land only occupies about 0.1% of the total land area (Table 12.4). It is clear that, unlike in other places in western China, cropping land is not a significant land degradation source responsible for eco-environmental deterioration in the region.

12.4.2 Rangeland status

Rangeland comprises the largest portion of land in the Alashan League. It occupies more than 60% of the total land area (Table 12.5). Rangelands in the league are mainly in the category of temperate arid rangeland with small shrubby plants as dominant and constructive species. Over 90% of the rangeland is classified as Grade III (49.44%) and Grade IV (42.83%), which indicates a moderate to low forage quality. Among the rangeland, unusable rangeland, or so-called gobi land, comprises more than half of the rangeland area (about one-third of the total land area). A large part of the rangeland in Alashan (73.91%) is in Grade VIII with very low productivity (less than 150 kg/ha DW of forage) and is considered as 'unusable' rangeland, which means of little value for animal grazing. In recent years, the areas of both rangeland and 'usable' rangeland have been reduced; in particular, usable rangeland has been reduced from 9,785,702 ha to 6,873,702 ha, being only about one-quarter of the total land area.

Desert comprises the second largest land portion in the Alashan League, occupying about one-third of the total land area (28%). Similar to cropping land, forestry land occupies only a marginal portion of the land area, although forestry land is much larger statistically than its actual distribution. By the Forestry Bureau's definition, any land with >30% woody canopy cover is classified into forestland categories. On the Alashan Plateau, xerophytic shrubs are frequently found in most of the rangelands. Real forestry of dense woody plants occurs only in the Helan Mountain valley in the south-eastern part of the plateau.

12.5 Causes of Land Degradation

Under increased human disturbance in the form of unrestricted animal production, exploitative utilization of other rangeland resources, land conversion for monocultural cropping and dryland cultivation operation and the influences of global climate change, Alashan rangeland is undergoing a fast degradation process.

Table 12.4. Pattern of land use in the Alashan League.

Attribute	1994	1999	2005
Total land area (ha)	27,024,400	27,024,400	27,024,400
Rangeland (ha)	24,021,000	24,021,000	24,021,000
Arable land (ha)	11,200	20,000	27,600
Cropping land (ha)	9,400	16,000	25,000
Grain production (t)	33,068	60,280	85,500

Table 12.5. Rangeland changes in the Alashan League in the past decades.

	1994	1999	2005
Total land area (ha)	27,024,400	27,024,400	27,024,400
Rangeland (ha)	24,021,000	17,534,358	17,534,358
Usable rangeland (ha)	9,785,702	9,785,702	6,873,702

12.5.1 Rangeland condition

As mentioned in the previous paragraphs, rangeland comprises the largest portion of the land cover on the Alashan Plateau. While the second largest portion (the desert) may remain more or less unchanged in terms of land cover, the rangeland condition has become the most pronounced indicator for assessing land degradation.

According to the survey made by an integrated eco-environment rehabilitation project on the Alashan Plateau, 30% of the rangeland in the league is in a severe degradation state. Vegetation coverage in the rangeland has decreased by 30–80%. Plant species number has been reduced from 130 to 30 in the rangelands. Most of the palatable plants have disappeared from the rangeland. Soil and water erosion has become obvious and desertification processes have accelerated in terms of expansion of the area of shifting sand dunes, and the oasis area has shrunk. For example, the area of the Ejina Oasis has shrunk from 650,000 ha in the 1980s to 340,000 ha at present.

12.5.2 Dependence of livestock on land resource

Animal grazing is the greatest disturbing influence on the land. As limited feed is provided from the cropping land, either as forage or as grain concentrates, grazing on the rangeland becomes almost the sole source of animal feed. With increased animal numbers and reduced usable rangeland, stocking rate on the Alashan Plateau increased substantially in the past decade. As shown in Table 12.6, in recent years usable rangeland decreased to an area of 6,873,702 ha in 2005, which supported 2,153,900 domestic animals (equivalent to 2,521,100 sheep units (SUs)). On average, the stocking rate reached an alarming high of 0.37 SU/ha in relation to the available forage in such an arid region. Figure 12.3 shows the rise in the livestock population over the period 1949–1998.

Rangeland primary productivities in different locations differ greatly in both space and time, so average values mean very little. In turn, animal carrying capacities vary enormously in different rangelands. Although a total rangeland area of 6,873,702 ha supported 2,153,900 domestic animals (equivalent to 2,521,100 SUs) in which the average carrying capacity for the whole rangeland was 4.84 ha/SU, 50.8 ha may be needed for an SU in some rangeland areas.

Table 12.6. Rangeland distribution and animal development in the Alashan League.

	1994	1999	2005
Usable rangeland (ha)	9,785,702	9,785,702	6,873,702
Total animal numbers	1,545,100	2,002,876	2,153,900
Total sheep units (SUs)	1,975,900	2,445,728	2,521,100
Sheep + goats	1,399,900	1,867,163	2,087,100
Large stock	135,200	115,713	86,800
Rangeland productivity[a]	366.3 kg/ha	N/A	N/A
Forage available to animal	165 kg/ha	N/A	N/A
Stocking rate (SUs/ha)	0.20	0.25	0.37

[a] Data quoted are from the Second Rangeland General Survey in 1984. No rangeland productivity data have become available since then.

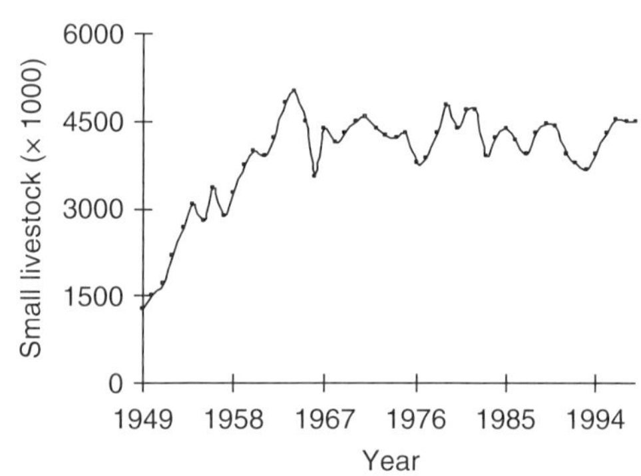

Fig. 12.3. Small livestock from 1949 to 1998. Livestock numbers reached a peak in the mid-1960s and fluctuated in response to policy directives, increasing the human population and the price of livestock and their products (wool, cashmere, skins).

12.5.3 Animal grazing and stocking pressure

Intensive livestock grazing started only some 50 years ago in the Alashan League. Earlier, grazing was characterized by a nomadic system in which rangeland was used in a natural manner and a low-level balance of animal demand–herbage supply was maintained. It is only in the past 50 years that the rangeland use pattern of nomadic grazing has been gradually replaced by the 'settle-down' animal husbandry manner, in which the herbage–animal imbalance became obvious, first in the settle-down area and then extending to a larger circle.

Based on a satellite survey in 1985, total forage production from rangeland in the Alashan League was 11,800,858 t, capable of providing feed for 3,101,163 SUs. In comparison with the animal number of 2,521,100 SUs in 2005, it seems that forage in the rangeland is well in excess of animal demand. However, the adverse fact is that rangeland productivity concentrates on the 4 summer months from May to August (over 90% of the production). While some 30% of the pasture production exceeds livestock demand during the summer months, pronounced feed deficits occur in the winter and early spring months. In most years, only in the wet and warm season from July to September can forage supply be in excess of animal demand. Most of the year, rangelands are heavily overgrazed. Late spring and early summer are the season most prone to feed shortage and land degradation due to grazing. Consequently, spring feed shortage became the most direct cause leading to fast degradation of the natural rangeland through overgrazing and heavy trampling by animals.

Moreover, the unreliability of the forage supply from the rangeland may worsen the overgrazing situation. Yearly and seasonal fluctuations in forage production in the rangeland greatly hinder the achievement of an actual feed–animal balance. Seasonal growth patterns of the plants on the Alashan Plateau are affected primarily by the unstable weather conditions in the early growth season. As the precipitation on the Alashan Plateau is unpredictable and changes greatly in the growing season from year to year, vegetation condition and primary productivity vary considerably both seasonally and annually. Consequently, grazing pressures on rangeland change dramatically under various seasonal and forage conditions. Feed shortage, even after a bumper forage year, in which animal numbers frequently increase, may often occur in the following spring.

12.5.4 Climate change

It was speculated that global climatic change, in terms of global warming, was one of the reasons for rangeland degeneration. However, analyses of the changes of natural precipitation and temperature over the past 30 years showed no convincing evidence that an accelerated desiccation has occurred in Alashan rangeland (Fig. 12.4).

12.5.5 Nutrient element balance and other factors

Insufficient replenishment of nutrient elements was considered the second reason for rangeland degeneration. It was estimated that annual net losses of nitrogen and phosphorus in a typical rangeland were 11.1 kg/ha and 0.085 kg/ha, respectively. Long periods of net offtake of nutrients from the rangeland system through animal production, with negligible input from outside, have depleted the reserves of key elements. Other factors, such as land conversion for annual cropping, the digging of herbs and firewood cutting

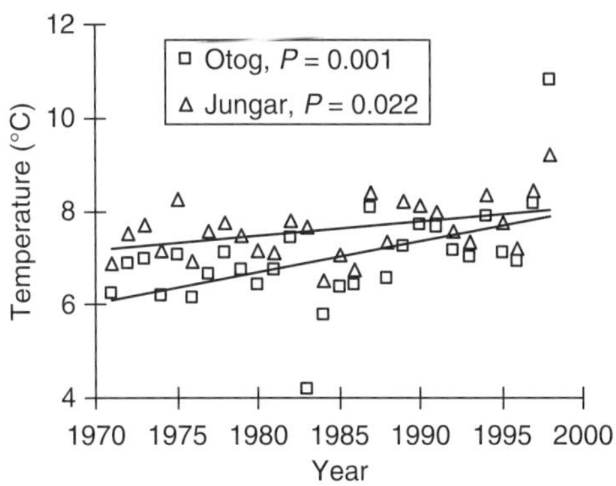

Fig. 12.4. Temperature changes over 30 years from two locations in Alashan.

may have accelerated the rangeland degradation process, but this pales into insignificance compared with the impact of overgrazing.

12.6 Analysis of Causes of Rangeland Degradation

12.6.1 Land use and tenure system

In the pastoral areas of China, a so-called 'Two Rights and One System' of land tenure has been applied since the late 1980s. The basic principles of the system are practised as 'the rangeland ownership remains with the state or the community, the user rights of the lands are given to private individuals for efficient management, and the user right can be transferred on a user-pays basis'. In this system, the key operation is the allocation of the formerly publicly owned, and managed, lands to individual households – grazing user right (GUR) transfer and, consequently, the responsibility for the use and care of the rangeland. As it is an important but sensitive issue, implementation of the GUR varied in different locations and situations, i.e. based on the number of persons per household or the number of animals owned (Williams, 2006).

The allocation of GUR in the Alashan League was implemented during 1995–1998, based on household number in a certain-sized community (normally a 'gacha' – the administrative village). Certain areas of rangeland were allocated to a household based on the rangeland available and a certificate for use of the land was issued to the household. The user right of the allocated rangeland was 'privatized'. However, in most actual situations, the rangeland was still used collectively by the community or by groups of households living in a neighbourhood.

In areas of scarce land resource and high land values, privatization of land may encourage the owner or the long-term user of the land to improve the land value and to protect it for longer-term benefit. Household-based rangeland use in some regions may also encourage the initiation of inputs into the rangeland by the land user. However, all this may not be applicable to the situation in the Alashan rangeland, where the land is plentiful but land quality is so poor. Therefore, the land-user right or the tenureship may have little influence on the rangeland degradation process (see Chapter 15).

12.6.2 Grazing method and animal production systems

As livestock grazing is the main land-use pattern in Alashan rangeland, overgrazing in recent years has been considered the most detrimental factor causing rangeland degradation, which has led to the fast reduction of primary productivity and loss of biological diversity in the ecosystem.

Traditionally, rangelands in the northern China pastoral areas had been used in a 'whole year-round grazing and forage feed on offer' manner in which animals were too heavily reliant on the herbage available on the rangeland. Seasonal fluctuations of the herbage production, plus frequent drought and snow disasters, often resulted in a shortage of herbage in winter and spring. Loss of up to 30% of the body weight of livestock was common in such production system and mortality rates were high at times.

For the grassland, ever-increasing grazing pressure, in particular during the spring season, caused large areas of grassland degradation by overgrazing and excessive trampling. Plant growth and vegetation recovery were severely constrained by grazing animals in the critical spring early-growth period. This stimulated a vicious circle of grassland deterioration, i.e. grazing pressure was heavy, herbage growth became less, so forage growth was reduced and grazing pressure became even heavier.

12.7 Strategy and Countermeasures for Combating Land Degradation

12.7.1 Releasing grazing pressure on rangeland

Since 2002, central and regional governments began a policy of 'subsidizing feed in compensation for grazing bans'. For the 2002–2010 period, a total of 20 million ha (Mha) in the whole Inner Mongolia Autonomous Region was put under the programme 'Grazing Bans and Rational Utilization for Rangeland Protection'. The main measures are whole-year bans, half-year bans, seasonal bans (3 months) and rotational grazing. In 2003, the 'Total Grazing Bans' policy was introduced to the Alashan League. About 30–50% of the rangeland (2,000,000–3,000,000 ha) was put under this policy. This countermeasure greatly relieved the grazing pressure on rangeland in the Alashan League and was effective for rangeland protection and for degenerated rangeland restoration, but the costs of livestock production in the rangeland regions rose enormously. The government provided a grain subsidy of 82.5 kg/ha for the whole-year bans in an effort to encourage the herders to adopt a stall-feeding animal production system. It is still too early to assess the long-term impacts of this policy.

12.7.2 Ecological migration of herders

In order to reduce grazing pressures, some herders were moved out of the rangeland areas. The government encourages this type of 'ecological emigration' by providing a house and some subsidy for resettlement. However, this policy has so far had mixed effects on the livelihood of the migrants, as well as on rangeland degradation. On the one hand, some of the herders were moved out of the harsh rangeland environment and found other ways to make a living. On the other hand, other herders were moved from their original pastures but did not adapt to new ways of making a living. The raising of livestock is still their major activity to generate an income and 'village-based herding' has become common, with the effect of large areas of rangeland in the immediate vicinity becoming stripped of vegetation (Yang *et al.*, 2007). In a newly settled site, herders are more concentrated on a much smaller rangeland area and therefore greater pressure is exerted on the land for feed production.

12.7.3 Rangeland re-seeding

Rangeland re-seeding has been a common practice in many rangeland areas in China. Re-seeding in the rangeland has two purposes. The first is to 'repair' the degenerated rangeland system, while the second is to increase the forage available for grazing animals. In the Alashan rangeland, sowing without irrigation is too risky and there are few successful cases. Sowing cultivated forages in cropping land became important in providing feed for stall-fed animals when the 'Grazing Ban' policy was introduced. However, compared with grain or vegetable crops, forages often give poor financial returns. Sowing with ecological species (e.g. *Artemisia* spp., *Caragana* spp. and *Hedysarum* spp.) is often a government-supported activity on a small scale, but the overall effects of such activities are insignificant, considering the vast area of the plateau.

12.7.4 Other countermeasures

Similarly to other pastoral areas in northern China, the Alashan League has also been given the policy of 'Grain for Green', which means that the government encourages the conversion of rain-fed cropping land back into the original rangeland or forestry land by giving food subsidies. In the Alashan League, such cropping lands are rare, so the measure of the land conversion policy has had little effect on the overall ecological condition of the rangeland.

Other indirect measures include the development of forage cropping, improvement of animal breeds, improvement of livestock-raising facilities and so on. Growth of forage crops is thought to be good for rangeland restoration by providing feed resources from the cropping land to reduce the grazing pressure on the rangeland. However, this measure is a two-edged sword. On the one hand, it might reduce the feeding pressure on the rangelands. On the other hand, it might encourage an increase in the number of animals, and therefore exert even greater pressure on the rangeland.

Improvement of animal breeds has similar indirect effects on the rangeland. It might reduce the grazing pressure by reducing the number of animals with the idea of 'raising fewer high-value animals instead of more inferior animals'. However, it is doubtful that herders will reduce, or even limit, their number of animals if they can see a more profitable result from raising more rather than fewer improved animals.

It is always a debate that improvement of livestock-raising facilities will have a positive impact on rangeland. The positive side presumes that improved livestock-raising facilities will keep the animals in place and facilitate stall-feeding, so as to release grazing pressure on the rangeland, while the negative side considers that improved facilities might encourage more livestock to be raised with, consequently, heavier pressure on land.

In some extreme situations, the herders were asked to reduce their number of animals regardless of their livelihood dependence on animal husbandry. Unless alternative income generation is found, this kind of operation is sure to be short-lived or ignored by both the local bureaux and the herders themselves.

Sowing or transplanting activities were assumed to have two purposes. First, 'artificial forage production' was regarded as a means to increase the forage available, support pen-feeding and thus reduce dependence on the rangeland. Secondly, it was considered as an ecological protection measure that would increase vegetation coverage and species richness in the rangeland.

The indirect countermeasures were considered to be good mostly for relieving the pressures of animal production on the rangeland. However, whether this is true or not remains unclear and questionable.

12.8 Lessons We Have Learnt and Further Thinking

12.8.1 Comprehensive understanding of the rangeland rehabilitation approach

In the past, the main countermeasures for direct rangeland restoration have included grazing bans and sowing or transplanting plants (herbage and shrubs). Some indirect measures to release the pressure of animal production on rangelands are also practised.

Among the measures, the whole-year grazing ban has been the most common one applied in the Alashan League rangelands since May 2003. The strengths and positive effects of whole-year grazing bans include:

- greatly improved natural vegetation; and
- the natural grazing animal production system has been shifted to a system more reliant on sowing.

The weaknesses and negative effects of the ban are:

- limited access to rangeland feed resources and increased difficulties of maintaining feed supply;
- increased cost of raising animals; and
- livelihoods have been affected due to these increased costs.

The practice of grazing bans has increasingly become a matter of finding alternative feed resources, or even an alternative livelihood, to reduce the grazing pressure on rangeland areas.

12.8.2 Overall consideration for harmony between ecological protection and socio-economic development

Although the economy in the Alashan League has been less reliant on agriculture (including animal production) in recent years, the raising of animals still plays an important role in the day-to-day life and livelihood of the local herders in the rural areas of the league.

It is critical to reduce the grazing pressure on the rangeland to allow vegetation to recover. Banning grazing has had the most direct effects in this regard. The 'Whole Year Round Total Grazing Ban' has had a profound influence on rangeland conditions, as well as on the livestock production system, in which animals were confined to 'pen-feeding'. Although this policy was effective for rangeland protection and degenerated grassland restoration, the costs of livestock production in the rangeland regions rose enormously. The livelihood of a considerable number of herders is certain to be adversely affected. Although some subsidy was provided for the grazing ban, the compensation was far less than the actual reduction in a herder's income. If the herders' interests cannot be protected, the ecological effect will hardly be achieved, as some 'illegal' activities, e.g. 'stealing grazing at night' and 'shifting herds to other pasture', are sure to happen sooner or later.

As the population increases, and particularly the rural population living on the land, reducing pressure on the land has increasingly become a matter of finding alternative ways of income generation from outside the rangeland ecosystem. Finding an alternative livelihood, therefore, has become another way to relieve pressure on the land. In the Alashan League, rural enterprises are not well developed as the herders often live in remote areas that lack exploitable resources. The mining industry in the league draws a considerable number of casual labourers but, so far, few other promising measures for supporting the shift from agricultural and pastoral activities to other sectoral activities have been found.

12.8.3 Animal grazing system restructuring

Undoubtedly, grazing natural rangeland is the most economic animal production practice in the short run if the detrimental effects on rangeland resources for long-term sustainable use are ignored. Long-term sustainable use is the goal, but it will take a lot of restructuring to bring this about.

It is clear that spring is the most critical time for both animals and the rangeland environment. Feed shortage becomes most acute as the winter storage of forage is exhausted. Even the animal's storage of energy is almost used up. Land in spring is so prone to deterioration due to an active surface from freezing–thawing weathering. The strategy of sustainable rangeland management must therefore lie in the principle of 'reducing the utilization pressure in the critical spring period'.

It is evident that, in the medium to long term, the traditional animal production system of 'whole year round natural rangeland grazing' should be replaced in order to allow the rangeland condition to recover. A seasonal utilization model from late summer to early winter on the rangeland would probably be suitable for the plant growth pattern on the Alashan Plateau (Fig. 12.5). In practice, seasonal grazing bans have shown the best results. Rangelands recovered effectively with minimum cost to livestock production. This measure has been readily accepted by herders as well the local governments in the Xilinguole League and should be extended to the Alashan League.

12.8.4 Feed production and supply system adjustment

With the introduction of grazing bans into the Alashan rangeland, animals will be more heavily reliant on feed production from outside the rangeland system. Crop residues, forage crops and feed grains will be the main feed sources for animal-raising activities. Even for the 'seasonal grazing ban', feeds are needed for a longer or shorter 'pen-feeding' period. Feed production and supply become the critical issues for restructuring the animal production system.

On account of the scarce water resources on the Alashan Plateau, the choice about using water for producing human food or animal feed is determined by a number of factors, concerning mainly economics and water-use efficiency. At present, the valuable water resource will be used for food crop production as a priority. In fact, very limited water is left for irrigated forage production. Conversion of food grain into feed concentrates seems a solution for pen-fed animals. However, the league has a total population of 158,300, which requires a total consumption of 31,660 t food annually (based on an average 200 kg per capita food consumption). Grain production in 2005 was 85,500 t. Technically, some 50,000 t grain could be used for animal consumption. On average, 20 kg grain/SU are available annually for animal raising based on animal numbers and grain production figures in 2005. Using this locally produced product might release

March–June: Total grazing ban	July–October: Rangeland grazing	November–February: Pen-feeding + rangeland browsing

Fig. 12.5. A schematic rangeland utilization model.

the grazing pressure on the rangeland partially, particularly in the winter season.

Crop straw, a by-product of cropping activities, is another feed resource as 5500 t crop straw is produced in the league. However, the straw is produced mainly in the cropping areas far away from the rangelands and, as it can only be economically consumed locally, it may have very little effect on relieving rangeland grazing pressure. Similarly, as forage crops are usually grown on irrigated lands, they have only a very limited effect on releasing animal-production pressures on remote rangelands. In considering these factors, animal-raising activities are moving gradually from the traditional 'pastoral areas' to 'cropping–pastoral interlacing areas', or even to the traditional cropping oases where village-based herding is becoming more common.

12.8.5 Feed–animal balance reconsideration

The biggest problem for feed–animal balance is the seasonal discrepancy in forage production and animal demand. Although there are positive balances (calculated!) between forage available and animal demand in the rangeland of the Alashan League and in many northern China pastoral areas, acute feed shortage in the late winter and spring season exerts great pressure on the natural rangeland. The currently practised feed–animal balance calculation method is based on the whole-year forage production, ignoring the big difference in seasonal pattern under the northern China climate. At best, the cold season and warm season difference has been recognized in the current calculation. The warm season normally starts in mid-April, which implies that vegetation by this date is good enough for grazing use. In reality, rangeland at this time is most vulnerable to grazing and it produces little, if any, forage before mid-May.

12.8.6 Deep thinking about the Alashan rangeland

What is the value of a piece of land? There are often conflicts between conservation and use of the land. For fertile land, its ecological value is appreciated first and the land merely becomes a resource asset. Its economic value is realized by the owner of the land. Fertile lands are often in the less ecologically sensitive regions and so their ecological values are almost negligible. For land such as the Alashan rangeland, what is more important, its ecological value or its economic value? The most likely answer is the former. However, we need to point out that, whatever its ecological value, the benefits accrue to the whole society in terms of the environmental services it provides. The individual land user can hardly realize any financial benefit from the protection of the land for ecological purposes, except for any improved productivity that might accrue. The incentive for rangeland conservation is therefore lacking. The use of the rangeland as a grazing resource, even if its value per unit area is small, is, in reality, a way of income generation for the individual land user. When the people need to make a living from the land resources, the wish to exploit its economic value will override any concern about protection. This thinking prevails not only for the 'communal land', but also for the 'privatized' or long-term contracted lands.

The compensation for ecological or environmental protection has become another hot topic in China. The practice of 'Grain for Green', in which food (even cash) is provided for the 'conversion of cropping land into forestry land or rangeland', may be a start. It shows that the government is paying more attention to the land degradation problem. However, in a heavily populated country like China, more than half its population is still farmers or herdsmen whose living is mainly from the land. Rural lands (not only rangeland) have too heavy a burden to support society's needs and ensure stability.

For the particular situation of the Alashan rangeland, perhaps the following questions should be answered before any concrete measures are introduced or implemented:

- What is the ecological value of the rangeland? And what is its economic value in the short to medium term and the long term?
- Should the rangeland be regarded as a productive resource, or as an eco-environmental asset?

- What is the compensation that should be given to the local people for giving up their user rights?
- Who should pay the cost of rangeland conservation/protection?
- What are the responsibilities of the people outside the rangeland?
- What are the mechanisms to stimulate the local people's incentives for eco-environmental protection?
- What are the alternatives for keeping the local people's living standard within the norm for the pace of national development?

References

Williams, A. (2006) Improving rangeland management in Alxa League, Inner Mongolia. *Journal of Arid Land Studies* 15(4), 199–202.

Yang, X., Zhang, K. and Ci, L. (2007) Vegetation-based analysis of *a priori* individual settlement-site-centered degraded gradient in semi-arid *Leymus chinensis* dominated steppe. *Journal of Vegetation Science* 19, 245–252.

13 Case Study 7: Qinghai–Tibetan Plateau Rangelands

Ruijun Long, Zhanhuan Shang, Xusheng Guo and Luming Ding
Lanzhou University, Lanzhou, Gansu, China

Synopsis
The rangelands of the Qinghai–Tibetan Plateau cover an extensive area on the roof of the world. Grazing animals, principally yaks and sheep, and grazing-based livestock production systems remain an important source of livelihood. The rangelands have developed under a continental climate that is one of the most severe in the world where pastoral livestock production continues to be practised. The traditional forage-based, extensively managed pastoral livestock production systems are showing a decline in overall productivity and about one-third of the rangelands exhibit severe degradation. Reasons for this situation are examined here.

Keywords: desert; steppe; alpine vegetation; animal production; yaks; pastoralism

13.1 Brief Statement of the Problem

In the past several decades, animal numbers have increased rapidly. This has, in turn, aggravated grazing pressures and accelerated rangeland degradation. Rangeland degradation is now one of the most serious environmental and socio-economic issues in the Qinghai–Tibetan Plateau region. Degraded rangelands of various types accounted for 32.1 million ha (Mha) in the 1960s, 42.5 Mha in the 1980s and over 88.48 Mha today. Although degradation of alpine rangeland has been occurring for decades on the plateau, it has become worse during the past decade, leading to increasing demands on the environmental services of the alpine rangelands.

By the 1990s, the seriously degraded rangeland area had reached about 40.25 Mha, accounting for 33% of the available rangeland area. The secondary bare lands have been appearing on the plateau since the 1980s; their occurrence and development are regarded as a 'cancer' of the alpine rangeland ecosystem because, once the sward formed over hundreds of years is removed, it will be impossible to rehabilitate. The secondary bare land (locally called 'black soil patch') covers an area of approximately 7.03×10^6 ha and accounts for about 16.5% of the total degraded land. Most of the bare lands are found in the headwaters area of the Yangtze, Yellow and Luancang Rivers of the Qinghai–Tibetan Plateau, which is well known as the 'water tower' of China. The consequences of soil erosion by water lead to dysfunction of the alpine rangeland system. The hillsides and hilltops of the secondary bare lands cannot grow any vegetation during the warm season as frequent rainfall induces soil erosion and washes away all the seeds and seedlings.

Also, 18.1% of the land on the Qinghai–Tibetan Plateau is affected by sandy desertification. The situation is particularly severe in the northern part of the plateau. This sandy desertification disaster has seriously damaged the ecological environment and socio-economic development in this region, and also seriously damaged ecological security and regional sustainable development in the whole region.

13.2 Introduction to the Tibetan Plateau Rangelands

The Qinghai–Tibetan Plateau extends from the southern slopes of the Himalayas in the south to the Altai in the north and from the Pamir in the west to the Minshan Mountains in the east. Covering an area of nearly 2.5 million km^2, it is the highest and largest alpine rangeland region in the world. Thus, it is referred to as the 'third pole' or 'the roof of the world' (Long, 2003). Situated in western China, its vast rangelands form the headwaters of Asia's most important rivers, including the Yellow, Yangtze, Mekong, Salween and Brahmaputra Rivers (Fig. 13.1).

These uplands are home for the internationally endangered Tibetan antelope (*Pantholops hodgoni*), wild yak (*Bos grunniens*), snow leopard (*Panthera uncia*), black-necked crane (*Grus nigricollis*) and other Central Asian wildlife. Given the high altitude and extreme harsh environmental conditions, agricultural cultivation is not possible on most alpine plateaus. The only way the land can be used is for livestock grazing. Therefore, the raising of ruminants (yak, Tibetan sheep and goats) plays the most important role in the socio-economic and environmental sectors of the plateau. These grazing animals provide over 90% of milk for human consumption and almost the entire meat requirement for the local people.

The available alpine rangelands of the plateau cover about 128.2 Mha, or approximately 30.7% of China's total area of rangelands. These alpine rangelands consist mainly of alpine meadow (49.3%), alpine steppe (including alpine meadow steppe and alpine desert steppe; 44.9%) and alpine desert (5.9%) (Bureau of Animal Husbandry and Veterinary Medicine, Ministry of Agriculture, China, 1994).

Currently, there are about 13 million yaks and 41.5 million sheep (Long, 1999), as well as large numbers of wildlife, on the plateau to support a human population of about 10 million. These animals are totally dependent on the native alpine rangelands for their survival (Long, 1999; Long *et al.*, 1999).

Fig. 13.1. The Qinghai–Tibetan Plateau is a large area of upland that is bordered by Sichuan and Yunnan in the south-east and by Gansu and Xinjiang in the north and north-west.

Continuous year-round extensive grazing (either transhumance grazing on the vast plain of the central plateau or seasonal rotation within certain mountain regions) is a unique land-use pattern on the Qinghai–Tibetan Plateau (Miller, 1990). However, before the 1950s, mobility and seasonal grazing were the main patterns used by pastoralists and mobility was probably the simplest and the most effective way of optimizing the use of alpine rangeland resources without harming the ecosystem. From the 1960s to the 1970s, most of the pastoral communities living in the eastern (Sichuan and Gansu) and northern (Qinghai) parts of the plateau changed from a migratory lifestyle to semi-sedentary or completely sedentary grazing practices.

The Household Responsibility System policy was implemented in China from the beginning of the 1980s (Chapters 2 and 15). Under this system, communal livestock are divided among every family, based on family size, and consequently some of the pastures used during the cold season (the so-called winter pastures near the herders' sedentary houses) are allocated to individual herders. The rest of the rangelands are normally situated in remote or alpine mountainous areas and are grazed mainly during the warm season (the so-called summer pastures). These are still used as communal lands and so engender less concern for grazers than the winter pastures.

The rangelands of the Qinghai–Tibetan Plateau are more than just a resource to sustain livestock and the livelihood of herders. Their diverse ecosystems of alpine meadows, shrub alpine meadows and forest alpine ecozones lead to extensive quantities of water being held underground, regarded as the 'underground reservoirs' of the plateau and also called the Chinese 'water tower'. Therefore, the alpine rangeland ecosystem provides three major functions, i.e. ecological service, productive service and livelihood service (Long *et al.*, 2007). The ecological service plays a fundamental role in supporting productive and livelihood services; a sustainable livelihood is based on a balance between the ecological and productive functions of the rangeland ecosystem. Xie *et al.* (2003) estimated the annual economic value of native rangeland in terms of ecosystem service on the Qinghai–Tibetan Plateau at RMB ¥2.57 × 10^{11}, accounting for 17.68% of the economic value of the grassland ecosystem services in China.

13.2.1 Climate

Air and solar radiation

Over the Qinghai–Tibetan Plateau, the atmospheric pressure and density are only about 50–60% and 60–70%, respectively, while ultraviolet radiation is much higher than at sea level. Annual sunshine is between 2000 and 3600 h and the value of solar radiation varies from 5000 to 8000 MJ/m^2/year, compared with only 2000–3000 MJ/m^2/year in the eastern lowland area of China at the same latitudes. Strong winds occur throughout the late winter and spring seasons, with a mean wind velocity of 3–4 m/s, even reaching over 5 m/s in spring.

Temperature and rainfall

Over the plateau, warming and watering occur synchronously. The warm season is from June to September, while over 60% of annual rainfall occurs within this period. The growing season of native plants varies from 90 to 120 days, but an absolutely frost-free season does not exist on the plateau. The average annual air temperature is generally below 0°C, while the average temperature in January drops below −10°C. The average in the hottest month (July) does not exceed 13°C. The rainfall decreases from south-east (<500 mm) to north-west (>300 mm). Disasters from hailstorms occur frequently in summer and from snowstorms in the spring.

In the past 50 years, however, global warming may have occurred over the Qinghai–Tibetan Plateau, which has led to an average temperature rise of about 0.25°C/year over Qinghai Province (Gong and Wang, 2002) and 0.40°C/year in the northern area of Tibet (Ding, 1996), but total precipitation has remained stable.

13.3 Alpine Soil Type and its Characterization

There are four different soil types on the Qinghai–Tibetan Plateau, namely: alpine meadow (including subalpine meadow, alpine scrubby meadow, subalpine scrubby meadow, peat-bog and peat), alpine steppe (including subalpine steppe), alpine desert (including subalpine desert) and alpine frigid soil. Table 13.1 summarizes the profile of

Table 13.1. Variation in organic matter (OM) and total nitrogen (N) contents of alpine soil types to a depth of 10 cm. Source: adapted from Xiong and Li (1987).

Soil type	OM (%)	Total N (%)	Sample no.
Alpine meadow soil	10.7	0.47	11
Subalpine meadow soil	15.7	0.69	13
Subalpine steppe soil	3.1	0.20	8
Alpine desert soil	0.49	0.04	2
Subalpine desert soil	0.76	0.06	2
Alpine frigid soil	0.79	0.06	7

chemical properties of these soil types and shows that the organic matter (OM) and total nitrogen (N) contents of alpine meadow soil types are much greater than those of other soil types. Similarly, the trend is for the amount of forage produced to increase with the OM and N content of the soil.

13.4 Alpine Rangelands

13.4.1 Vegetation

The available alpine rangelands area covers 128.2 × 10^7 ha, including 49.3% of alpine meadow, 44.9% of alpine steppe and 5.9% of alpine desert, on which about 13 million yaks and 41.5 million sheep and large numbers of wildlife graze all year round.

The alpine meadows are distributed mainly in the eastern and southern parts of the Qinghai–Tibetan Plateau, extending from 27° to 39°N latitude and from 82° to 103°E longitude. Due to the greater rainfall on alpine meadows than on alpine steppes and deserts, its primary productivity, in terms of biomass production and diversity of vegetation, is much higher. The vegetation on alpine meadow contains sedge species, which include *Kobresia pygmaea*, *K. microglochin*, *K. humilis*, *K. bellardii*, *K. capillifolia*, *K. royleana*, *K. tibetica*, *K. setchwanensis*, *K. kansuensis*, *Carex moorcroftii*, *C. prewalskii*, *C. scabrirostris*, *C. ivanoviae* and *Elymus sinocompresus*; grass species, which include *Elymus nutans*, *Stipa* and *Festuca*; forb species, which include *Polygonum viviparum*, *P. sphaerostachyum*, *Potentilla anserina* and *Artemisia frigida*; and shrub species, which include *Hippophae tibetana*, *Lonicera tibetica*, *Dasiphora fruticosa* and *Salix*. Following the uneven annual rainfall and changing elevation, forage yield of alpine meadows with a groundcover canopy of 80–100% varies widely, from a low of 1500 kg DM/ha in eastern Tibet to a high of 4000 kg DM/ha in south-western Sichuan and Gannan Prefecture of Gansu Province. At present, about 8.5 million yak (60.7% of total yak on the plateau) and 15–20 million Tibetan sheep are sharing those 63.2 Mha of alpine meadows. However, the degradation of meadows is developing extensively across the plateau, particularly in the headwaters of the Yellow, Yangtze, Luancang/Mekong Rivers in the southern part of Qinghai.

Alpine steppe is widely distributed in the centre of the Qinghai–Tibetan Plateau and on most of the southward-facing slopes of the mountains, with a total area of about 55.5 Mha. Water shortages lead to poor vegetation diversity and an open canopy ranging from 40 to 70% groundcover. In consequence, its yield varies from 350 kg to 1000 kg DM/ha/year. Its community consists of grasses and sedges, such as *S. purpurea*, *S. glareosa*, *S. capillacea*, *S. bungeana*, *S. breviflora*, *S. krylovii*, *S. aliena*, *F. ovina*, *Poa annua*, *K. parva*, *K. tibetica*, *C. scabriolia* and *C. trofusca*, on which the largest proportion of livestock raised is sheep, followed by goats and then yak.

13.4.2 Ecological functions of alpine rangelands

Long *et al.* (2007) indicate that the rangelands of the Qinghai–Tibetan Plateau have at least three major functions, i.e. ecological services, productive services and livelihood services (Fig. 13.2). Maintaining and developing the Tibetan grassland ecosystem depend mainly on its intrinsic ecological services; therefore, the ecological services play a fundamental role in maintaining and supporting the productive and livelihood services. The productive services act as a trigger to alter the rangeland ecosystem. The balance between ecological and productive services affects the quality of the livelihood services, which reflects the level of integrated Tibetan grassland ecosystem management.

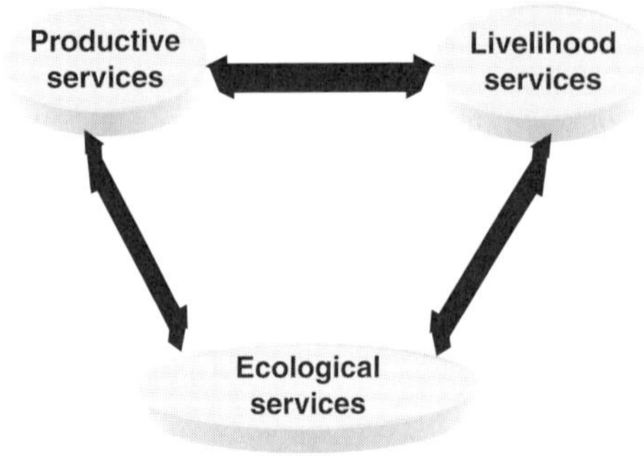

Fig. 13.2. Ecosystem services in the rangelands of the Tibetan Plateau (Long *et al.*, 2007) (with permission).

Xie *et al.* (2003) estimated the annual ecological value of native rangeland in terms of ecosystem services on the Qinghai–Tibetan Plateau was RMB ¥2.57 × 10^{11} (about 3.95 × 10^{10} US\$), which accounts for 17.68% of the ecological value of the grassland ecosystem services in China. However, as indicated in Table 13.2, the different types of alpine rangeland contribute different levels of ecological values; therefore, the weighting of ecological value in alpine meadow (62.52%) is much higher than that in montane meadow (14.14%) and high-cold steppe (12.92%). Following the climate change from warm and humid in the south-west to cold and droughty in the north-west part of the plateau, the ecological value accordingly decreases swiftly. This suggests that global warming could impact ecological services through altering the biodiversity of the rangelands. Human activities also lead to alteration of ecological value in certain rangeland types; for example, conversion of native rangeland to croplands in Qinghai and Tibet has led to loss of ecological values of up to RMB ¥2.29 × 10^8, which accounts for 0.9% gross domestic product (GDP) of these two provinces together (Xie *et al.*, 2003). This indicates that rangeland degradation would result in reduction of its ecological service value. Furthermore, the plateau, as a natural protective screen for China in the south-west, plays a tremendous role in driving and regulating the climate of western and south-western China, even the whole northern hemisphere.

Table 13.2. Ecological service value contributed by different rangeland vegetations on the Qinghai–Tibetan Plateau. Source: Xie *et al.* (2003).

Rangeland vegetation	Area (10^4 Mha)	Value (RMB/Mha/year)	Value (10^8 RMB/year)	Proportion (%)
Temperate meadow steppe	21.10	3,702.72	9.68	0.38
Temperate steppe	171.50	4,585.36	47.72	1.86
Temperate desert steppe	43.20	2,782.52	6.15	0.24
High-cold meadow steppe	558.60	1,424.12	53.68	2.09
High-cold steppe	3,737.40	960.89	332.22	12.92
High-cold desert steppe	867.9	888.90	52.97	2.06
Temperate steppe desert	10.70	610.34	1.56	0.06
Temperate desert	4.50	1,455.42	0.46	0.02
High-cold desert	596.80	1,029.75	1.85	0.85
Tropical herbosa	1.00	366.20	0.51	0.02
Tropical shrub herbosa	35.40	5,142.49	19.60	0.76
Warm-temperate herbosa	2.70	5,536.87	2.23	0.09
Warm-temperate shrub herbosa	27.60	8,272.43	21.83	0.95
Lowland meadow	7.90	7,909.36	4.28	0.17
Temperate montane meadow	705.00	5,414.80	363.65	14.14
Alpine meadow	5,824.70	5,158.14	1,607.97	62.52
Marsh	37.20	2,760.61	25.42	0.99
Total	12,653.20	6,832.66	2,571.78	100.00

13.5 Animals

13.5.1 Wild herbivores

The Qinghai–Tibetan Plateau rangeland is home not only to many species of wild animals, but also to billions of livestock such as yaks, Tibetan sheep and goats. There are 125 species of wildlife found across the plateau, which accounts for almost 36% of rare and protected wildlife species in China. World-famous large herbivores such as Tibetan antelope (*P. hodgoni*), wild yak, Tibetan gazelle (*Procapra picticaudata*), Tibetan argali sheep (*Ovis ammon*), blue sheep (*Pseudois nayaur*) and Tibetan wild ass (*Equus kiang*) inhabit the high plains and mountains of north-west Tibet, along with their predators, which include wolf (*Canis lupus*), snow leopard (*Uncia uncia*), brown bear (*Ursus arctos*) and lynx (*Lynx lynx*). Besides the larger herbivores, there are populations of rodents such as black-lipped pika (*Ochotona curzoniae*), zokor (*Myo spalax fontanieri*) and the Tibetan woolly hare (*Lepus oiostolus*). The Himalayan marmot (*Marmota himalayana*) increased rapidly in the areas where the rangeland condition declined, which, in turn, led to degraded lands and, in severe cases, to secondary bare land.

13.5.2 Livestock

Grazing management on the plateau is carried out by a seasonal rotation system, whereby the animals are moved to a different area with the approach of a new season. Livestock production on the plateau is strongly influenced by climatic conditions. Continuous year-round extensive grazing is a unique land-use pattern on the Qinghai–Tibetan Plateau; thus, these animals are totally dependent on the native alpine rangelands for their survival. Yaks are one of the most important domestic animals as they provide milk products, meat, wool, hides, draught power, riding and fuel energy (dung). Yaks also play an important role in many pastoral rituals and religious festivals. The yak makes life possible for humans in one of the world's harshest environments. All of these factors have caused a continuous increase of yak numbers on the Tibetan Plateau during the past 50 years, while the population of sheep and goats has shown a similar trend in the western regions (Fig. 13.3). Koch (2003) suggested that in the Xinghai County, Qinghai Province, significantly increasing livestock numbers during the past 50 years have had an effect on pasture productivity (Fig. 13.3).

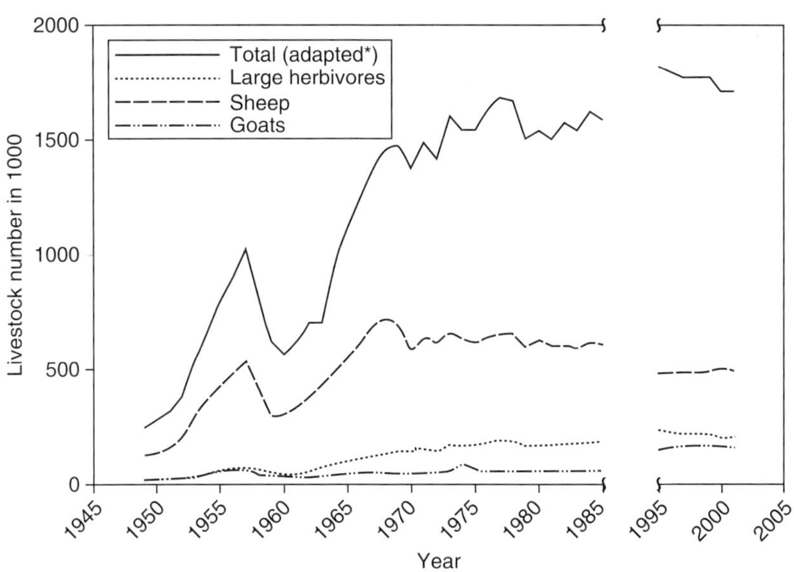

Fig. 13.3. The development of livestock numbers shows a marked increase (Xinghai County, Qinghai Province, 1949–2001, *weighted according to FAO).

Overall, overgrazing is recognized as the most fundamental cause of degradation. Across Qinghai, in the 1980s, one sheep unit shared 0.91 ha of grassland area, by the 1990s it had decreased to 0.74 ha, by 2003 it had dropped to 0.5 ha and, currently, the figure ranges from 0.45 to 0.50 ha. The overgrazing rate is 16% for Qinghai and 78% for Tibet, with the average for both being 45% (Qian *et al.*, 2007). Overgrazing has resulted in degradation of the grassland ecosystem; in alpine meadows, it will cause the structure of the plant community to deteriorate and finally result in the degradation of grasslands (Brockway *et al.*, 2002).

Many experiments have shown that grazing caused the change in the structure of the community. Overgrazing and unsuitable grazing can lead to a decrease of the regenerative ability of meadow vegetation and the degradation of soil and meadow (Wang *et al.*, 1989; Yang *et al.*, 1989; Zhou, 2001; Su *et al.*, 2004). Long-term grazing may also decrease the biomass of meadow vegetation and seriously reduce the return of litter to the soil. At the same time, the degradation of meadows can result in the decrease of amination, nitrification and nitrogen fixation. The decrease of microorganism diversity may lead to the rapid decrease of soil fertility (Li *et al.*, 1989; Zhou, 2001). These factors will weaken the flow of energy and substance circulation in the ecosystem, finally resulting in the disruption of the ecosystem, as well as maladjustment of its function.

Death of livestock

Snowstorms are one of the main natural disasters that result in a large loss of animals on the Qinghai–Tibetan Plateau rangeland. The statement 'a heavy snowstorm disaster occurs every 10 years and low-grade snowstorms occur every 3 years' is a good description of snowstorm disasters across the Qinghai–Tibetan Plateau. From 1954 to 2008, 11 major disasters caused by heavy snow were recorded in Qinghai Province, particularly in southern Qinghai, including Yushu, Guole and Huangnan Prefectures and Qinghai lake areas. These led to the death of over 8 million animals (yaks and sheep) (Table 13.3). Under the traditional livestock farming system, the lack of forage in the winter and early spring always leads to a big seasonal weight loss in animals (20–30% of body weight between November and March). The animals suffer a lot from both starvation and cold weather. Mortality rates can be high during harsh winters. The animal husbandry system on the Tibetan Plateau faces higher risks as compared with other regions in China, where animals can find alternative grassland for the winter months.

13.5.3 Rodents

Two species of rodent, the plateau pika (*O. curzoniae*) and zokor (*M. baileyi*), are major small native

Table 13.3. Areas covered and animals killed by heavy snow in Qinghai Province. Source: Long and Ma (1997).

Year	Time	Prefecture[a]	No. of counties	No. of counties	Animals involved[b]	Animals lost[b]
1954	Winter	HAN, YS, GL	–	–	–	31.30
1960	Winter	Around lake	–	–	–	61.30
1975	Jan.–Feb.	YS, GL	19	72	730	86.72
1982	Mar.–Apr.	HN, HAN, YS, GL	13	51	370	95.73
1983	March	HN, HAN, YS, HB, GL	17	45	350	65.20
1985	May	YS, GL	16	37	390	136.80
1987	Apr.–May	HN, HAN, YS, GL	10	43	201	19.26
1989	Feb.–Apr.	HAN, HB, GL, HN	10	58	616	113.84
1993	Feb.–Apr.	YS, GL, HAN	20	123	–	60.52
1996	Feb.–Apr.	YS, GL	–	–	–	129.33
2008	Feb.–Apr.	YS, GL, HN	15	–	509	31.91
Total						831.61

[a] HAN, Hainan; YS, Yushu; GL, Guole; HB, Haibei; HN, Huangnan.
[b] × 10,000.

mammals that are found on the Qinghai–Tibetan Plateau. These rodents play a role as main contributors to maintaining the alpine grassland ecosystem functional through the transport of organic matter between the subsoil and the surface; their burrows provide channels to transport water, air and essential nutrients to underground plant stems and fibrous root systems; burrows also provide a nesting environment for birds and small reptiles. The pika and zokor are the main contributors to redistribution of soil and vegetation in alpine degraded meadow as they have higher survival rates in heavily grazed sites (Zhou *et al.*, 2003a).

However, the population of rodents increases rapidly following degradation of alpine meadow, and even breaks out more frequently in some areas of the plateau than in others. They destroy subhealthy rangelands much more through their overburrowing and gnawing behaviour (Zhou *et al.*, 2003a,b) (Table 13.4). Also, the rodents compete with livestock for herbage. The activities of the rodents accelerate erosion and the degradation rate by loosening the *Kobresia* sod and killing its roots (Limbach *et al.*, 2000), and stop the succession of vegetation.

13.6 Degradation and Recovery

13.6.1 Degradation

Over the past 50 years, the Qinghai–Tibetan Plateau has been regarded as one of three major pastoral production areas in China; hence, the most important role of alpine rangelands was raising livestock, particularly after the Household Responsibility System policy was implemented in China. From the beginning of the 1980s, the herders themselves were able to decide their own animal-farming business model, including investment, herd size, livestock and rangeland management, marketing, etc. However, almost each family's herd size has been increasing significantly compared with the animal numbers that were allocated to the family in the 1980s, as herders always try to maximize the size of their herds (Fig. 13.3). During this period, both the herders and the government were concerned most about income and GDP obtained from livestock farming sectors. Ecological services of alpine rangelands received less care; therefore, overgrazing activities and rangeland degradation occurred across the plateau. Although herders and the government were able to gain more benefits (income or GDP) from livestock overload on the rangelands, this was unsustainable for the long-term livelihood of local herders and the environment (Fig.13.4). Once the alpine rangeland ecosystem is damaged, both herders and animals would lose their homes/habitats and it would be very difficult and costly to recover. Such an issue has existed in the headwaters of the Yellow, Yangtse and Luancang Rivers.

In the past several decades, animal numbers have increased rapidly. This has, in turn, aggravated grazing pressures and accelerated rangeland degradation (Table 13.5). Rangeland degradation is now one of the most serious

Table 13.4. The alpine meadows damaged by rodent activities in the headwaters region of the Yangtze and Yellow Rivers.

County	Area of land affected by rodents ($\times10^4$ ha)	Proportion of total land area damaged by rodents (%)	Carrying capacity decreased ($\times10^4$ sheep units)
Maduo	13.98	6.08	27.78
Dari	29.73	21.23	55.74
Maqing	16.35	16.19	20.00
Qumalai	14.12	6.67	26.47
Zhiduo	16.83	8.95	8.07

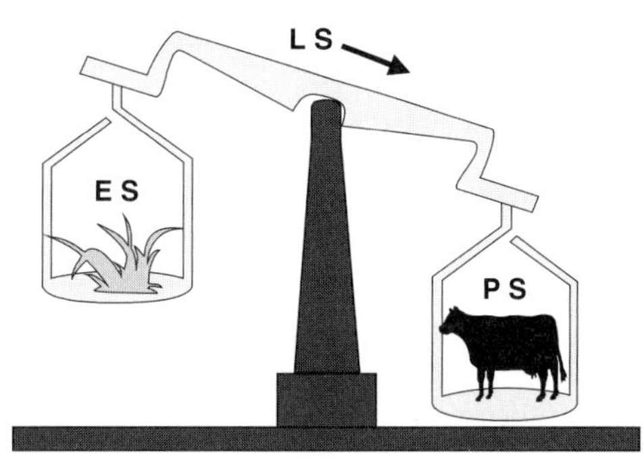

Fig. 13.4. Interactions of the three services in the alpine rangeland ecosystem (ES, ecological service; PS, productive service; LS, livelihood service).

Table 13.5. Trend of degraded rangelands on the Qinghai–Tibetan Plateau (unit: ha × 10,000). Source: after Long and Ma (1997).

Region	Available rangeland	Percentage of degraded rangeland		
		1980s	1990s	2000s
Tibet	6,636.1	18.1	30.0	>70
Qinghai	3,161.0	28.8	31.8	>80
Sichuan	1,416.0	27.3	33.0	>50
Gansu	1,607.2	44.4	49.0	>60
Total	12,820.4	25.1	33.2	>69

environmental and socio-economic issues in the Qinghai–Tibetan Plateau region, accounting for 32.1 Mha, 42.5 Mha and over 88.48 Mha in the 1980s, 1990s and 2000s, respectively. The degradation of the Qinghai–Tibetan Plateau grassland system is continuing. Although degradation of alpine rangeland has been occurring for decades on the plateau, it has become worse during the past decade due to a rapid increase in human and animal populations across the plateau. This has led, in turn, to increasing demands on the alpine rangelands.

Thus, rangeland degradation is often manifested by decreased diversity of plant species, reduced sward height and vegetation cover, increased undesirable and unpalatable grass species and even the occurrence of toxic species harmful to animals. Above all, there is a sharp reduction of acceptable biomass production. If the vegetation density is insufficient to cover the ground surface, wind erosion and desertification take place. In general, the alpine rangelands of the Qinghai–Tibetan Plateau are suffering degradation, wind erosion of the soil and desertification. These problems make the sustainable management and use of the rangeland resources more difficult and, in addition, make the alpine ecosystem even more fragile and unstable than before.

13.6.2 Desertification

About 18.1% of the land area on the Qinghai–Tibetan Plateau has been classified as sandy desertification, and the situation is worse in the northern part of the plateau. It will soon rank third in China as a sandy desertification zone (Tables 13.6 and 13.7) (Dong, 2001). The desertification is mostly discontinuously distributed in spots, strips and patches; only part of the desertified land is concentrated over a relatively large area. Desertification exhibits an evident regional differentiation. The sandy desertification disaster has seriously damaged the ecological environment and socio-economic development in this region, and has also seriously damaged the ecological security and regional sustainable development in the peripheral region (Zhou *et al.*, 2003a).

Wetland and alpine meadow play important roles in animal production and the conservation of water resources in the upper basin of the

Table 13.6. Area of sandy desertification lands in the northern plateau (10^4 Mha) (Dong, 2001).

Region	Sand-desertification lands				Gravel-desertification lands			Total	
	Moving dune	Semi-stable dune	Stabilized dune	Subtotal	Exposed sandy	Semi-exposed sandy	Subtotal	Area	%
Anduo	3.75	13.94	6.32	24.01	18.65	20.88	39.52	63.53	4.14
Bange	2.04	19.00	1.15	22.18	90.93	122.31	213.24	235.42	15.36
Shenzha	0.00	8.84	0.00	5.84	7.33	26.44	33.77	39.62	2.58
Nima	1.11	21.30	0.00	25.41	472.09	145.28	617.37	642.78	41.93
Ritu	1.62	1.05	0.00	2.68	95.18	54.87	150.05	152.72	9.96
Geji	0.65	1.57	0.00	2.22	46.66	33.80	80.46	82.68	5.39
Gaize	0.68	0.60	0.00	1.28	126.07	146.75	272.82	274.10	17.88
Cuole	1.15	0.45	0.00	1.60	0.12	40.35	40.46	42.07	2.74
Total	14.00	63.76	7.46	85.22	857.02	590.68	1447.70	1532.92	100.00
%	0.91	4.16	0.49	5.56	55.91	38.53	94.44	–	100.00

Table 13.7. The degree of sandy desertification lands in the northern plateau (10^4 Mha) (Dong, 2001).

	Heavy desert land		Medium desert land		Light desert land		Total	
Region	Area	%	Area	%	Area	%	Area	%
Anduo	3.75	5.90	32.59	51.30	27.19	42.80	63.53	100.00
Bange	2.04	0.87	109.92	46.69	123.46	52.44	235.42	100.00
Shenzha	0.00	0.00	13.17	33.25	26.44	66.74	39.62	100.00
Nima	4.11	0.64	493.39	76.76	145.28	22.60	642.78	100.00
Ritu	1.62	1.06	96.23	63.01	54.87	35.93	152.72	100.00
Geji	0.65	0.79	48.23	58.34	33.80	40.88	82.68	100.00
Gaize	0.68	0.25	126.67	46.21	146.75	53.54	274.10	100.00
Cuole	1.15	2.73	0.57	1.36	40.35	95.90	42.07	100.00
Total	14.00	0.91	920.78	60.07	589.14	39.02	1532.92	100.00

Yellow River. In recent decades, desertification of these areas has not only threatened the ecosystem functions but also resulted in changes in vegetation and soil features. Furthermore, where hygrophytes were replaced gradually by mesophytes, xerophytes and annual psammophilous plants, their cover and biomass decreased along with the progressive desertification. Also, species diversity declined, suggesting that desertification of wetland and alpine meadow contributed to species loss and biomass reduction. Soil texture becomes coarser and the water holding capacity is reduced. Soil organic matter and nutrients (total N, P and K) content reduce dramatically with progressive desertification. Overall, desertification of wetland and alpine meadow has decreased plant diversity, reduced dominance of palatable species and changed soil physical and chemical properties (Wang *et al.*, 2007).

In the headwaters of the Yangtze and Yellow Rivers of southern Qinghai Province, areas of sandy and gravel desertification lands reach 64.37 $\times 10^4$ Mha; this leads to a significant influence on local ecological and economic sustainable development. Desertification has led to a significant negative impact on functions of the wetland ecosystem, particularly reduction of water buffering capacity.

13.6.3 Secondary bare land (black soil patch) in the headwaters region

In the 1990s, the degraded rangeland area reached about 4.25×10^7 ha, accounting for 33% of the available rangeland area, in which the secondary bare land (locally called 'black soil patch') covered an area of approximately 7.03×10^6 ha, accounting for about 16.5% of the total degraded land. Most of the bare lands are found in the headwaters area of the Yangtze, Yellow and Luancang Rivers of the Qinghai–Tibetan Plateau, which is well known as the 'water tower' of China. The headwaters of the three rivers are located in the centre of the Qinghai–Tibetan Plateau and extend from 89°24′ to 102°15′N latitude and from 31°32′to 36°16′E longitude. Its administrative district, situated in the southern part of Qinghai Province, includes the entire Yushu and Guole Prefectures and two counties from Huangnan and Hainan Prefectures, respectively, as well as the Tanggula Township of Geermu City.

The headwaters area covers 36.31×10^4 km^2, with elevation between 3500 and 4500m above sea level. Rangeland covers 2634.2×10^4 ha, including 13.8% of non-degraded rangelands, 67.6% of degradation lands and 18.6% of secondary bare land (Ma *et al.*, 2002). The secondary bare lands have been appearing on the plateau since the 1980s and their occurrence and development are regarded as a 'cancer' of the alpine rangeland ecosystem because, once the sward formed over hundreds of year is removed, it will be impossible to rehabilitate it under such a harsh environment. Vegetation developed on the secondary bare land existing in the valley or relatively flatter areas is dominated by annuals or biennials, by inedible and poisonous forbs and grasses with a low coverage of about 30% in the growing season. Furthermore, the aboveground dry biomass is always blown away after heavy trampling by livestock during deep winter and spring seasons. However, the hillside and hilltop secondary bare lands cannot grow any vegetation during the warm season as frequent water flow and concomitant soil erosion wash away all the seeds and seedlings. The consequences of water and soil erosion and other forms of land degradation lead to dysfunction of the alpine rangeland system.

In Dari County of the headwaters area, alpine rangeland converted to secondary bare land due to overgrazing and climate warming extended from 0.17 Mha in 1985 to 0.58 Mha in 1994 and 0.78 Mha in 2006. A similar trend also occurs in other counties of the area. Therefore, the ecological and grazing environment and livelihood are facing great threats due to the speeding up of expansion of the secondary bare lands. So far, the most difficult problem is inadequate effective approaches to slow down and stop development of the secondary bare lands (Shang and Long, 2007; Shang *et al.*, 2007).

13.6.4 Recovery of alpine degraded rangelands

The severe degradation of alpine rangelands on the Qinghai–Tibetan Plateau has gone on for decades, and so has the recovery effort by academia and the government. However, due to complexities of landscape, variation of climates, factors that lead to rangeland degradation as well as constraints imposed by culture, there is no single way to deal effectively with alpine rangeland degradation. Enclosure (grazing ban) is used in many areas to recover the degraded lands; it works, but in some cases it in turn speeds up the land degradation when the area enclosed is not large enough to support a family's herd size. Re-seeding, fertilizing and watering, although effective, are rarely used to improve the native rangeland condition. Reduction of herd size to stop overgrazing is encouraged by governments, but leads to loss of a herder's income. Therefore, the recovery of degraded lands is limited.

Herders themselves have limited finances and lack awareness of the long-term consequences of failure to address the problems now. The government has released some policies to deal with rangeland degradation, such as 'returning grazing land to pasture' to encourage the growth of plants. This has shown quite a positive impact in small-scale trials, but needs replication and scaling up. How to deal effectively with the 'cancer' of the alpine rangeland ecosystem, secondary bare land, still needs a lot of work. Lack of fundamental work in the past on the processes and mechanisms of degradation has hampered present-day efforts. However, the establishment of artificial and semi-artificial pastures on secondary bare land has been used widely to recover the bare land (Peng *et al.*, 1980; Li, 1996, 1999; Li and Huang, 1996; La and Liang, 2000), but efforts to re-establish natural vegetation in these vegetation types have generally been unsuccessful. To be successful, attention must be paid to controlling rodents, such as the pika, insect pests and poisonous weeds by adhering to the principles of integrated pest management (Wang *et al.*, 1995; Ma and Lang, 1998; Wang and Cheng, 2001). On the Qinghai–Tibetan Plateau, the recovery work and efforts to mitigate impacts on degraded rangelands lag far behind the rate of degradation spread.

13.7 What Lessons Have We Learnt?

Rangeland ecosystems on the Qinghai–Tibetan Plateau are complex, not only in the ways that physical forces shape the landscape, but also in the ways that socio-economic, political and institutional forces interact and impact the people who use the rangeland resources (principally, forage, fuel and water). As stocking rates continue to increase, restoration of Tibetan ecosystems to a higher productive state will have a low probability of success.

Herding families are becoming settled, not only as a result of government policies designed to promote privatization of production resources, but also because overpopulation and overuse of natural resources are causing fundamental social and economic changes among the herding households themselves. Among these factors, the decline of natural resources capacity to support animal production is the major stress on the cultural and social identity of Tibetan animal production households. With attempts to transform pastoral livestock production towards a market economy, increased livestock offtake has often been the goal. This has been promoted through privatization of herds and land, settling of herders, production of rain-fed forage and introduction of less mobile intensive grazing management. Stocking density on rangelands has risen to unprecedented levels. While many of these interventions have improved the delivery of social services, in many instances they have conflicted with the goal of maintaining rangeland health and stability because they limit the critical factor of mobility (Sheehy *et al.*, 2006). Movements between seasonal pastures, a mainstay of traditional practice, are being eliminated or reduced, herd composition is being restructured along commercial lines and herders are being compelled to become livestock farmers. The environment and pastoral cultures are under threat where mobility has been eliminated or substantially reduced (Humphrey and Sneath, 1999).

When we trace back to try to discover the reasons leading to the accelerated degradation of alpine rangeland, the impact of policy should be a key factor. During the past 50 years, management of these rangelands has undergone major shifts from feudalism to collectivism to privatization of livestock with individual grazing rights. Many of the policy initiatives had unintended deleterious consequences. Unless future government policy favours the preservation of ecological services as the first objective instead of the GDP, the trend of alpine rangeland degradation will continue downward.

References

Brockway, D.G., Gatewood, R.G. and Paris, R.B. (2002) Restoring grassland savannas from degraded pinyon-juniper woodlands: effects of mechanical overstorey reduction and slash treatment alternatives. *Journal of Environment Management* 64, 179–197.

Bureau of Animal Husbandry and Veterinary Medicine, Ministry of Agriculture, China (1994) *Data on the Grassland Resources of China*. Agricultural Technological Publishing House, Beijing, pp. 2–5 (in Chinese).

Ding, Y.J. (1996) Response of cryosphere to climatic warming since 1980 over the northern hemisphere. *Journal of Glaciology and Geocryology* 18(2), 131–138 (in Chinese).

Dong, Y.X. (2001) Driving mechanism and status of sandy desertification in the northern Tibet plateau. *Journal of Mountain Science* 19(5), 385–391.

Gong, D.Y. and Wang, S.W. (2002) Uncertainties in the global warming studies. *Earth Science Frontiers* 9(2), 371–376 (in Chinese).

Humphrey, C. and Sneath, D. (1999) *The End of Nomadism? Society, State and the Environment in Inner Asia*. Duke University Press, Durham, North Carolina.

Koch, K. (2003) www.pik-potsdam.de/avec/peyresq2003/posters/katja_koch.pdf, accessed 12 August 2008.

La, Y.L. and Liang, Z.Y. (2000) Investigation in effect of control of degradation of 'Black Soil' by artificial seeding. *Qinghai Prataculture* 9(4), 32–34 (in Chinese).

Li, J.Z., Yang, T. and Zhu, G.R. (1989) Studies on soil biology and its decomposing function in alpine meadow. In: Northwest Institute of Plateau Biology, Chinese Academy of Sciences (eds) *The International Forum Corpus of the Alpine Cold Meadow Ecosystem*. Science Press, Beijing, pp. 25–28 (in Chinese).

Li, Q.Y. (1999) The discussion on building artificial grassland in 'Black Soil Land' degraded alpine meadow. *Qinghai Environment* 9(2), 64–66 (in Chinese).

Li, X.L. (1996) Effects of resowing grasses on resuming vegetation of 'Black Soil Patch'. *Pratacultural Science* 13(5), 17–19 (in Chinese).

Li, X.L. and Huang, B.N. (1996) A preliminary report on seeding and reseeding grasses and *Kobresia* species on 'Black Soil Patch' grassland. *Chinese Qinghai Journal of Animal and Veterinary Sciences* 26(4), 9–11 (in Chinese).

Limbach, W., Davis, J., Bao, T., Shi, D. and Wang, C. (2000) The introduction of sustainable development practices of the Qinghai Livestock Development Project. In: Zheng, D. and Zhu, L. (eds) *Formation and Evolution, Environment Changes and Sustainable Development on the Tibetan Plateau*. Academy Press, Beijing, pp. 509–522.

Long, R. (1999) Feed value of native forages of the Tibetan Plateau of China. *Animal Feed Science and Technology* 80, 101–113.

Long, R. (2003) Alpine rangeland ecosystems and their management in the Qinghai–Tibetan Plateau. In: Wiener, G., Jianlin, H. and Long, R. (eds) *The Yak*, 2nd edn. FAO/RAP, Bangkok, pp. 359–388.

Long, R. and Ma, Y. (1997) Qinghai's yak production systems. In: Miller, D.J., Craig, S.R. and Rana, G.M. (eds) *Conservation and Management of Yak Genetic Diversity*. Proceedings of a Workshop on Conservation and Management of Yak Genetic Diversity at ICIMOD, Kathmandu, pp. 105–114.

Long, R.J., Dong, S.K., Hu, Z.Z., Shi, J.J., Dong, Q.M. and Han, X.T. (1999) Effect of strategic feed supplementation on productive and reproductive performance in yak cows. *Preventive Veterinary Medicine* 38, 195–206.

Long, R.J., Dong, S.K. and Shang, S.K. (2007) *The Qinghai–Tibetan Plateau Rangelands: From Productive Services Oriented to Functional Integration.* International Centre for Tibetan Plateau Ecosystem Management (ICTPEM), Lanzhou University, Gansu Province, China, 13 pp.

Ma, M., Jiao, Y. and Cheng, G. (2002) Change in land coverage in Northwest China during the past decade monitored by NOAA CHAIN. *Journal of Glaciology and Geocryology* 24(1), 69–71 (in Chinese).

Ma, Y.S. and Lang, B.N. (1998) Establishing a pratacultural system – a strategy for rehabilitation of 'Black Soil' on the Tibetan Plateau. *Pratacultural Science* 15(1), 5–9 (in Chinese).

Miller, D.J. (1990) Grasslands of the Tibetan Plateau. *Rangelands* 12, 159–163.

Peng, L.M., Yan, X.W., Zhou, J.S. and Lu, Y.X. (1980) Bare land of alpine meadow and their rebuilding in Qumalai region of Qinghai province. *Chinese Journal of Grassland* 4, 7–17 (in Chinese).

Qian, S., Mao, L., Hou, Y., Fu, Y., Zhang, H. and Du, J. (2007) Livestock carrying capacity and balance between carrying capacity of grassland with added forage and actual livestock in the Qinghai–Tibet Plateau. *Journal of Natural Resources* 3, 389–397.

Shang, Z. and Long, R. (2007) Formation causes and recovery of the 'Black Soil Type' degraded alpine grassland in Qinghai–Tibetan Plateau. *Frontiers of Agriculture in China* 1(2), 197–202.

Shang, Z., Long, R. and Ma, Y. (2007) Review on environmental problems in the headwater areas of Yangtze and Yellow rivers in Qinghai–Tibetan Plateau. *Pratacultural Science* 24(3), 1–7 (in Chinese with English abstract).

Sheehy, D.P., Miller, D. and Johnson, D.A. (2006) Transformation of traditional pastoral livestock systems on the Tibetan steppe. *Sécheresse* 17(1–2), 142–151.

Su, Y.Z., Zhao, H.L., Zhang, T.H. and Zhao, X.Y. (2004) Soil properties following cultivation and non-grazing of a semi-arid sandy grassland in northern China. *Soil and Tillage Research* 75(1), 27–36.

Wang, G.X. and Cheng, G.D. (2001) Characteristics of grassland and ecological changes of vegetation in the source regions of Yangtze and Yellow rivers. *Journal of Desert Research* 21(2), 101–107 (in Chinese).

Wang, H., Guo, Z., Xu, X., Liang, T. and Ren, J. (2007) Response of vegetation and soils to desertification of alpine meadow in the upper basin of the Yellow River, China. *New Zealand Journal of Agricultural Research* 50, 491–501.

Wang, Q.J., Yang, F.T. and Shi, S.H. (1989) The pilot study on the *Kobresia humilis* regeneration in alpine meadow. In: Northwest Institute of Plateau Biology, Chinese Academy of Sciences (eds) *The International Forum Corpus of the Alpine Cold Meadow Ecosystem.* Science Press, Beijing, pp. 83–93 (in Chinese).

Wang, Q.J., Zhou, X.M., Shen, Z.X. and Chen, B. (1995) Analysis on benefit about restoration and rebuilding degraded alpine meadow under different control strategies. *Alpine Meadow Ecosystem* 4, 345–352 (in Chinese).

Xie, G.D., Lu, C.X., Xiao, Y. and Zheng, D. (2003) The economic evaluation of grassland ecosystem services in Qinghai–Tibet Plateau. *Journal of Mountain Research* 21(1), 50–55 (in Chinese).

Xiong, Y. and Li, Q.K. (1987) *Soils of China.* Science Press, Beijing, 236 pp.

Yang, F.T., Wang, Q.J. and Shi, S.H. (1989) *Kobresia humilis* meadow biomass variation, seasonal and annual. In: Northwest Institute of Plateau Biology, Chinese Academy of Sciences (eds) *The International Forum Corpus of the Alpine Cold Meadow Ecosystem.* Science Press, Beijing, pp. 61–71 (in Chinese).

Zhou, H., Zhou, L., Zhao, X., Yan, Z., Liu, W. and Shi, Y. (2003a) The degraded process and integrated treatment of 'black soil beach' type degraded grassland in the source regions of Yangtze and Yellow rivers. *Chinese Journal of Ecology* 22, 51–55 (in Chinese with English abstract).

Zhou, H., Zhou, L., Liu, W., Zhao, X., Lai, D., Cai, R., Zhao, B. and Li, Y. (2003b) The study on the reason of grassland degradation and the strategy of sustainable development of animal husbandry in Guoluo Prefecture, Qinghai Province. *Pratacultural Science* 20, 19–25 (in Chinese with English abstract).

Zhou, X.M. (2001) *Chinese* Kobresia *Meadow.* China Science Press, Beijing, pp. 131–206 (in Chinese).

14 Case Study 8: Northern Xinjiang

Jin Gui-li and Zhu Jin-zhong
College of Pratacultural and Environmental Sciences, Xinjiang Agricultural University, Urumqi, China

Synopsis

This case study examines the situation in northern Xinjiang, a vast and important pastoral rangeland region in far north-west China. The climate and topography combine to create a complex suite of rangeland types from alpine meadows to harsh desert margins. Human impacts have been severe and massive changes have been wrought over the past 50–60 years. Land degradation has been accelerated as the population of humans and their livestock increases. Recovery has been slow and difficult. A new approach to rangeland-based livestock production is outlined.

Keywords: overgrazing; dust and sandstorm; artificial oases; desertification; water diversion; alpine meadows; desert steppe; land conversion; stocking intensity

14.1 Brief Statement of the Problem of Rangeland Degradation in Xinjiang

Xinjiang is situated in a fragile ecological region and the area of desertified land is becoming larger and larger each year. The desertified area has expanded to 1.044 million km^2 in Xinjiang, accounting for 48% of the total area in Xinjiang, of which the sandified area accounts for 54%. Xinjiang is listed first among the top 18 provinces/autonomous regions in China that are affected by desertification. The expansion of desertification is generally at a speed of around 400 km^2 annually, accounting for approximately 15% of the total incremental area of desertification, and each year there are about 8 million ha (Mha) of rangeland stricken by sandification in China. The rate of expansion is nationally, and even globally, significant. Rangeland vegetation has been destroyed and soil erosion by both wind and water has worsened. For example, mobile sand dunes are moving southward at a speed of 0.5–2.6 m/year in the south of the Kuerbantonggute Desert. In the Shihezi reclamation area, 20,000 ha of cropland and rangeland have been threatened.

The desert area has increased three times in the past 20 years on the modern delta of the Wulungu River in the east of Buluntuohai. The rangeland in downstream reaches of the Tarim River has suffered loss of vegetation since water diversion upstream and large areas of reed and native poplar trees (*Populus diversifolia*) and others have wilted and died. The desertified and potentially desertified areas midstream and downstream in the Tarim River account for 87% and 93%, respectively, of the total land area there.

Rangeland coverage, height and forage yield have decreased on average by 30–70%; for example, compared with the 1960s, vegetation coverage has decreased from 90% to 30–50%, height from 24.6 cm to 14.2 cm and fresh grass yield is now less than 50% on the rangeland in the Yuerdousi Basin in the Tianshan Mountains.

Both the frequency and intensity of dust and sandstorms show an increasing trend, which seriously deteriorates the local environment and even influences other regions of China, East Asia and the Pacific region, bringing more and more threats to global ecological security. The frequency of sandstorms has increased by 45% since the 1990s. Among the top ten special class sandstorms that have occurred in China since 1949, seven occurred in Xinjiang. Sandstorms are much stronger and more frequent in Turpan, Hami and Hetian Prefectures, and the accumulated economic loss is estimated as high as several billion dollars.

Eco-environmental deterioration leads to more frequent disastrous weather. According to local meteorological data, the total frequency of meteorological disasters in the 1980s was 4.54 times higher than that in the 1950s, droughts are seven times more frequent than in the 1950s and the frequency and severity of floods have increased by 3.52 times in Xinjiang as a whole. Along with the expansion of rangeland degradation, strong windy weather and floating dust weather are becoming more and more frequent. For example, floating dust weather in Jinghe County averaged 0.4 days in the 1960s and in the 1990s was 44.7 days, a 100-fold increase.

In Xinjiang, rangeland degradation has become one of the important causes of water and soil losses and more frequent sandstorms. Moreover, such deterioration is increasingly expanding, which has a tremendous negative impact on the local eco-environment and socio-economic development.

14.2 The Setting of Xinjiang Rangeland

14.2.1 Climate

Xinjiang is located in the hinterland of the Eurasian continent and on the north-west frontier of China (Fig. 14.1).

Being isolated from the oceans and surrounded by high mountains, the moist oceanic airflow cannot reach Xinjiang; therefore, it is a typical arid region and the climate there belongs to the continental temperate zone and warm temperate zone. Xinjiang has a typical temperate desert climate characterized by abundant sunlight and heat resources, dramatic changes of temperature and scarce rainfall. Mean annual precipitation is 156.5 mm and mean annual evaporation is over 1000 mm. The aridity index is as high as 7 in Xinjiang and the ecosystems there are extremely fragile.

The topography is exceedingly complex, which makes for great vertical zonality in the north and south of Xinjiang. Additionally, because the basins are surrounded and divided by high mountains, the climates are diversified and some local climates are quite special. A series of high mountain ranges run from west to east, such as the Altai Mountains on the northern edge of the Junggar Basin, the Tianshan Mountains in the middle and the Kunlun Mountains on the south edge of the Tarim Basin, which has led to a geographical environment that is relatively closed. Overall, the rangeland eco-environment in Xinjiang is complex and diversified as a result of interactions between climate and topography.

All these physical conditions have constrained the zonal distribution and regional characteristics of the rangeland in Xinjiang, which are ecologically fragile and diversified, and impacted their generation, development and utilization significantly (Dai *et al.*, 2007).

14.2.2 The rangeland soil

The development and distribution of rangeland soils are constrained by the terrain, parent material, climate, hydrographic condition and biological activity, which give rise to many and varied rangeland soils in Xinjiang. They can be classified into *zonal temperate desert soil* and *warm temperate desert soil* generated under natural conditions, a series of *azonal aqueous soils* soaked by phreatic water, or partially by surface water, a series of *salinized–alkalinized soils* during salinization and desalinization and a series of *mountain soils* stratified according to elevation.

Different rangeland vegetations grow on the various soil types, for example: *desert rangeland* on grey-brown desert soil, desert brown soil and carbonate soil; *steppe desert* on brown soil; *mountainous steppe* and *desert steppe* on chestnut soil and light chestnut soil, respectively; *mountainous meadow* on black earth; alpine steppe and alpine meadow on

Fig. 14.1. Geographical location of Xinjiang, China. The focus of this case study is on the northern area.

alpine steppe soil and alpine meadow soil; and *azonal lowland meadow* on azonal meadow soil.

14.2.3 The rangeland flora and type

There is a total of 3270 species in 687 genera and 108 families of higher plants (including subspecies and varieties) on the rangeland in Xinjiang, accounting for 30.5% of the total families, 21.6% of the total genera and 12.1% of the total species in China. Forage species account for 91.9% of the total rangeland species in Xinjiang and 382 species are superior-quality forages. The rangeland quality is relatively good in some areas because of the many superior forage species there, which are unique and important germplasms in the world. Xinjiang is one of the regions with the most abundant forage resources in northern China.

The rangeland types in Xinjiang are diversified. The special topographical structures and great altitudinal difference from 6000 m of the Kunlun Mountains down to −154 m of the Turpan Basin, the high mountains, hills, basins, valleys, plain oases, gobi and desert, and the Pamir Mountains, the so-called 'roof of the world', present the complex and diversified topographical attributes that have resulted in the different types of rangeland from the plains to the mountains (Table 14.1).

14.2.4 The rangeland population

Xinjiang is one of the major areas for nomadic habitat in China. In Xinjiang, the ethnic minorities such as Kazakh, Khalkhas, Mongolian, Tajik and a few Uzbek and Tatar are traditional nomads. They have been practising nomadism

Table 14.1. The area and percentage of different rangeland types in Xinjiang.

	Total area		Available area		
Rangeland type	Area (10,000 ha)	Percentage	Area (10,000 ha)	Percentage	Percentage of available area
Temperate meadow-steppe	116.60	2.04	108.62	2.26	93.15
Temperate steppe	480.77	8.40	442.25	9.21	91.99
Temperate desert-steppe	629.86	11.00	580.97	12.10	92.24
Alpine steppe	433.19	7.56	386.09	8.04	89.13
Temperate steppe-desert	441.85	7.72	356.63	7.43	80.71
Temperate desert	2133.19	37.26	1609.99	33.54	75.47
Alpine desert	111.75	1.95	80.48	1.68	72.01
Lowland meadow	688.58	12.03	603.62	12.57	87.66
Temperate mountain meadow	287.06	5.01	265.70	5.53	92.56
Alpine meadow	376.37	6.56	341.90	7.12	90.84
Marsh	26.66	0.47	24.44	0.52	91.67
Total in Xinjiang	5725.88	100	4800.68	100	83.84

for centuries. Ever since the New China was founded, some people of these traditional nomads have transferred to other livelihoods, e.g. cropping, commerce and industry, etc. However, the majority of Kazakh, Khalkhas, Mongolian and Tajik in Xinjiang still engage mostly in pastoral livestock as their major livelihood. In 2001, the total population of these four major traditional nomadic groups in Xinjiang was 1.692 million, of which nearly three-fifths still depended on pastoralism (Table 14.2). Livestock production in Xinjiang is very important in China's pastoral development (Table 14.3).

Table 14.2. Populations of the four major nomadic groups in Xinjiang (2007).

Ethnic group	Population (10^4)	Proportion of total population of Xinjiang (%)	Proportion of four major nomadic groups (%)
Kazakh	143.50	7.00	80.16
Khalkhas	17.59	0.86	9.83
Mongolian	17.46	0.85	9.75
Tajik	4.47	0.02	0.26
Total	179.02	8.73	100

14.3 Causes of Rangeland Degradation in Xinjiang

The causes of rangeland degradation in Xinjiang could be summarized as physical and anthropogenic, but mostly the latter. The fragile rangeland eco-environment in Xinjiang is the external cause of the degradation. The change of watercourse, reduction of precipitation and increase of air temperature and evaporation, etc., are all the physical causes of the rangeland degradation (Xu and Zhao, 2007). According to relevant surveys, annually averaged hay production decreased by 60.16 kg/ha because of the reduced precipitation. Accordingly, it is estimated that the hay production from winter pastures of northern Xinjiang (including Yili, Tacheng, Bole, Altai and Changji Prefectures, total area approximately 10.3884 Mha) will be reduced to 625 million kg. It is evident that reduction of precipitation is one of the most important causes accelerating the degradation of natural rangeland in these areas (Guo *et al.*, 2001; Eli *et al.*, 2002; Gao, 2002; Ju *et al.*, 2004; Jin *et al.*, 2007).

14.3.1 Climate change and rangeland degradation

Since the middle of the 20th century, Xinjiang has experienced a gradually fluctuating increase of

Table 14.3. Major economic indicators of livestock in Xinjiang and their comparison in China.

	1990		1995		2000		2005	
Indicator	Quantity	Proportion in China (%)	Quantity	Proportion in China (%)	Quantity	Proportion in China (%)	Quantity	Proportion in China (%)
Year-end animal stock (10^4 head)	575.9	3.7	580.0	3.1	632.9	4.2	739.7	4.6
Meat production (10^4 t)	30.5	1.1	52.4	1.0	90	1.5	141.5	1.8
Milk production (10^4 t)	30.8	7.4	45.2	7.8	72.5	7.9	152.2	5.3
Wool production (10^4 t)	4.9	20.6	5.5	19.7	7.0	23.9	9.4	23.9
Production value of livestock (10^8 yuan)	29.5	1.5	82.3	1.4	114.5	1.5	183.5	1.4

temperature and precipitation (Figs 14.2 and 14.3). For the past 20 years, it has been characterized by a warmer, rainier climate (Dai *et al.*, 2007).

From the 1950s to the 1980s (1950–1984), the weather went through a cold period and then a warm period. The temperature in the 1950s was comparatively lower, in the 1960s higher and in the 1970s moderate. In Xinjiang, the difference between maximal and minimal annual means among each decade is 1–2°C. In eastern and southern Xinjiang, the difference between mean temperature of each decade over 30 years was less than 0.2°C, showing only a slight change, whereas there was a significant change after the 1990s. The annual precipitation also changed with alternating wet and drier periods. Precipitation in the 1950s and after the mid-1980s was on the high side, and in the 1960s and 1970s it was on the low side. Both temperature and precipitation demonstrate a decadal-scale periodicity and a linear increment trend, while the temperature is more significant than the precipitation.

There is no evidence that the climate gradually became more arid in terms of either temperature change or precipitation change. The present rapid rangeland degradation cannot be attributed primarily to climate change. Climate plays an important role in rangeland 'fluctuation', but this is not the primary factor for the large-scale rangeland degradation. Xinjiang is

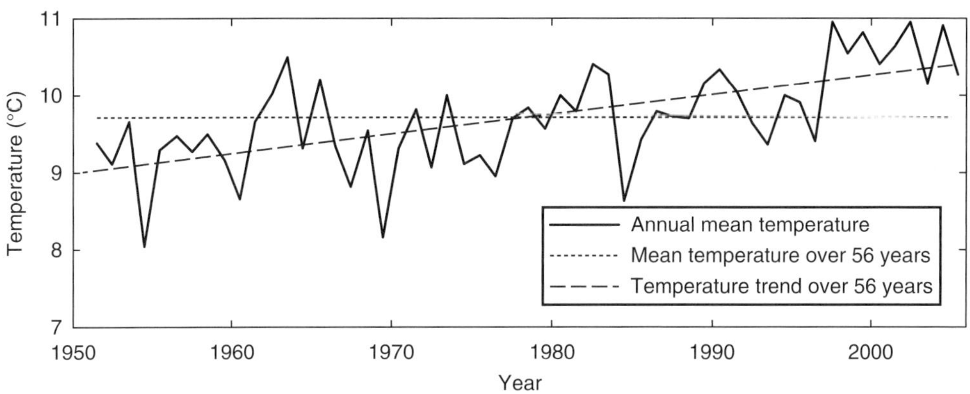

Fig. 14.2. Temperature changes in Xinjiang from 1951 to 2005.

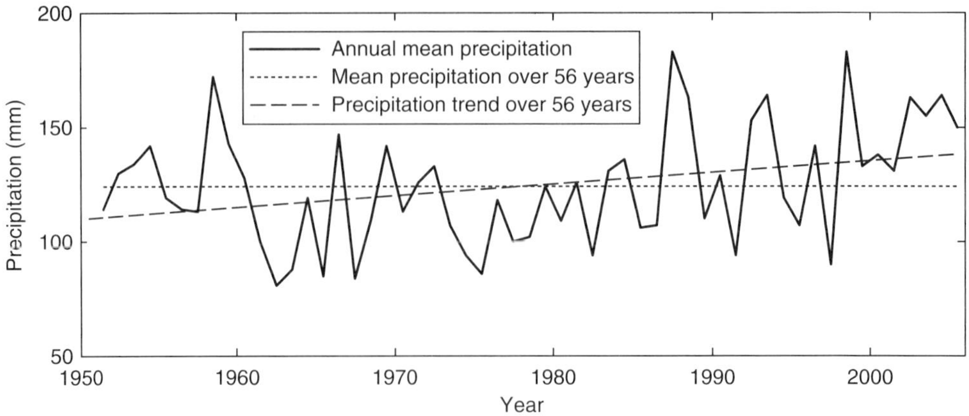

Fig. 14.3. Precipitation changes in Xinjiang from 1951 to 2005.

located in an arid desert region, the rangelands that cover both mountains and plains are the outcomes of natural selection and environmental adaptation. The climate is merely one factor contributing to rangeland degradation.

14.3.2 Glacier shrinkage and rangeland degradation

Along with the global warming, the glaciers are melting and shrinking in Xinjiang. From 1963 to 2000, the area of glaciers on the Tianshan Mountains was reduced on average by 12.5%, e.g. No. 1 Glacier in the headwater region of the Urumqi River in the Tianshan Mountains (Table 14.4) (Li *et al.*, 2003; Lan *et al.*, 2007).

Glacial retreat at No. 1 Glacier during 1959–1985 averaged −94.5 mm/year, whereas during 1986–2000 it increased to −358.4 mm/year (a 2.8-fold increase). Accordingly, melting runoff from the glacier increased greatly as well; during 1958–1985, it averaged 508.4 mm/year from No. 1 Glacier, whereas during 1985–2001 it was 936.6 mm/year by the same calculation method (Fig. 14.4). Thus, it can be seen that the temperature has increased quickly since the 1980s and, consequently, this has accelerated the melting of glaciers.

The melted glaciers increased the rangeland area and increased water recharge to the regional river systems. For example, the area of alpine meadow in Changji District, Xinjiang, increased over 20 years from 29,005 ha in 1980 to 49,292 ha in 2000. Currently, some rangelands have benefited from glacier melting but, in the long term, this will have an important impact on rangeland degradation.

Table 14.4. Changes of the ice area of No. 1 Glacier in the headwater region of the Urumqi River in the Tianshan Mountains.

Date	Ice area (km^2)
Aug. 1962	1.950
Oct. 1964	1.941
Aug. 1986	1.840
Aug. 1992	1.833
Aug. 1994	1.742 (1.115 in west; 0.627 in east)
Aug. 2000	1.733 (1.111 in west; 0.622 in east)

14.3.3 Population expansion and rangeland degradation

During the period of 1949–2001, the total population in Xinjiang increased from 4.5 million to 18.76 million, in which the population of the four major nomadic groups, i.e. Kazakh, Khalkhas, Mongolian and Tajik, and others increased from 0.5758 million to 1.6922 million, an increase of 193.9%. The remaining 17 million was from inward migration of, principally, Han Chinese and/or their descendants.

As the population continues to grow, the demand for livestock products is increasing year

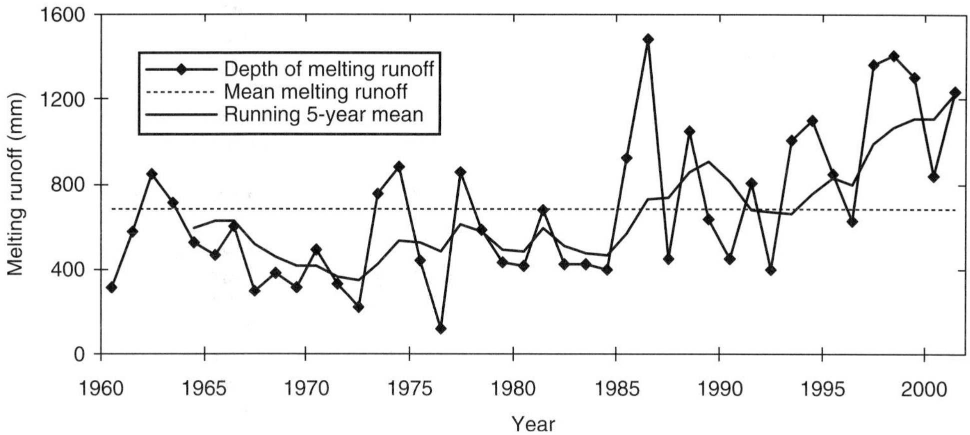

Fig. 14.4. Changes in melting runoff and running 5-year mean from No. 1 Glacier in the headwater region of the Urumqi River in the Tianshan Mountains.

after year and livestock numbers are increasing continuously, which increases the pressure on the rangeland progressively and results in feed imbalance. Moreover, the lifestyle of nomads and their subsistence economy are impacted. Conversion of rangeland for crop production has increased the animal loading rate and has exacerbated rangeland degradation.

14.3.4 Laggard livestock production manner

Pastoral livestock production in Xinjiang is based on the notion of 'depending on heaven'; such a laggard conventional production and operational manner has lower efficiency, lower output and lower economic benefit. In recent years, livestock numbers have increased because of the expanding population. Additionally, poor management and unwise use of the rangeland resources, without a view to the long term and with insufficient awareness of the limited availability of the resources, are leading to increasing exploitation of the rangeland and blind pursuit of quantity rather than quality. While making positive progress in terms of total output, many negative consequences have been triggered. In the past three decades, the livestock quantity in Xinjiang has increased rapidly (Fig. 14.5). The animal stock numbers overall increased from 2476.98 × 10^4 in 1978 to 5206.37 × 10^4 at

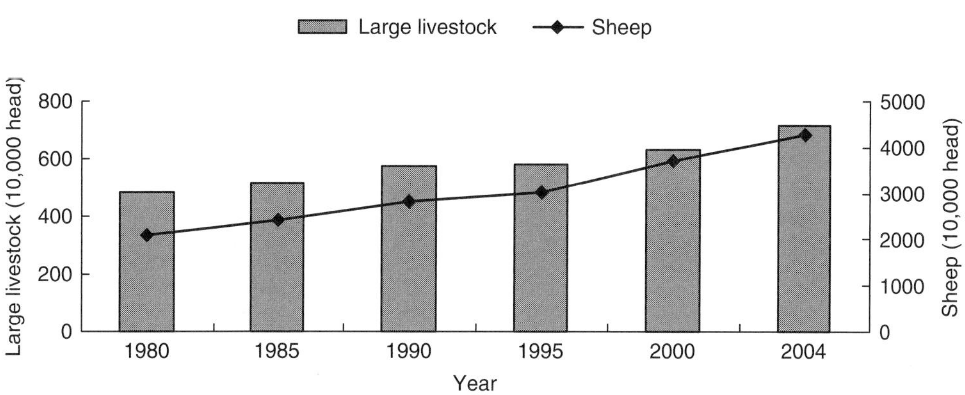

Fig. 14.5. Growth of livestock populations in Xinjiang (1978–2004).

the end of 2004 (more than double). In some prefectures where the livestock production developed rapidly, the livestock numbers showed a five- to six-fold increase over that in 1949. Grass and livestock are now severely unbalanced, resulting in accelerated rangeland degradation. Growing demand for animal products and decreasing forage supply are increasingly incompatible and the eco-environment has become more and more deteriorated.

Historically, the nomadic economy was the dominant production model in the pastoral areas, which was characterized by 'chasing water and grass for grazing animals', periodic migration and transhumance (seasonal rotation). The nomadic economy was low cost and high benefit, and naturally it was the right choice for the herders (Adilhan, 2004). However, in this transhumance model, the herders and their animals had to move from season to season over a large geographical area. The rapid increase in livestock numbers upset the balance and this impacted strongly on the rangelands. Overgrazing under this new circumstance is an important cause of rangeland degradation. The herders' capacity to resist disasters is now weaker and their animal mortality is high (Mansur *et al.*, 2002).

14.3.5 Irrational rangeland development and use

The irrational development and use of rangeland, such as land conversion, firewood cutting and so on, are the major reasons for its degradation. Results from Wu *et al.* (2005; see also Zhao, 2002a) show that, in the decade of the 1990s, 672,921 ha of rangeland was converted into other land uses in Xinjiang (Table 14.5).

Since the 1950s, the government has launched unprecedented land conversion projects. There have been several episodes; a 'legionary' phase (1950–1957), a 'full-scale conversion' phase (1958–1970), a 'sporadic conversion' phase (1971–1985) and a 'project development' phase (1986–1997). Currently, the accumulative area of converted land is 4.093 Mha in the whole of Xinjiang, that is, 3.38 times the area of cropland in 1949. The Manasi River Irrigation Area, the Kuitun River Irrigation Area, the Urumqi River Irrigation Area and the Yeerqian River Irrigation Area, etc., were founded one after the other. Originally, these irrigation areas and the downstream areas were natural rangelands with superior forage species for grazing animals and/or for hay fields, which could provide large amounts of forage for the animals year after year and where, generally, surface water and groundwater could be recharged during flooding periods.

Along with the expansion of cropland, the high-quality natural pastures were gradually decreasing, which reduced forage supply from the pastures. In addition, irrational land conversion resulted in large areas of the land being wasted. Less than half of the area converted from rangeland became high-yield farmland. The area of abandoned cropland and wasteland is 1.3 Mha, accounting for approximately one-third of the total farmland in the whole of Xinjiang (Fig. 14.6). These areas are generally located on the outskirts of the desert oases, where the virgin vegetation has been destroyed completely and cannot be rehabilitated easily. Firewood cutting and digging for herbal medicines have also destroyed rangeland vegetation and worsened water and soil losses. Disorderly wood logging led to forest resources being reduced by over one-third, and river valley woodlands suffered even more. The native poplar trees on the banks of the dried-out rivers could not avoid their fate of complete collapse. The area of desert shrubland decreased by over 50%. The ecological benefit of

Table 14.5. Dynamic conversion of rangeland into other land use types in Xinjiang in the 10-year period from the end of the 1980s to the end of the 1990s (Xu, 2004).

Items	Arable land	Woodland	Water body	Land for construction	Unused land	Total
Area (ha)	365,222.01	19,113.91	47,634.08	87,438.02	153,513.34	672,921.36
Proportion (%)	54.27	2.84	7.08	12.99	22.81	100.00

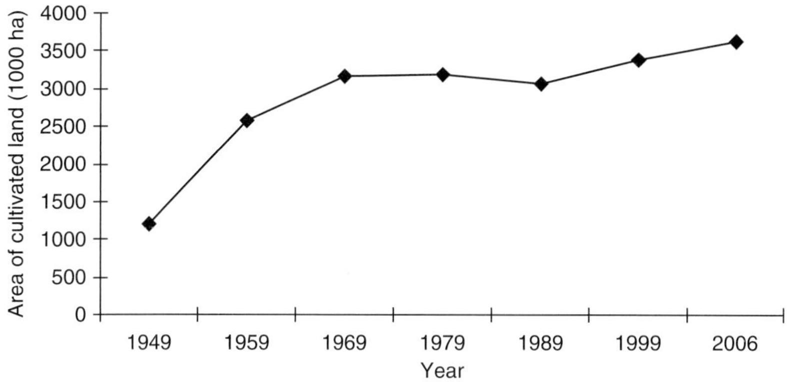

Fig. 14.6. Changes of cultivated land area in Xinjiang.

the newly planted young woodland was very much lower than the native vegetation.

Xinjiang is rich in medicinal herb species. Since the 1980s, being encouraged by high prices, the collection of herbal plants on the rangeland has been active. Statistics showed that the area of liquorice (*Glycyrrhiza uralensis*) was reduced from 1.9333 Mha in the 1950s down to around 666,000 ha and, annually, over 300 million kg have been dug and collected and roughly 15,000 ha of rangeland destroyed because of this. Cutting shrubs and semi-shrubs on the rangeland for firewood is an especially serious problem. It is estimated that nowadays firewood-consuming households account for 60% of the total rural households in the whole of Xinjiang and annual demand for firewood is 780 t, mostly from desert vegetation.

14.3.6 Unsound policy and regulatory systems

The policies and regulations for the administration/management of rangeland resources are insufficient and, even under the present rangeland laws and regulations, there is poor supervision. Various forms of an output-linked contracting liability system have been implemented in over 90% of Xinjiang (Chapters 2 and 16). However, most areas have not yet practised the family-based contracting system completely. Some rangelands have been used in a disorderly way and nobody has been responsible for their maintenance. This is disadvantageous for the sustainable utilization and development of the rangelands. In China, a dual contracting system on both rangeland and livestock has been put in place in pastoral areas. This endows the contracting herders with the right to exploit rangeland resources (even misusing and overgrazing the resources) without any responsibility for rangeland conservation. Another factor is that, currently, the contracting period of rangeland in China is relatively short, which is disadvantageous for encouraging steady and persistent contractor investment in their rangeland resources or the conservation and improvement of rangeland productivity. On the contrary, the rangeland might be destroyed by such patterns of use (Chapter 16).

The legal and regulatory framework and level of awareness of the majority of herders and farmers of the need for rangeland protection are weak. There is a lack of law enforcement in the local communities, the rangeland supervision and monitoring systems have not been extended broadly and their functions have not been developed fully.

14.3.7 Insufficient investment

Starting from the 1950s, under support from the central and local governments, infrastructure development programmes have been carried out on a large scale in pastoral areas. This includes herders' settlement projects and other projects related to developing 'reasonable utilization' of rangeland resources through the construction of water conservancy facilities, road systems, warm sheds and herders' residential sites. All these have played an important role in promoting the development of productivity. However, because of the underdeveloped local economy and extreme lack of investment over the long term, most areas are still poor in terms of infrastructural scale and level, which cannot satisfy the requirement for the development of modern prataculture (a set of practices related to scientific agropastoral integration).

Modern prataculture was derived from the traditional pastoral livestock industry. The laggard infrastructures will be a 'bottleneck' for modern pratacultural development in some areas. The herders' settlement programme, the establishment of new-type industries and the infrastructural development in rural energy, water conservancies, transportation, information services and telecommunications and productive facilities all need central, local, collective and private inputs for increasing investment and accelerating development.

14.4 Ecological Processes of Rangeland Degradation

Rangeland degradation results in the succession of vegetation and a series of changes of the rangeland eco-environment, including the soil. A typically degraded spring–autumn pasture in Xinjiang – *Seriphidium transillense* desert – was studied to understand the mechanism of degradation.

The study area (43°49'–43°56'N, 87°02'–87°05'E), which is an open, flat alluvial plain belonging to the spring–autumn pastures of Ashili Village, is located on the middle part of the northern slopes of the Tianshan Mountains. It is approximately 40 km east of Urumqi City and 32 km south-east of the Asian continent geographical centre (ACGC). The elevation ranges between 750 and 950 m.

The study area has an extremely arid climate which belongs to the middle Asian desert climate, with a mean annual rainfall of 180–190 mm, but a potential annual evaporation of >1760 mm and a drying index of 4–10. Unlike most other arid zones, this area has an average annual snow cover period of about 103 days, starting from late November and ending in mid-March of the following year. The study area has a mean annual temperature of 6.5°C, but with a hot summer and a cold winter; the frost-free period is 160–190 days. The soil is classified as grey desert soil and the soil parent material is loess-like.

The study area was subdivided into four sites, each representing one of the four successive stages of degradation – non-degradation (ND), mid-degradation (MD), heavy degradation (HD) and overdegradation (OD) – for the study on rangeland vegetation, soil seed bank and soil properties, respectively.

14.4.1 Process of rangeland degradation

Change of vegetation

There were obvious species compositional shifts between different degradations. Twenty-five plant species belonging to 15 genera were found in the study area and species of *Chenopodiaceae* and *Compositeae* accounted for 40% of the total species number. The species names and their important value in each degraded area are listed in Table 14.6.

The composition of plant community changes from *S. transillense* + *Petrosimonia sibirica* + *Ceratocarpus arenarius* (in ND) to *S. transillense* + *Gagea bulbifera* + *C. arenarius* (in OD) to *G. bulbifera* + *Geranium pratense* + *S. transillense* (in HD) to *P. sibirica* + *G. pratense* + *Trigonella arcuata* (in OD) under grazing pressure.

In non-degraded areas, *S. transillense* is the dominant species. Due to domestic livestock grazing, the important value of *S. transillense* drops dramatically in degraded rangeland and consequently it loses its dominant status and releases more resources to other plants, which leads to an increase of species number (Jin *et al.*, 2007). In the overdegraded stage, *P. sibirica* replaces *S. transillense*, becoming the dominant species. There are two to four geophytes in each degraded stage in comparison with non-degraded sites because the geophytes can successfully avoid the trampling of animals.

S. transillense became sparse and lower in the degraded areas because of grazing pressure and a lack of sufficient nutrients accumulated in the previous autumn. The frequency, coverage, yield and density of *T. arcuata* increased with degraded degree, while its height became lower and lower. These changes attributed to the increasing of space

Table 14.6. Species name and important value in each degraded area.

Species name	ND	MD	HD	OD
Seriphidium transillense	0.55	0.36	0.06	<0.01
Petrosimonia sibirica	0.16	0.05	0.04	0.41
Trigonella arcuata	0.03	0.03	0.05	0.12
Ceratocarpus arenarius	0.07	0.08	0.03	0.03
Altaicum schischk	0.04	0.02	0.05	0.01
Malcormia africana	0.03	0.01	0.02	–
Ceratocephalus orthoceras	0.07	0.01	–	0.02
Gagea bulbifera	–	0.33	0.55	0.05
Tulipa iliensis	–	0.01	0.01	0.02
Geranium pratense	–	–	0.08	0.27
Rapeseris sp.	–	0.01	0.01	–
Astragalus vicarious	–	0.03	0.02	–
Kochia prostrata	–	–	0.01	–
Salsola collina	–	0.01	–	0.05
Koelpinia linearis	–	–	0.04	0.01
Tragopogon kasahstanicus	0.01	–	–	–
Alyssum sibiricum	0.02	–	–	–
Convolvulus arvensis	<0.01	–	–	–
Lepidium perfoliatum	0.01	–	–	–
Anchusa ovata	–	0.01	–	–
Chenopodium album	–	–	<0.01	–
Ceratoides latens	–	–	0.01	–
Plantago lessingii	–	–	–	0.01
Allium chrysanthum	–	–	–	<0.01
Poa bulbosa	–	–	–	0.01

Important value (IV) = (RC + RY + RD + RF)/4. RC, RY, RD and RF mean relative coverage, biomass, density and frequency, respectively. –, the species does not exist in the stage and its IV is zero.

and nutrients released by *S. transillense* and the low height may help the plants being eaten. *P. sibirica* adopts an *r* strategy, which produces innumerable seeds and forms a large population and, consequently, this protects the rangeland from complete collapse, while *T. arcuata* and *P. sibirica* are 'defensive fighters' in overdegraded rangeland.

The rangeland feeding value becomes worse with the rangeland degradation degree (Table 14.7). Although two to five palatable species increased in degraded areas, the total palatable yield decreased dramatically because the domestic livestock imposed heavy grazing pressure on the palatable plants. The unpalatable vegetation yield in overdegradation sites is five times higher than that of other degradation sites.

Soil seed bank

Composition of persistent soil seed bank: ND has 6 species, MD has 8 species, HD has 7 species and OD has 8 species, Seed numbers were 683, 455, 374 and 611 seeds/m^2, respectively, in four degradation stages (Table 14.8). In ND, the proportion of useful species seeds, e.g. *S. transillense*, is 50%, but 5% in MD, 0% in HD and OD. Constructive species cannot complete their life histories under higher grazing stress, so seed numbers decrease. But in OD invasive species gradually replaced constructive species and their reproductive strategy made seed numbers increase. To the three vertical layers, the distribution rate of seeds is, respectively, 62%, 14% and 24% in ND, 72%, 14% and 14% in MD, 44%, 35% and 21% in HD and 83%, 6% and 11% in OD. The total seed numbers decreased gradually with soil depth, with 66% of total seeds distributed in 0–5 cm soil depth. Grazing stress makes soil compaction increase, so it is difficult for seeds to get buried and most of the seeds exist in the surface layer of the soil, where they are vulnerable to trampling and wind and water erosion.

Table 14.7. Changes of rangeland feeding value in different degradation regimes (unit: g/m², %).

	ND		MD		HD		OD	
Forage selection rating	Yield	Cover	Yield	Cover	Yield	Cover	Yield	Cover
Palatable	106.1	35.8	27.7	22.9	32.5	33.0	7.7	4.9
Seriphidium transillense	102.1	34.7	20.6	11.5	4.4	1.4		
Trigonella arcuata	0.9	0.4	0.5	0.7	1.4	1.3	6.3	3.2
Altaicum schischk	2.5	0.6	0.7	0.9	4.6	1.8	0.2	0.5
Ceratocephalus orthoceras	0.6	0.1	<0.1	0.4			<0.1	0.1
Gagea bulbifera			3.9	9.2	15.7	25.7	0.9	0.9
Tulipa iliensis			0.2	0.1	0.2	0.5	0.2	0.1
Koelpinia linearis					3.5	2.0	0.1	0.1
Astragalus vicarious			1.8	0.5	2.1	0.2		
Kochia prostrata					0.6	0.1		
Unpalatable	4.9	2.2	1.2	1.1	2.7	2.7	24.3	27.2
Petrosimonia sibirica	3.9	1.7	0.3	0.3	0.1	0.4	13.4	13.9
Ceratocarpus arenarius	0.8	0.5	0.5	0.7	0.2	0.2	0.6	0.9
Malcormia africana	0.2	<0.1	0.3	0.1	0.8	<0.1		
Geranium pratense					1.6	2.1	8.9	11.3
Salsola collina			0.1	<0.1			1.4	1.1

The seed bank in the soil of both suffrutex *Artemisia* and ephemeral/ephemeroid species is more than the number of their aboveground plant seedlings, but for annual species the reverse is true. Grazing stress makes the two types of species suffer more and they have no chance to produce seeds, so perennials convert to an agamogenesis strategy. Furthermore, most of the seeds remain dormant in the soil, to give annual plants an opportunity to grow and propagate.

Changes of soil

BULK DENSITY AND SOIL TEXTURE. Soil bulk density and soil textual differences of 0–10 cm, 10–20 cm and 20–30 cm depths between different degradations are shown in Table 14.9. The soil bulk density of each layer in degraded stages is lower than the counterpart layer of the non-degraded stage. Soil bulk density of 0–10 cm decreased gradually, although it did not differ significantly; nevertheless, the average bulk density of 0–30 cm increased with degradation. From non-degradation to overdegradation, the sand content increases steadily from 42.94% to 62.40%, which is a common phenomenon with desertification. The silt content is higher in degraded areas than in non-degraded areas, while the clay content is lower. With rangeland degradation, the surface soil becomes coarse as finer material is washed or blown away.

CHEMICAL PROPERTIES. The results in Table 14.10 indicate the change in the chemical properties in different degraded stages. The soil pH did not differ significantly ($P > 0.05$) at either 0–10 cm or 20–30 cm; the pH at 10–20 cm in overdegradation differed significantly ($P < 0.05$) from other degradations. They were not significantly different ($P > 0.05$) from each other in total salt (TS) content at different levels of degradation.

Soil organic matter (SOM) content had an increasing trend with degradation in each layer; the content in overdegradation was 1.5 times that in the non-degradation site and, in the soil profile, the content decreased steadily with depth. The SOM content of 20–30 cm in non-degradation and 0–10 cm in overdegradation differed significantly ($P < 0.05$) from other degradations within their counterpart layers. The total nitrogen (TN) content of 10–20 cm and 20–30 cm in non-degradation differed significantly ($P < 0.05$) from other degradations in the same layers. In profile, the TN content decreased gradually. The available nitrogen (AN) content of 0–10 cm in overdegradation and 20–30 cm in non-degradation differed significantly ($P < 0.05$) from other degradations in the same layers.

Table 14.8. Species composition of soil seed bank in different layers at different degradation stages.

Degradation phase	Species name	0–5 cm	5–10 cm	10–15 cm	Total
ND	*Eragrostis pilosa*	65	0	0	65
	Seriphidium transillense	293	65	33	390
	Tetracoma quadricornis	0	0	98	98
	Chenopodium album	33	0	33	65
	Amaranthus retroflexus	0	33	0	33
	Malcormia africana	33	0	0	33
Subtotal		**423**	**98**	**163**	**683**
MD	*Ferula ferulaeoides*	0	0	22	22
	Petrosimonia sibirica	0	0	22	22
	Amaranthus retroflexus	130	0	0	130
	Trigonella arcuata	22	0	0	22
	Eragrostis pilos	108	22	0	130
	Seriphidium transillense	22	0	0	22
	Chenopodium album	0	22	0	22
	Tetracoma quadricornis	65	22	22	108
Subtotal		**347**	**65**	**65**	**477**
HD	*Amaranthus retroflexus*	49	0	0	49
	Trigonella arcuata	16	16	0	33
	Eragrostis pilos	49	49	33	130
	Chenopodium album	33	33	16	81
	Peganum harmala	0	16	0	16
	Tetracoma quadricornis	0	16	33	49
Subtotal		**163**	**130**	**81**	**374**
OD	*Petrosimonia sibirica*	39	0	0	39
	Trigonella arcuata	52	0	39	91
	Eragrostis pilos	234	26	0	260
	Peganum harmala	26	0	0	26
	Kochia prostrata	52	0	0	52
	Malcormia africana	39	0	26	65
	Tetracoma quadricornis	52	13	0	65
	Salsola collina	13	0	0	13
Subtotal		**507**	**39**	**65**	**611**

Table 14.9. Comparison of bulk density and soil texture (means ± SE) between different degradation stages.

Item	Depth (cm)	ND	MD	HD	OD
Bulk density (g/m^3)	0–10	1.41 ± 0.01^a	1.39 ± 0.04^a	1.35 ± 0.04^a	1.34 ± 0.02^a
	10–20	1.59 ± 0.01^a	1.44 ± 0.02^a	1.49 ± 0.01^a	1.44 ± 0.02^a
	20–30	1.60 ± 0.02^a	1.34 ± 0.03^b	1.52 ± 0.03^{ab}	1.49 ± 0.01^a
Soil texture					
Sand (%)	0–10	48.06 ± 0.20^a	48.14 ± 0.79^a	46.69 ± 0.57^a	36.88 ± 0.59^b
	10–20	36.14 ± 0.08^b	47.78 ± 1.08^a	49.39 ± 0.35^a	29.07 ± 0.67^c
	20–30	44.32 ± 0.29^b	44.03 ± 0.09^b	63.67 ± 0.23^a	30.63 ± 0.33^c
Silt (%)	0–10	17.33 ± 0.13^c	22.55 ± 0.33^a	17.60 ± 0.62^c	19.50 ± 0.18^b
	10–20	16.82 ± 0.17^c	22.40 ± 0.86^a	17.97 ± 0.43^{bc}	19.03 ± 0.80^b
	20–30	10.92 ± 0.17^d	28.57 ± 0.39^a	12.59 ± 0.33^c	22.12 ± 0.37^b
Clay (%)	0–10	34.62 ± 0.32^b	29.31 ± 0.58^c	35.71 ± 0.95^b	43.32 ± 0.56^a
	10–20	47.04 ± 0.25^b	29.82 ± 0.37^d	32.64 ± 0.14^c	51.37 ± 0.31^a
	20–30	44.76 ± 0.16^b	27.41 ± 0.32^c	23.74 ± 0.32^d	48.12 ± 0.85^a

^{a,b,c,d}Values with the same superscript letter are not significantly different.

Table 14.10. Soil chemical properties (means ± SE) between different degradation stages.

Soil parameters	Depth (cm)	ND	MD	HD	OD
pH	0–10	8.70 ± 0.02[a]	8.80 ± 0.02[a]	8.87 ± 0.06[a]	9.18 ± 0.10[a]
	10–20	8.90 ± 0.01[b]	8.78 ± 0.05[b]	8.91 ± 0.06[b]	9.64 ± 0.05[a]
	20–30	8.92 ± 0.01[a]	8.89 ± 0.10[a]	9.01 ± 0.08[a]	9.57 ± 0.11[a]
TS (g/kg)	0–10	1.06 ± 0.01[a]	0.97 ± 0.08[a]	0.72 ± 0.05[a]	1.57 ± 0.05[a]
	10–20	1.42 ± 0.02[a]	2.13 ± 0.03[a]	1.55 ± 0.6[a]	2.45 ± 0.09[a]
	20–30	0.40 ± 0.01[a]	1.43 ± 0.04[a]	1.22 ± 0.03[a]	2.65 ± 0.06[a]
SOM (g/kg)	0–10	9.17 ± 0.14[b]	12.22 ± 0.06[ab]	12.02 ± 0.42[ab]	18.04 ± 0.10[a]
	10–20	7.86 ± 0.17[a]	9.17 ± 0.07[a]	10.30 ± 0.18[a]	10.66 ± 0.03[a]
	20–30	5.78 ± 0.08[b]	8.77 ± 0.02[a]	10.10 ± 0.03[a]	9.37 ± 0.06[a]
TN (g/kg)	0–10	0.95 ± 0.01[a]	0.78 ± 0.03[a]	0.75 ± 0.03[a]	1.22 ± 0.03[a]
	10–20	0.44 ± 0.01[b]	0.61 ± 0.04[a]	0.77 ± 0.03[a]	0.74 ± 0.01[a]
	20–30	0.35 ± 0.01[b]	0.51 ± 0.01[ab]	0.71 ± 0.03[a]	0.67 ± 0.02[a]
AN (mg/kg)	0–10	23.7 ± 1.0[b]	29.1 ± 1.9[b]	29.7 ± 1.2[b]	52.1 ± 4.0[a]
	10–20	16.1 ± 0.9[a]	19.8 ± 3.6[a]	20.1 ± 1.5[a]	23.1 ± 2.9[a]
	20–30	13.2 ± 0.8[b]	23.5 ± 1.0[a]	31.7 ± 0.7[a]	21.9 ± 3.9[a]

[a,b]Values with the same superscript letter are not significantly different.

14.4.2 Process of rangeland restoration by enclosing and grazing reduction

The result of 4 years of enclosure and reduced grazing pressure on *S. transillense* desert degraded rangeland (Jin and Zhu, 2007; Jin *et al.*, 2007) shows that the dominant plant species change is from *G. bulbifera* + *S. transillense* to *C. arenarius* + *P. sibirica*.

The quantitative characteristics of the main plant species change obviously, except *S. transillense*. After enclosure, the rangeland vegetation coverage increased from 33.4% to 81.2% and aboveground biomass rose from 40.1 g/m^2 to 111 g/m^2. The density, coverage and biomass of ephemeral/ephemeroid plants decreased significantly after enclosing, whereas those of annual plants such as *C. arenarius* and *P. sibirica* increased greatly. The semi-shrubs such as *S. transillense* showed no obvious change (Table 14.11).

After enclosing and grazing reduction, the composition and quantity of persistent seeds in the soil seed bank were more abundant than before. In the unenclosed sampling sites, seeds of five species were identified, with a total density of 359/m^2. In the enclosed sampling sites, seeds of six species were identified, with a total density of 845/m^2, showing an increasing tendency of species diversity and density in the soil seed bank after enclosing (Table 14.12).

After enclosing and grazing reduction, soil total nitrogen, total phosphorus, total potassium and pH value decreased to different extents and organic matter increased (Table 14.13). Without grazing interference, the vegetation was flourishing and absorbing more nutrients from the soil. The salt content also decreased, which was beneficial to the rangeland plants.

Adopting enclosing with grazing bans is suggested for recovery of key perennial vegetation. Enclosing relieves grazing pressure, providing opportunity for the degraded rangeland to rehabilitate, thus promoting natural restoration of the rangeland, which is a simple and effective measure and has been applied widely in the restoration and reconstruction of degraded rangeland. This enclosing with reducing grazing pressure mode may restore community production through renovating annuals, but it does not restore perennials (Guo *et al.*, 2001).

14.5 Ecological Processes of Degraded Rangeland Restoration

14.5.1 Optimizing rangeland use

Xinjiang is vast; great differences exist in the conditions for prataculture from one area to another; therefore, development is unbalanced. Based on the distribution of rangeland resources and the difference in the pratacultural economic structure

Table 14.11. Changes of species composition between enclosing and non-enclosing in *Seriphidium transillense* desert rangeland.

Item	Main plant species	IV	Height (cm)	Coverage (%)	RF	Density (plant/m^2)	Aboveground biomass (g/m^2)
Enclosing	*Ceratocarpus arenarius*	0.53	1.4	57.4	50	3794	78.7
	Seriphidium transillense	0.12	2.6	7.4	49	20	13.4
	Gagea bulbifera	0.10	3.7	6.6	50	164	5.1
	Geranium pratense	0.09	1.1	5.4	46	158	6.1
	Petrosimonia sibirica	0.03	0.6	2.2	13	145	2.1
	Ceratocephalus orthoceras	0.03	0.9	0.0	25	14	0.2
	Altaicum schischk	0.03	1.0	0.8	14	5	2.9
	Trigonella arcuata	0.02	0.8	0.0	20	18	0.4
Non-enclosing	*Gagea bulbifera*	0.28	1.9	14.8	50	330	8.5
	Seriphidium transillense	0.23	2.3	8.1	34	13	12.4
	Astragalus vicarious	0.15	0.7	4.4	20	17	9.5
	Ceratocarpus arenarius	0.08	0.8	0.8	38	35	0.6
	Kochia prostrata	0.06	2.0	1.8	20	6	4.2
	Geranium pratense	0.05	0.5	1.0	21	49	1.4
	Trigonella arcuata	0.05	0.7	1.0	23	10	0.9
	Petrosimonia sibirica	0.03	0.6	0.4	18	21	0.3
	Altaicum schischk	0.03	0.2	0.6	10	2	1.1

IV, important value; RF, relative frequency.

Table 14.12. Comparison of soil seed bank (0–15 cm).

Species	Enclosing quantity (no./m^2)	Non-enclosing quantity (no./m^2)
Petrosimonia sibirica	65	33
Eragrostis pilosa	260	98
Tetracoma quadricornis	195	130
Seriphidium transillense	65	
Trigonella arcuata	195	
Kochia prostrata	65	
Ferula sp.		33
Chenopodium glaucum		65
Total	845	359

and productivity level, a diversified pratacultural production system that is adapted to local situations should be established to take advantage of the regional resource mix (Zhao, 2002b; Xu, 2004; Zhu, 2006).

In accordance with the different physical, economic, social and resources attributes, forest resources, rangeland resources, water and land resources are being managed comprehensively and optimized for use, and different developmental models incorporated with farming and livestock have been generated for the best cost-effective use of the resources. The use of rangeland has been adjusted in terms of the spatial and temporal pattern of the rangeland ecosystem to mitigate the pressure on seasonal pastures, especially the spring/autumn pastures. To restore the damaged rangeland ecosystem, maintenance of ecosystem equilibrium and improvement of pastoral livestock production are theoretically and realistically significant.

Implementing optimization of the rangeland reallocation and operational model is very significant for rehabilitation of the deteriorated rangeland ecosystem. For instance, in a three-season grazing rotation system implemented in Bayinbuluke, the ratio of rangeland area in the cold season (winter pasture):warm season (summer pasture):intermediate season (spring/autumn pasture) is 1:2:1. Specific to the extreme overgrazing problem on summer and winter pastures on the north slopes of the Tianshan Mountains, the original four-seasonal transhumance system of summer–spring/autumn–winter–spring/autumn pastures was adjusted to a three-season rotation system of spring/autumn–summer–spring/autumn, which has mitigated

Table 14.13. Change of soil content in different layers.

Item	Soil layer (cm)	Total N (g/kg)	Total P (g/kg)	Total K (g/kg)	Organic matter (g/kg)	pH	Salt (g/kg)	Bulk density (g/cm³)
Fenced	0–10	0.80	1.20	11.80	13.50	8.73	0.57	1.36
	10–20	0.62	0.96	10.50	8.95	8.93	0.38	1.52
	20–30	0.58	0.93	9.40	8.50	9.08	1.66	1.43
Unfenced	0–10	0.74	1.30	12.95	12.48	8.75	1.18	1.37
	10–20	0.78	1.11	13.00	8.43	8.97	1.02	1.51
	20–30	0.58	1.06	11.80	8.40	9.14	0.92	1.46

pressure on the natural pastures and optimized the operational model for pastoral livestock production. Through restructure of the cropping industry and animal population, increased unit production and reallocation of the natural rangeland, the ecological, economic and social benefits have been increased considerably. In recent years, along with the progress of the herders' settlement programme, a new pastoral operational model of warm-season grazing and cold-season pen-feeding plus grazing on the north slopes of the Tianshan Mountains was proposed specifically for the post-settlement regime. In this way, local grass–animal balance was realized and the eco-environment was protected and improved.

14.5.2 Undertaking rangeland enclosure and improvement

After entering the new century, the projects of converting farming into grass, rangeland fencing and reconfirmation of basic rangeland, etc., have been implemented broadly. For slightly degraded rangeland, seasonal enclosure was conducted to create the opportunity for self-regeneration. For seriously degraded rangeland, in addition to enclosure, the measures of re-seeding, fertilizing, ploughing and reconstruction of vegetation have been undertaken. For those fragile rangeland ecosystems, e.g. desert rangeland, grazing bans have been implemented. It is expected that the implementation of these measures will restore the large-scale degraded rangeland (Liu, 2002; Liu *et al.*, 2007a,b).

The evidence showed that, after the enclosure and improvement, coverage, height and biomass were increased significantly, species composition changed greatly and high-quality forage species and their yield increased as well.

14.5.3 Implementing the herders' settlement programme

The objectives of the herders' settlement programme are to change the conventional pastoral production model to restore the deteriorated rangeland ecosystem fully. The Xinjiang regional government proposed in 1986 that the pastoral practice had to be changed from transhumance towards settlement. Again, in May 1996, it declared specifically its commitment to herders' settlement as the focal task to change the conventional livestock production model. Since 1987, the settled and semi-settled herders' households account for over 80% of the total households. After the settlement, each herder's household was allocated a certain area of artificial forage land and certain infrastructure and service facilities were constructed in the settlement communities. A new production model of grazing in the warm season and pen-feeding in the cold season has been established. It is expected that the herders' living environment will improve, and their scientific and cultural attributes and living standard will improve also. Resettlement symbolizes a historical initiative of converting traditional transhumance pastoral practice to the modern livestock production model in Xinjiang.

Although the herders' settlement programme has made certain progress in Xinjiang, there are still some aspects that must be improved and perfected. Currently, the major issues concerning the herders' settlement are lack of consolidated awareness of the essentials of the herders' lifestyle, the quality of settlement is low in general, productivity is poor and the phenomenon of overgrazing on the natural pastures has not yet been changed fundamentally. Poor infrastructure development is often blamed. According to an investigation conducted by Xinjiang's Department of Animal Husbandry in 2000, less than 50% of settled herders' households satisfied the basic standards, which were:

- 'three accesses – access to water supply, road and electricity';
- 'four possesses – possess forage land, house, warm shed and forestland'; and
- 'five availabilities – school, grocery, training centre, clinic and technical service centre'.

Constraints under which herders operate are:

- limited available land for artificial forage production;
- poor soil fertility and limited water;
- low level of technical knowledge and skill required to plant forage crops; and
- low operational efficiency and poor returns.

All of these constraints lead to a situation where the people are settled but their animals are not. The situation of rangeland degradation has not been changed fundamentally. Therefore, the herders' settlement programme needs to be improved further (Sulayman and Mansur, 2002).

14.5.4 Implementing ecological displacement project

Based on the reality of rangeland degradation in Xinjiang and in view of the sustainable use of resources, Professor Xu Peng of the Xinjiang Agricultural University proposed the Rangeland Ecological Displacement Theory in 2000 to combat rangeland degradation and pastoral renaissance (Xu, 2002). It aimed at controlling rangeland degradation and restoration of the rangeland ecosystem through developing artificial pastures, implementing *grazing exclusion, resting and reduction* and other measures on the natural pastures to enhance self-regeneration and rehabilitation capacities of the natural rangeland. If successful, eventually it would protect and improve the rangeland eco-environment, promote great development of livestock production and realize harmonious and comprehensive development of rangeland resources–livestock production–eco-environmental development. The government of Xinjiang has incorporated the Ecological Displacement Project into Xinjiang's future economic and ecological programming as a key project. The specific measures of the Ecological Displacement Project are below.

Ecological restoration and maintenance of natural rangeland centre

Implement *converting grazing into grass* and *grazing exclusion, resting and reduction* on the natural rangeland. Actualize a measurable orderly reallocation system for rangeland use, displace degraded rangeland for rehabilitation and secure the rangeland ecosystem for long-run maintenance in order to realize long-term use of rangeland resources (Li *et al.*, 2004).

Settlement–grazing rotation system

An alternative to the transhumance system is to allow grazing in the warm season and confine the animals to pens in the cold season. The objective is to increase livestock products and herders' income by improving the nutritional efficiency (transforming forage to animal products such as meat, wool/mohair or milk). Moreover, a modern pastoral livestock production system can be established to facilitate restoration of the natural rangeland ecosystem and maintenance of the system to promote development of the pastoral area.

Novel productivity backup

Implement *converting grazing into grass*, develop artificial forage production for better supply. Focus on artificial forage plus hay field improvement for high yield and agricultural by-product processing and utilization for establishing a new forage/feed production system and supply capacity for compensating the lost rangeland productivity due to converting grazing into grass, ensuring the implementation of *convert grazing into grass* and *settlement–grazing rotation system*, and realizing grass–livestock balance.

14.5.5 Implementing ecological migration programme

For those pastoral areas where there is a shortage of rangeland resources and their ecological, productivity and living environments and conditions have deteriorated, including some areas in Hetian, Kashigar and Kezilesu Prefectures, pastoral business activities will be abandoned. Alternative economic options will be found through implementing an ecological migration programme to help solve the issues of environment deterioration of the rangeland and poverty reduction (Li *et al.*, 2004).

14.5.6 Monitoring rangeland loading rate

A dynamic rangeland resources monitoring and comprehensive management information system that is 3S (RS, GPS and GIS) technology-based is established for providing a basis for the examination and ratification of loading rate and reasonable utilization of the seasonal pastures (Zou *et al.*, 2003). Such a system has been established already in some counties with a large area of rangeland resources, and monitoring of the rangeland loading rate has been highly regarded and broadly attended by the relevant agencies.

14.5.7 Implementing advantageous policy adjustment

Establishing compensation system for rangeland displacement

To encourage herders to establish their artificial pastures by providing fiscal support for the displaced rangelands through loans, tax reductions and so on for maintaining grass–animal balance and mitigating the pressure on the natural pasture.

Establishing rational population management system

To formulate a policy for management of the populations engaged in pastoral business in terms of quantity, quality and reallocation. First, to control the populations engaged in pastoral business to a certain limit to reach a harmonious proportion among the quantity of people, livestock and forage; realized livestock quantity is decided by the loading capacity of the pasture and quantity of the people is decided by the available pasture size. Secondly, to popularize public education of the herders broadly to improve their productive and operational capacity; meanwhile, that will also benefit the herders with more alternative business opportunities during the industrial development. And, thirdly, to divert the population and handle the issue of ecological migration properly (Guo *et al.*, 2001).

References

Adilhan, Y. (2004) From nomadic life to settlement: a great change in the traditional life and productive way of nomadic people. *North West Ethno-National Studies* 4, 132–140, 166.

Dai, X.G., Ren, Y.Y. and Chen, H.W. (2007) Multi-scale feature of climate and climate shift in Xinjiang over the past 50 years. *Acta Meteorologica Sinica* 65(6), 1003–1010.

Eli, E., Aximu, A. and Rehemudula, A. (2002) Over-used land desertification in Xinjiang and its prevention and harness measures. *Territory and Natural Resources Study* 1, 26–27.

Gao, N. (2002) Study on ecological replacement of grassland in Xinjiang. *Journal of Xinjiang Agricultural University* 25, 70–72.

Guo, X.Z., Zhao, D.Y. and Zhang, J.L. (2001) The counter measure of comprehensive treatment to recover environment of Xinjiang grassland. *Grass-feeding Livestock* 3, 43–45.

Jin, G.L. and Zhu, J.Z. (2007) Discussion on rangeland degradation. *Grassland and Turf* 5, 79–82.

Jin, G.L., Zhu, J.Z. and Chen, L.N. (2007) The analysis of plant community evolvement tendency on degraded *Seriphidium transillense* desert rangeland in Yili. *Prataculture Science* 24(10), 26–30.

Ju, Q., Nurbay, A. and Pan, X.L. (2004) Degeneration and strategies for management of grassland of Xinjiang. *Environmental Protection of Xinjiang* 26(3), 43–46.

Lan, Y.C., Shen, Y.P. and Wu, S.F. (2007) Changes of the glaciers and the glacier water resources in the typical river basins on the north and south slopes of the Tianshan Mountains since 1960s. *Journal of Arid Land Resources and Environment* 21(11), 1–8.

Li, H.M., An, S.Z. and Zhu, J.Z. (2004) Optimization model for stock raising management on grassland after settlement of herdsmen. *Pratacultural Science* 21(5), 58–62.

Li, Z.Q., Han, T.D. and Jin, Z.F. (2003) A summary of 40-year observed variation facts of climate and glacier No. 1 at headwater of Urumqi River, Tianshan, China. *Journal of Glaciology and Geocryology* 25(2), 117–123.

Liu, H.L., Zhu, J.Z. and Jin, G.L. (2007a) Division on degraded successional series of *Seriphidium transillense* desert grassland. *Xinjiang Agriculture Sciences* 44(2), 137–141.

Liu, H.L., Zhu, J.Z. and Jin, G.L. (2007b) Analysis on plant diversity in different degraded stages of *Seriphidium transillense* desert grassland. *Xinjiang Agriculture Sciences* 44(56), 632–636.

Liu, Y.L. (2002) Thinking and constructive countermeasure of grassland ecological replacement in Xinjiang. *Xinjiang Agricultural Sciences* 39(4), 195–199.

Mansur, S., Sulayman, A. and Zhou, J. (2002) The sustainable development of grassland resources and livestock husbandry of Xinjiang. *Pratacultural Science* 19(4), 11–15.

Sulayman, A. and Mansur, S. (2002) The sustainable development of grassland resources and stock farming of Xinjiang. *Journal of Xinjiang Normal University (Natural Sciences Edition)* 21(1), 62–65.

Wu, S.X., Zhou, K.F. and Liu, Z.X. (2005) Study on the temporal and spatial dynamic changes of land use and driving forces analyses of Xinjiang in recent 10 years. *Arid Land Geography* 28(1), 52–58.

Xu, P. (1993) *Xinjiang Grassland Resource and Utilization.* Xinjiang Health Technology Publisher, Urumqi, China.

Xu, P. (2002) Developing superiority of regional environment and strengthening protection and construction of ecology. *Journal of Xinjiang Agricultural University* 25, 1–3.

Xu, P. (2004) Characteristics of Xinjiang grassland resource and treatment of degeneration. *Environmental Protection of Xinjiang* 26, 34–37.

Xu, Q. and Zhao, J.B. (2007) The relations between sandstorm activities and climatic conditions in Xinjiang. *Journal of Arid Land Resources and Environment* 21(12), 116–120.

Zhao, W.Y. (2002a) Deterioration, causes and control strategies of grassland resources in Xinjiang, China. *Pratacultural Science* 19(2), 19–22.

Zhao, W.Y. (2002b) The desert steppe degeneration and strategy for improving in Xinjiang. *Grassland of China* 24(3), 68–72.

Zhu, J.Z. (2006) A reflection on the development of prataculture in Xinjiang (2). *Xinjiang Agricultural Sciences* 43(4), 281–285.

Zou, Y.R., Zhang, Z.X. and Zhou, Q.B. (2003) Spatial pattern and its analysis of China's grassland change in recent ten years using remote sensing and GIS. *Journal of Remote Sensing* 7(5), 428–432.

Part IV

The Future – How to Prevent the Next Major Degradation Episode

The three chapters in this part examine the importance of land tenure arrangements, property rights and institutional arrangements in the cycle of rangeland degradation and recovery. The initial nationalization of China's grasslands undermined the legitimacy of local customary rights systems over the use of the range. As the central and local government failed to encourage mutual cooperation, the management of grasslands evolved into an open-access system. Two sets of factors were most important in creating this situation – property rights structures and institutional arrangements, including effective mechanisms to monitor changes and react accordingly.

Rangeland degradation in China's vast pastoral region cannot be blamed solely on population growth, overgrazing or reclamation of marginal land. Rather, it has its roots in the failure of successive Chinese governments to create conditions under which collective management could be effective.

The last chapter assesses whether we can prevent the next degradation episode. The historical episodes described in the book represent a failure to manage for the extreme climate variability that characterizes north and west China's vast arid rangelands. Thus, they represent a historical 'test bed' for our current scientific understanding of rangelands and government and land-user responses. The challenge is to overcome decades of neglect of rangelands as ecosystems and their exploitation for economic gain, of inappropriate policy interventions (with unintended consequences) against a background of negative attitudes towards herders.

We outline some views about the benefits of using pastoral rangelands in a sustainable way and discuss some likely consequences of allowing further degradation to occur. The long-term future of traditional nomadic pastoralism is discussed in the light of environmental, economic, social and political changes.

There are still a few options open to dryland inhabitants, but these are being reduced as each year passes. This chapter tries to set some parameters for future generations to ensure intergenerational equity, while at the same time allowing the present generation to prosper, or at least survive.

15 Land Tenure Arrangements, Property Rights and Institutional Arrangements in the Cycle of Rangeland Degradation and Recovery

Adrian Williams,[1] Meiping Wang[2] and MunkhDalai A. Zhang[3]
[1]*Formerly Centre for the Management of Arid Environments, Australia;*
[2]*Gansu Agricultural University, Lanzhou, China;* [3]*Chinese Academy of Social Sciences, Hailar, China*

Synopsis

This chapter examines the importance of land tenure arrangements, property rights and institutional arrangements in the cycle of rangeland degradation and recovery. Rangeland degradation in China's vast 'Three Norths' region cannot be blamed solely on population growth, overgrazing or reclamation of marginal land. Rather, it has its roots in the failure of successive Chinese governments to create conditions under which collective management can be effective. The initial nationalization of China's rangelands undermined the legitimacy of local customary rights systems over the use of the range. This chapter examines options for the future and weighs into the debate about privatization of rangelands.

Keywords: privatization; grazing user rights; responsibility system; legal; regulatory environment; enforcement; cultural; nomads; policy; customary rights

15.1 Land Tenure Arrangements on Rangelands (Grasslands, Steppe and Shrublands) in China

Prior to 1949, nomadic pastoralism had been practised sustainably on the arid and semi-arid rangelands of Inner Asia for over 2000 years. It is believed that yaks were domesticated on the Tibetan Plateau around 4500 years ago. Building of the Great Wall commenced some 2000 years ago, as much to control nomadic societies as military incursions (Miller, 2002). The concepts and practices of nomadic cultures, held by the whole community, were to respect the natural ecology, to be systematic in the use of rangeland resources, to protect the rangeland and to give higher priority to rangeland condition than to livestock productivity (Qingwu, 2001). For Mongolians, the core idea of the nomadic culture was centred on an awe of life, a reverence and respect for nature and harmonious coexistence between humans and nature (Gegenguva, 2002). Such was the feeling for their environment that herdsmen in the 1930s were prepared to lay down their lives to protect rangeland from being sold to warlords for use under cultivation (Zhang *et al.*, 2007).

Livestock mobility was crucial to maintaining rangeland condition and being able to minimize the effects – on rangeland, livestock and people – of variability in seasonal conditions. The control of the land was vested in tribal leaders or Buddhist monasteries, with strict codes of acceptable practices and seasonal use of rangeland areas.

Life was hard. The frequent movements of yurt (*ger*), livestock and essential goods, with the concomitant search for and investigation of water sources, were labour and energy sapping. Diet and nutrition were poor. Overcoming natural disasters (e.g. rainstorms, floods, sandstorms) was part of a life survived solely on self-reliance, as was overcoming threats from wolf packs, plagues of small mammals and the plague ('Black Death') itself. Livestock management required avoiding areas where poisonous plants were prevalent and settling conflicts amicably with neighbouring people, while avoiding getting livestock mixed. The average lifespan of herders in many areas in those days was 36 (Wang *et al.*, 2006). Yet here was a system of land management used sustainably across Inner Asia for thousands of years, and still used in some countries today.

Following the foundation of the People's Republic of China (PRC) in 1949, the State Council in 1953 considered pastoral production in Inner Mongolia, Suiyuan, Qinghai and Xinjiang Province and called for a 'rational utilization' of the rangelands. The Council regulated that pastures should be protected and prohibited the clearing of any more land for cultivation. However, this positive step in the protection of rangelands was accompanied by the edict that livestock development should be the main focus in semi-agricultural or semi-pastoral areas. This was a shift from putting the condition of the rangeland first and foremost.

During the period 1953–1957, agriculture was organized into collectives with the aim of improving productivity in order to feed the growing population and to raise capital for the country. In this time, when nearly everything was nationalized, all the pastures formerly vested in tribes and Buddhist monasteries were converted to the ownership of state cooperatives. Throughout the country, as in areas like Su'nan County in Gansu Province, the rangeland's carrying capacity was thought to be greater than the number of grazing animals and grazing pressure was not particularly heavy. However, that was to change.

The implementation of the 'Three responsibilities – one reward' system in the late 1950s and early 1960s gave herders the incentive to invest in production. The three responsibilities were to: (i) follow instructions within the required time frame; (ii) contract for a fixed output or production quota; and (iii) provide all the investment. The reward was granted by local government to herders who exceeded their production quotas. The simplest way to guarantee to exceed the production quota was to keep more livestock. Thus, livestock numbers rose sharply.

Government focus remained on improving production while protecting the rangeland. In 1958, the national programme for agricultural development placed priority on rangeland protection, but also on improving grass productivity through livestock breeding and finding new water resources. The stage was set for the Ministry of Agriculture and related departments dealing with pastoral matters at the provincial level to set up institutions to undertake rangeland protection, technological innovation and extension to open up the rangelands to modern planned management.

In 1963, the State Council ratified the 'Regulations on policies regarding the minority ethnic groups and people's communes in pastoral areas'. Once again, rangeland protection was seen to be a most important aspect, involving prevention and control of moving sand, rodent control and water resource protection. Developing water conservancy facilities was seen as a priority for improving rangeland. All of this was to be paid for by the central government, indicating how important it judged pastoral development to be. The increase in water resulted in further increases in livestock numbers.

In Inner Mongolia, according to a figure by Dalintai (2008), the livestock population (converted to and measured in sheep units) rose from around 10 million in 1949 to a peak of around 54 million in 1969, and remained around 50 million until 1989.

Problems were surfacing throughout the 1970s. Some collectives experimented with rotational or seasonal grazing, but found that they needed formalized and demarcated boundaries with neighbouring collectives to prevent them from grazing the areas that were being rested. Fences started to appear and subdivide the rangelands.

In 1979, the central government passed the 'Resolution on several problems related to the speeding up of agricultural development'. This stressed the need to improve rangeland management. The resolution included the building of more watering facilities, improvement to livestock breeds and more rational use of rangeland through employing rotational grazing in order to *increase* carrying capacity. As realized by experience (see above), such rotational grazing relied on the need for no neighbour to graze the rangeland a collective was resting, and thus the grazing user rights (GUR) for communes and production teams came into being and were defined. In so far as these measures permitted the practice of seasonal grazing and seasonal resting of rangeland within the tenure structure of the time, it was probably a good thing, a first official sign of acceptance of the wisdom inherent in the nomadic system of rangeland and livestock management.

In the 1980s, land tenure changed dramatically. In 1982, the central government changed the rules of pastoral tenure, with the aim of improving rangeland management and protecting rangeland condition, by handing over responsibility for rangeland management to individual herding households. Households were given a form of leasehold tenure over areas of natural rangeland through the Household Responsibility Contract System. Thus, livestock became privately owned by the herders and then areas of rangeland were allocated to individual herding households as GUR. Land allocation was based on family size and the number of livestock the household owned at the time. Boundaries were mapped and recorded, with a copy presented to the household. The leasehold tenure was set to last for 30 years. This move was unprecedented across the whole of the rangelands of Inner Asia. It led to further subdivision of the rangelands and further loss of livestock mobility. As we shall see later in this chapter, while the Household Responsibility Contract System may have been well considered and developed at the central government level, there were inequalities in its implementation at the local level – for a range of reasons – which were likely to have contributed to the degradation of the rangeland.

In 1983, it was the turn of areas of unsustainable cultivated land in the pastoral zone. Central government policy required that such cultivated land be returned to rangeland, or converted to forestland or, in some suitable sites, to fish farming. This policy had little or no effect at the local level, where the importance of the environmental degradation caused by land clearing was not recognized – or for other reasons. Land fertility continues to be degraded by clearing for cropping, and wind erosion on cleared land continues to cause deflation, leading to desertification, mobile sand dunes and dust storms.

Severe penalties were promised by the central government and State Council in 1984 for people damaging forestry or rangeland. This had an impact on rangeland rehabilitation in western China, but had little effect on the removal of edible or medicinal plants from rangelands, or on the use of plant material for fuel for cooking and heating.

The Sixth Standing Committee of the National People's Congress promulgated the *Rangeland Law* in June 1985. This law was 6 years in the making by relevant departments. It made provisions for the protection of the rangelands (grasslands, steppe and shrublands) against deliberate destruction and unsanctioned clearing of vegetation and farming the rangelands. Culprits were responsible for the restoration of the rangeland and were liable to fines in the case of serious damage. The difference from the legislation of the 1970s was that the Rangeland Law was given implementation power, through the establishment of Rangeland Monitoring and Management Departments at different levels of government to enforce the law. This had some effect on curtailing activities that led to rangeland degradation. However, it was a number of years before natural rangeland protection procedures were defined by law, along with monitoring and measuring techniques and departmental responsibility at each level of government. These were passed into law on 1 March 2003, following adoption as the amended version of the 1985 Rangeland Law by the Standing Committee of the National People's Congress in December 2002.

Despite the new legislation, Brown (2001) reported a decision in Beijing in 1994 that required that all cropland consumed by urban development in fast-growing provinces should be offset by land reclaimed (cleared for cropping) elsewhere. Hence, provinces that were losing cropland could pay other provinces to clear and cultivate new land. Such a policy would exacerbate the problems associated with the clearing of rangeland for cropping and would be at odds with the government's position on land clearing for cropping in rangeland areas.

It is clear that the changes to land tenure and land management since 1949 have allowed massive increases in livestock populations in the rangelands of northern China. In Inner Mongolia, for example, the livestock numbers (as sheep units) increased fivefold over 1949 numbers and maintained the fivefold increase from 1967 until 1989 (Dalintai, 2008). This occurred in the formative years of the PRC and helped to feed and clothe the rapidly expanding human population, as well as generate income from abroad, as in the case of cashmere fibre export. These achievements should not be forgotten.

But what has been the cost of this so-called 'rational utilization' of rangelands?

15.2 Some National Effects of Property Rights Structures and Institutional Arrangements

The 2004 Report on the State of the Environment in China (SEPA, 2004) stated that 90% of the available natural rangelands of China were suffering various degrees of degradation, with an annual growth in degraded area of 2 million ha (Mha). Reasons for this were given as: (i) the trend in overgrazing, which has not been curbed; (ii) rangeland damage resulting from irrational development, industrial pollution and plagues of insects and rodents; and (iii) frequent illegal collection of herbs and other commercial plants. Understandably, no mention was made of the effects of underlying policies and legislation.

In the same year, Hawley (2004) reported that, from the mid-1990s to 2000, 3559 km^2 of China had turned into desert each year (compared with 2180 km^2/year in the 1980s and 1616 km^2/year during the 1970s). In total, China has lost 418,000 km^2 of land to desert since the 1950s.

Spreading deserts are also displacing people. Brown (2004) reported that in China, where the Gobi Desert is growing by 10,400 km^2/year, the refugee stream is swelling. Chinese scientists report that there are now desert refugees in three provinces – Inner Mongolia, Ningxia and Gansu. An Asian Development Bank preliminary assessment of desertification in Gansu Province has identified 4000 villages that face abandonment. Previously, Brown had reported that the northern half of China was literally drying out as rainfall declined and overpumping depleted aquifers. Water tables were falling almost everywhere, altering the region's hydrology. Based on 30 years of US satellite data, it appeared that thousands of lakes in northern China had disappeared (Brown, 2001). (In some areas, groundwater levels are falling and salinity is rising due to clearing and reduction in evapotranspiration.)

Clearly, the land-use policies and practices since 1949 have provided only a short-term, unsustainable windfall in production, but have 'mined' the rangeland resources to do it. Today, livestock grazing numbers are little different from what they were under traditional nomadic herding management. A key factor that has been lost in the non-traditional ways of managing the rangeland is livestock mobility. At no stage in the development of rangeland policies were the land managers, the herders, consulted.

15.3 Some Effects of Property Rights Structures and Institutional Arrangements at the Grass-roots Level

This chapter has dealt so far with the macro settings of livestock and land tenure and management. By going down to the village level, we can gain an understanding of the effects of these settings on individual households and learn much about the system and how it might be improved.

15.3.1 Inequitable land allocation contributing to high stocking rates, degradation and poverty

The government changed livestock and land tenure systems in an attempt to increase the productivity of rangelands. In addition, the centrally controlled economy changed to a market-based economy from 1978. Goals of satisfying burgeoning domestic consumption and increasing exports (e.g. cashmere fibre), while praiseworthy, led to significant increases in livestock numbers and a concomitant increase in rangeland degradation.

The change from commune and team-based production to the Household Contract Responsibility System in the 1980s, in line with the change to a market economy, might have proceeded logically in the order:

1. Extension personnel trained and in place to advise on individual responsibility for rangeland and livestock management.
2. Estimates made of rangeland carrying capacity according to landscape, soils, vegetation condition, mean annual rainfall and economic factors such as distance from markets and suppliers. Rangeland monitoring sites installed.
3. Rangeland allocated to herding households according to estimates of rangeland carrying capacity and livestock numbers required to make a satisfactory income.
4. Livestock allocated to households based on the number of household members.

However, these steps tended to proceed in reverse of the order outlined above. The change from team ownership to household ownership of livestock required that livestock be allocated to households as the first step. Skewed distributions or redistribution of livestock through trade led to skewed distributions of grazing land under the GUR system.

The land area required by a household varies according to the productivity of the land – affected by climate, soils and vegetation. However, these were not the criteria on which initial land allocations were based. According to Taylor (2006), when households were issued with 30-year user rights leases, GUR areas were calculated for households on the basis of household population and livestock numbers at the time. In some cases, land allocation was based solely on current livestock numbers. Households with both large families and a large number of animals received larger initial allocations. Smaller households that were politically well-connected and owned large herds received priority. Small households with few stock received small allocations. Yet, in time, households with small land allocations that initially had low livestock numbers increased the size of their flocks or herds in attempts to rise out of poverty and respond to local government calls to fulfil production quotas.

Figure 15.1 provides an example from Hatuhuduge Gacha, Alxa League, Inner Mongolia. The data were obtained from the Economic Management Station of the Bureau of Agriculture and Animal Husbandry, and from herders. In the analysis, households were grouped according to the stocking rates they employed in 2001 (*x* axis). The mean allocation of land to households within each of these categories of

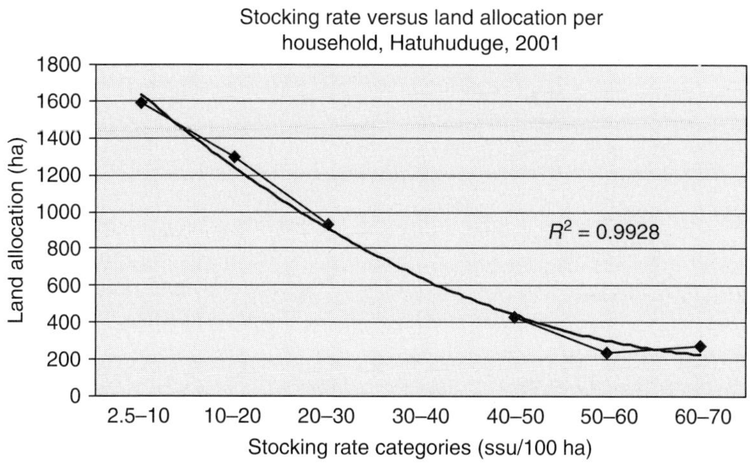

Fig. 15.1. Effect of land allocation on stocking rates (expressed as small stock units/100 ha) in Hatuhuduge Gacha.

stocking rate was calculated (*y* axis). (There were no households within the 30–40 small stock units/100 ha category.) Clearly, herding households with smaller land allocations now employ the highest stocking rates.

When categories of stocking rate were compared with land allocation per individual within the households at Hatuhuduge, a similar pattern emerged to that in Fig. 15.1. Changes in household populations since the land was allocated mean that household population as a basis for land allocation no longer holds and, while included initially for no doubt laudable reasons, is not a long-term basis on which to base land allocation.

It is well accepted that uncontrolled high stocking rates lead to overgrazing, which in turn leads to rangeland degradation. Hence, the system used to allocate land can be seen as a contributor to current degradation problems where land allocations have been inadequate to avoid the need to employ high stocking rates.

Under the current GUR system, the financial returns of households within a community could be used as a guide to what might be an adequate land allocation. Figure 15.2, also from Hatuhuduge, shows the mean financial figures for the households in each of seven categories of GUR allocations. Clearly, households with low GUR allocations have low net returns, while households above a GUR allocation of 1000 ha have higher net incomes. (The highest land allocation category contains a sample size of one and the high costs in this category are thought to be due to major purchases in the sampling year (e.g. fencing).)

Mean household annual net incomes double between land allocations of 200–500 ha and 1000–1500 ha and rise by 50% between land allocations of 500–1000 ha and 1000–1500 ha.

A similar pattern appears when the financial returns per individual within households are compared with land allocation on a per capita basis. Increasing the land allocation per person from 100–200 ha to 200–300 ha increased annual per capita net income by around 500 RMB, or 25%. Larger land allocations maintained this financial advantage.

Thus, apart from affecting stocking rates and grazing pressure, land allocation can be seen as influencing household financial returns and levels of poverty. Similar examples were found in other gachas within Alxa League.

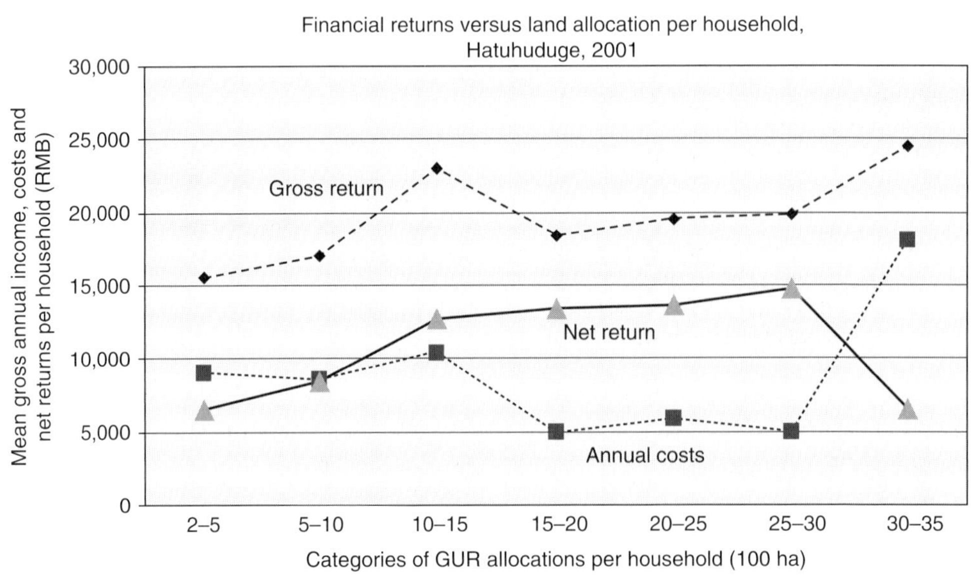

Fig. 15.2. Effect of land allocations on household financial returns, Hatuhuduge Gacha.

If the current system of land tenure is to continue, the data presented from Hatuhuduge support the notion of a minimum size for GUR allocations. The sizes of such allocations will vary according to livestock carrying capacity.

Financial returns will vary with market prices and seasonal conditions. However, in the Hatuhuduge example, a minimum household land allocation of 1250 ha and/or a minimum per capita land allocation of 300 ha would have a strong influence on reducing stocking rates (Fig. 15.1) and on increasing mean household and individual incomes (Fig. 15.2).

Such a system for calculating a first approximation of adequate land allocation in different communities with different climates, soils and vegetation could be used across the rangelands of China. The calculations are based on the household returns collected by the government every 6 months and held at the Banner/County Economic Management Stations of the Bureau of Agriculture and Animal Husbandry. Thus, the required data are already available.

It is important to note that there is sufficient land at Hatuhuduge to make it possible to reallocate a minimum of 1250 ha to each household. If the GUR system is to continue, all land allocations need to be of sufficient size to allow households the ability to run 'viable' herding enterprises without resorting to overstocking the land. Where land allocations are inadequate, there needs to be a fair and equitable reallocation of land. Unless these conditions are met, continued land degradation is inevitable. The question of land reallocation is addressed later in this chapter.

15.3.2 Communities rather than households as the administrative unit – little incentive for good land management

As in the days of the collectives, communes and production teams, the administrative unit in terms of rangeland management remains the administrative or natural village. This has serious consequences for the condition of the rangelands today. In attempts to overcome rangeland degradation or to be seen to react quickly to orders from higher levels of government to address degradation problems, local governments are applying grazing bans across whole communities. A grazing ban was imposed on the whole of Ningxia Province by its provincial government in the early 2000s. In such communities, not all households have degraded their rangeland to the point where a grazing ban is warranted or justified. (It appears that it is often the households with inadequate land allocations that have the worst of the degraded land.) Thus, while the community rather than the individual household remains the administrative focus for government, there will be little or no incentive or reward for managing rangelands well.

15.3.3 Grazing bans and fragmentation of communities

Grazing bans usually result in each household having to sell all their livestock, often at below market prices to opportunistic traders. While the herding households may be allowed to remain on their allocated land, and the government provides an allowance of around 18,000 RMB/year (depending on which part of the country) for one member of the household to manage and improve the rangelands, this is a much lower income than that to which many households would have been accustomed. Thus, family members need to travel away to seek employment, since employment opportunities on the rangelands are few. Thus, grazing bans result in the fragmentation of communities, and particularly communities of ethnic minorities.

Compare this with a situation where the condition of the rangeland of each individual household is assessed and grazing bans, fines or orders to rest or repair areas of country are imposed as required on a household-by-household basis. Thus poor land management could be addressed and good land management fostered and rewarded (through a lack of grazing bans, etc.). With modern satellite analysis, GIS and GPS positioning in the field, assessment of the household rangeland condition is highly practical (Chapter 16). The increased cadre of well-trained and motivated extension staff that would be required would cost a fraction of the economic losses caused by present degradation.

15.4 Other Factors that Have Affected Rangeland Condition

15.4.1 The loss of livestock mobility – a crucial factor in sustainability of rangeland grazing

A comparison between pastoral systems and land degradation across sites in Inner Asia that were broadly similar in terms of landscapes and climates identified that high stocking rates did not lead automatically to rangeland degradation. It appeared that, in steppe case study sites, the amount of livestock mobility was a better guide to the severity of reported degradation of pastures than stocking rate (Sneath, 1999). Areas that had retained livestock mobility within the pastoral practices suffered much lower levels of rangeland degradation, even at higher stock numbers. Case study sites in Inner Mongolia (Haragant in the north-east of Hulunbeier and Chingelbulag in northern central Xilingol) that supported lower stocking rates but had no livestock mobility (data collected at the banner level) exhibited worse rangeland degradation than other study sites with higher stock numbers but retained livestock mobility. Sneath (2000) goes on to say that the results of the study support a widespread perception among herders that a low amount of livestock movement is a major cause of pasture degradation.

The allocation of grazing land to individual households in China through a Household Contract Responsibility System (GUR) has never been attempted anywhere else in Inner Asia. Let us reiterate, livestock mobility appears to be a crucial factor in reducing or avoiding rangeland degradation. Livestock mobility is hampered or removed completely under the GUR system.

15.4.2 The change to a market economy and using financial returns to assess carrying capacity

Since 1978, livestock production on rangelands has been changing from a system of self-sufficiency and barter to a market-driven economy. Quota systems have been employed by local governments, along with rewards to households for reaching quotas as a way to boost production (under the 'three responsibilities – one reward' system). Quotas have been monitored according to the numbers of livestock turned off or amount of fibre (cashmere and wool) produced. In order to minimize the risk of not meeting quota requirements for livestock or fibre, herders have tended to take the easier option of carrying large flocks rather than seeking methods to increase productivity. The lack of an active extension service in some areas has not helped this situation.

Faced with lower net income and decreasing margins at high stocking rates, herders are inclined to increase livestock numbers further, thereby making the situation worse. Data from Gong Dalai Gacha illustrate this point.

But, first, we must explain the apparent differences between Hatuhuduge and Gong Dalai Gacha. At Hatuhuduge, we have seen that stocking rates decrease and financial returns increase with increasing land allocation. At Gong Dalai, financial returns also appear to increase with increasing land allocation. However, at Gong Dalai, we see a contrary trend of increased returns at higher stocking rates. This is explained by the facts that land at Gong Dalai Gacha has been allocated more equitably and the rangeland is of a more productive type. The differences in minimum, maximum and mean land allocations at Hatuhuduge and Gong Dalai and the differences in land type are given in Table 15.1.

Figure 15.3 compares stocking rate categories with gross annual income and annual costs. As the stocking rate increases, the separation between gross income and costs increases to a maximum, achieved at the second highest stocking rate (20–24 sheep equivalents/100 ha). Beyond this, at the highest stocking rate (16–20 sheep equivalents/100 ha), this separation between income and costs decreases. The separation is referred to more normally as net income and the fall in net income at the highest stocking rate is seen more easily in the graph of net income in Fig. 15.4, to which a trend line has been fitted.

This effect of diminishing net returns with increasing stocking rate is to be expected (Wilson, 1986). As stocking rates increase, pasture utilization increases, ultimately to a point where there is insufficient pasture to satisfy the nutritional requirements of the grazing animals. When that point is reached, individual animal productivity starts to fall. However, there is a trade-off between individual animal performance and productivity

Table 15.1. The differences in minimum, maximum and mean land allocations and the differences in land type in Hatuhuduge and Gong Dalai Gacha.

	Hatuhuduge Gacha	Gong Dalai Gacha
Number of households	22	21
Minimum GUR allocation (ha/household)	240	764
Maximum GUR allocation (ha/household)	3180	3938
Mean GUR allocation (ha/household)	1247	1701
Major rangeland types	Semi-stable sandy desert Hummock sandy rangeland (small areas of desert rangelands)	Desert rangelands (small areas of gobi and low mountains)

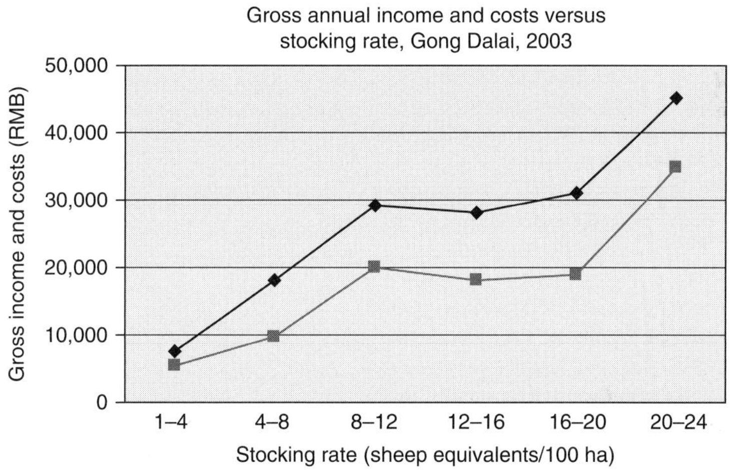

Fig. 15.3. The mean gross annual income and annual costs for households within different stocking rate categories in Gong Dalai Gacha.

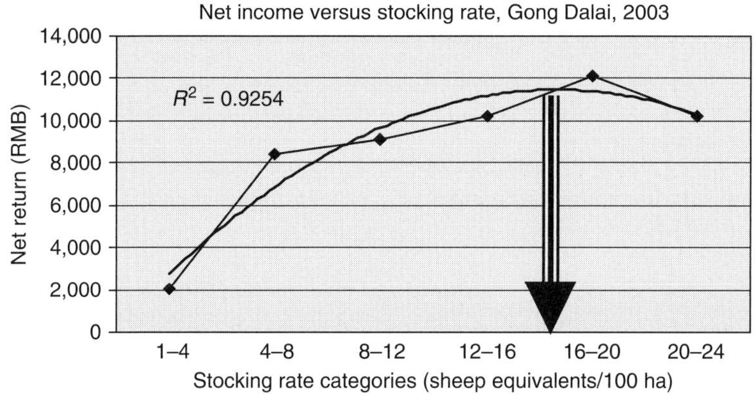

Fig. 15.4. Mean net returns for households within different stocking rate categories in Gong Dalai Gacha.

per unit area of pasture. While individual animal productivity might be falling, the productivity per 100 ha as stocking rate increases will increase to a certain maximum value. However, once this point is passed, continued increase in pasture utilization will so reduce individual animal performance that the overall production per unit area will fall. In Fig. 15.4, we see this happening as net return declines at the highest stocking rates. Here, the herders have pushed pasture production too far.

The message from Fig. 15.4 is that those herders at the highest stocking rates would earn more net income – and improve the condition and productivity of their rangeland – by *reducing* stocking rates.

Based on the trend line drawn over the net income values in the graph, the ideal stocking rate for maximizing net returns would be around 16 sheep equivalents/100 ha (indicated by the arrow). This is equivalent to 1 sheep equivalent/6.25 ha or 1 sheep equivalent/94 mu. This figure is close to the stocking rate recommended by staff of the local Rangeland Management Station (personal communication) of 1 sheep equivalent/110 mu following their studies of pasture production.

Carrying capacity has always been estimated on some basis of biomass produced in a growing season compared with individual livestock needs. Article 18 of the Law on Combating Desertification (LCD) in the PRC stipulates this method (an unofficial English summary of the law is provided by ADB, 2001). However, the concept of carrying capacity is really a lot more complex than this, incorporating not only pasture production, but also seasonal conditions, allowance for poor seasons, livestock type, grazing system and distance from markets and suppliers.

A system to allocate carrying capacity based solely on pasture production has a number of drawbacks:

- The analysis is based on a small amount of point data.
- It requires specific fieldwork, requiring time and money.
- The results are sensitive to the season when the data are collected and either require revision each year as data become available or wait for a poor season to use as a benchmark.
- The system does not consider the other ingredients that contribute to an assessment of carrying capacity.

In contrast, a system based on a comparison of net financial returns versus stocking rates: (i) is likely to be as accurate as the biomass system; (ii) can integrate all aspects of the production system; (iii) requires no additional fieldwork; (iv) can be based on a chosen recent year; and (v) can be performed rapidly.

An important advantage of the 'net financial return' method of estimating carrying capacity is that it can involve the herding community and, rather than developing an ultimatum on stocking rate, the recommendations include a welcome extension message, showing how it is possible for many herding households to make more money by adjusting their livestock numbers and stocking rates.

The method is based on the premise that rangeland condition affects livestock nutrition and livestock performance directly (Wilson, 1986), which in turn affects financial performance.

Any carrying capacity or stocking rate recommendation, however derived, should be seen as a 'first approximation' of sustainable stocking rate and should not be thought of as the absolute answer. Ongoing rangeland monitoring (e.g. biodiversity, soil cover, species frequency and plant sizes) will be required to prove or refine stocking rate decisions.

15.4.3 Changes in grazing species – the effect of markets

Populations of large grazing species (cattle, horses, camels) have declined over a number of years in preference for sheep, and particularly goats. Reasons for this include the relatively strong markets for cashmere fibre and sheep meat and the fact that smaller livestock species are easier to manage. When livestock are viewed as wealth, then the sale of one sheep or goat may be seen as a smaller and more appropriate withdrawal from the 'bank' than the sale of a large animal. A larger number of small livestock provides more financial flexibility than a smaller number of large livestock. Markets for camel products are presently poor and domesticated Bactrian camel numbers have fallen significantly over recent years as a result.

The change in grazing species has affected the height at which vegetation is grazed and has resulted in more soil disturbance, compaction and erosion due to the action of sheep and goat hooves. The only way to redress the balance in species is through improved marketing of cattle and camel products.

15.4.4 Damage by small mammals and insects

Brandt's vole (*Microtus brandti*) is endemic to the rangelands of central Inner Mongolia, while the plateau pika (*Ochotona curzoniae*) is endemic on the Tibetan Plateau. These burrowing herbivores are a normal part of the ecology and food chains in these areas (Pech *et al.*, 2005), but can cause rangeland degradation when their numbers reach plague proportions, as in 2004 when damage was reported to 38.93 Mha (SEPA, 2004).

The same year saw a plague of insect pests (rangeland locust, meadow webworm, white puncture vine noctuid and rangeland caterpillar) on 39.22 Mha of the rangelands of China. Such cyclical explosions of populations cause rangeland degradation and financial loss (SEPA, 2004).

15.4.5 Extreme weather events

China's rangelands suffer periodically from four extreme weather events. These are torrential rain, strong winds, drought and exceedingly heavy snowstorms. All affect rangeland condition.

Reduced soil cover increases the effect of direct raindrop impacts in intense rainstorms, splashing soil particles and the fertility they hold into runoff water. Denuded and degraded hillsides and heavily grazed rangelands permit increasing proportions of rainstorms to become runoff, thereby carrying away plant nutrients, as well as putting built structures at risk. Erosion, reduced soil fertility and reduced soil moisture recharge limit plant growth and a downward spiral of poor plant growth, less soil cover and increased erosion of soil and fertility develops (Chapter 5). Increased runoff also results in less groundwater recharge in the area where the rain falls, contributing to falling groundwater reserves.

Strong winds that occur mainly in spring whip up fine soil particles from bare soil and carry these for several thousand metres to become dust storms. At the same time, heavier soil particles are bounced along the soil surface, causing impacts with other particles in the soil surface, which, in turn, are released. This saltation process causes sand dunes to form and advance (Yang *et al.*, 2008). The saltation process may be reduced by the growing of windbreaks of suitable trees and shrubs – where there is sufficient water to maintain the trees. However, windbreaks have little or no effect on dust storms.

There are perceptions of climate change and decreased precipitation. Drought conditions are certainly a periodic feature of the rangelands of China. Increased runoff and decreased soil moisture recharge will decrease plant growth and provide the perception of increasing drought. In addition, an average year can provide drought conditions if the rainfall is concentrated in one part of the growing season, whereas a drier than average year may produce better pasture growth because the little precipitation there is has been well distributed across the growing season.

Is a long-term trend in falling precipitation likely? The short answer is yes. There is a growing body of evidence that, within large-scale global weather patterns, local land-use practices and land condition can influence local precipitation (Otterman, 1974; Charney *et al.*, 1975; Sagan *et al.*, 1979). First, livestock reduce the rate at which water penetrates the soil surface and this leads to increased runoff by reducing vegetative and litter cover and by compacting the soil. Then, lower soil moisture and increased water stress reduce plant productivity and vegetative cover, thereby worsening the problem of runoff further, reducing soil moisture and decreasing plant growth (Belsky and Blumenthal, 1997). Increased albedo from bare soil increases the reflectance of the sun's energy and reduces the absorption of energy and its conversion to sensible heat. Heat is required to generate thermals to carry moisture from evapotranspiration to altitudes where such moisture could assist in cloud formation and precipitation under suitable atmospheric conditions. A study by Huang *et al.* (1995) in Western Australia demonstrated the connection between high albedo and lack of rising thermals when comparing wheat fields (high albedo) with neighbouring *Eucalyptus* woodland (low albedo). Further, an American National Academy of Sciences study, quoted in Brown (1985), found that between one-third and two-thirds of all rainfall in the Sahel came from local evapotranspiration rather than from the oceans. Hence, loss of vegetation from overgrazing and loss of water-holding capacity due to soil erosion produces less evaporation, and hence less rainfall. This could be a factor in China's arid and semi-arid rangelands, which are hundreds of kilometres from the nearest ocean. The foregoing would support the view that rangeland degradation will ultimately lead to declining rainfall and increasing drought, and that such conditions will become more severe as time goes on.

Climate change has another pernicious effect, that of increasing the rate at which glaciers in the high mountains are melting (Chapter 3). While this represents a short-term windfall in terms of water availability for irrigators, herders and towns and cities downstream of glaciers, increases in other forms of water reserves are urgently needed in preparation for the day when the supply of meltwater from glaciers ends.

Heavy winter snows can lead to disasters for both herding households and the rangelands. In 2004, for example, areas of Inner Mongolia, Xinjiang, Qinghai, Gansu and Tibet experienced snow and freezing disasters, resulting in the deaths of 99,300 cattle, sheep and goats and direct economic losses of over 100 million RMB (SEPA, 2004). When livestock are unable to graze or browse due to thick layers of snow, herders are forced to provide conserved fodder. Should the fodder be exhausted before the snow has gone, livestock face starvation and herders can face disastrous losses. In springs following heavy snow, livestock are likely to be turned out to graze on the rangeland as soon as weather conditions allow, due to shortages of conserved fodder. Thus, early plant growth is likely to be grazed before annuals have established sufficient root systems to support productive aboveground growth, or before any plant has produced an active photosynthetic area sufficient to translocate energy reserves to the root system. A system of delaying grazing in spring and keeping livestock in their pens until plants are well established (known as 'deferred grazing' or 'spring grazing rest') is being promoted to improve total growing season pasture production (Li *et al.*, 2003). But this relies on herders acquiring sufficient conserved fodder and grain to see them through winter and the deferred grazing period in spring.

15.5 Lessons Learnt from the Preceding Information

Here are some of the key lessons that have been learnt from observations of the past 60 years of rangeland policies and management in China.

There are many influences on the condition of the rangelands in China, but, of these, precipitation and grazing are the most important. While precipitation is largely beyond human control, grazing management is very much controlled by human decisions and government policies.

Traditional nomadic grazing placed more importance on the condition of the rangeland than on livestock productivity. Since the early 1950s, that philosophy has been reversed, with the result that today up to 90% of the grazed natural rangelands of China are degraded to some extent and the area of degradation is growing at 2 Mha/year (SEPA, 2004). Numbers of grazing livestock on the rangelands of China now are little different from those pre-collectivization, when nomadism was still a major form of grazing management (Sneath, 2000; Dalintai, 2005).

Comparison with other areas of steppe rangeland in Inner Asia has shown that livestock numbers alone are not the cause of overgrazing. Other areas within Inner Asia have higher stocking rates than on the rangelands of China, but suffer low levels of rangeland degradation (Sneath, 2000). Instead, the mobility of livestock seems to be the key determinant of the level of degradation.

Inadequate and inequitable land allocations to households under GUR and the slow development and extension of livestock-carrying capacities on different rangeland types have contributed to rangeland degradation.

Herders have always been seen as illiterate and the cause of the (overgrazing) problem, rather than being embraced and consulted as people with unique rangeland knowledge and insights and being on the front line in fighting and solving the problem (Miller, 2002).

Attempts at improving the rangelands through researching the introduction of more productive species over a period approaching 60 years have yielded little other than a realization that local plants are well adapted to their environments and will be productive if managed appropriately.

Livestock production has been based on livestock numbers rather than on a serious attempt to raise productivity. While livestock breeding programmes have been in vogue, a holistic approach to increasing productivity is required that includes:

- improved market intelligence;
- turn-off strategies aimed at selling to market requirements;
- breeding (genetics to satisfy market requirements);
- breeding season (to coincide with pasture growth);
- nutrition (energy, protein, minerals, vitamins, water of adequate quality);
- grazing management and sustainable stocking rates;
- adequate land on which to practise profitable herding enterprises;
- adequate winter shelter;
- disease control and improved health (linked to nutrition); and
- rodent pest control by environmentally sensitive means.

To this list might be added the development of herder cooperatives to coordinate sales and purchase of supplies. In irrigation oases, farmers produce fodder crops or fodder residues for their own livestock and for sale to herders. Apart from cornstalks and maize, turnips can be grown as a second crop following maize, utilizing stored soil moisture, and they require only one irrigation. If prevented from freezing, turnips can provide a nutritious winter and spring fodder.

Cultivating rangeland to grow crops may produce an acceptable crop yield for 1 or 2 years, but it depletes soil fertility, organic matter and the soil reserves of the seed of rangeland species. Following abandonment after a few years, the site will take many years to redevelop its soil cover. During that time, it will be subject to wind and water erosion. Rangeland should be cultivated only where this can be undertaken sustainably. If that judgement on sustainability includes supplies of groundwater for irrigation, the sustainability of the water source itself should be checked, particularly if it originates from glacier meltwater.

15.6 Possible Futures

The condition of the rangelands of China is at a crossroads. Whether they take the route to rehabilitation and sustainable production or the road to further degradation hangs largely on three factors:

1. The willingness of policy makers to embrace change, to make strategic decisions based on the best available scientific and socio-economic information and to facilitate change through new policy settings.
2. The amount government is prepared to pay to repair the rangelands, develop sustainable rangeland production (albeit at a lower level than previously), reduce dust storms and slow the movement of sand dunes.
3. The rebirth of a sense of community.

The chairman of China's Environment and Resources Committee of the National People's Congress estimates that remediation of degraded land in the areas where this is technically feasible would cost US$28.3 billion. With such large sums of money required for remedial work of a technical nature, it is imperative to consider the entire picture of rangeland degradation and its causes. Unless the underlying cause of rangeland degradation is understood and removed, remediation work will be unsuccessful in the long term. A full appreciation of the causes of rangeland degradation may indicate a less expensive and yet effective means of repairing the rangelands.

Rehabilitating the rangelands will be an enormous undertaking that will require the widest multidisciplinary approach. Within this short, technocentric chapter, we would not presume to provide a blueprint for development. However, we would like to offer some recommendations.

The importance of livestock mobility has been stressed several times above. In order to allow for increased livestock mobility, policy makers would need to rethink critically the household GUR. One of the best of possible futures for China's rangelands would be to develop a system of scientific, ecologically sound, semi-nomadic summer and autumn grazing linked to short-duration grazing. Under the system envisaged, herders and their livestock would return from summer and autumn grazing areas to areas of home range and winter housing to overwinter (in some less degraded areas, this occurs already). Livestock would be pen-fed for as long as possible in spring, while spring growth becomes established on the rangelands. Such a system should be trialled in a small number of willing herding communities over sufficient area so that the influence of poor seasonal conditions can be minimized when such seasons arise.

Should such a system prove practical and acceptable, its implementation on a broader scale would likely require the vesting of GUR in the community or administrative area government. Individual herders would retain a reduced household GUR area for their winter quarters and family home. No one system would be effective over such a vast and diverse area as northern China. For instance, seasonal grazing in mountainous areas is likely to be practised differently from the way it might be practised in areas of extensive rangeland plain. Therefore, it would be important for the government to set general policies, an enabling framework and annual implementation budget and allow local scientists, administrators and herders to develop local practices according to local conditions.

The engagement of herders in the process is seen as imperative to its success, since it is they who are the day-to-day managers of the rangelands. Further, developing community 'ownership' of aspects of rangeland administration (such as which areas need resting from grazing in any particular season) and 'ownership' of the role to settle grazing disputes will save local government resources, but, more importantly, it will empower the community to practise sustainable rangeland management.

If the decision is to retain GUR similar to their current form, then there is a need to ensure that all herding households have sufficient land allocation to run a profitable herding enterprise and have the provision to sell or agist livestock in drought years. In some communities, there may be insufficient land to provide all herders with a profitable GUR allocation. There may be a number of ways to overcome this problem. The first, already practised in some areas, is for senior herders to exchange their GUR for a government pension and move to the nearest town. This gives government the opportunity to reallocate that land to neighbouring herders to increase their GUR areas. A further option may be for a herding household to keep few livestock of its own, but derive income from agisting a neighbour's livestock. This effectively increases the neighbour's grazing area. Providing herders with the option to trade their GUR (i.e. to sell their leasehold rights to the grazing land) could be a further option that would allow herders to leave the herding industry with dignity. Such transactions require that land is provided with a monetary value, based on carrying capacity, and would also require policies and frameworks such that land sale will not be likely to cause rangeland degradation. Principal among potential purchasers of the leasehold right would be neighbours seeking to enlarge the size of their GUR, or government seeking to reallocate the land to neighbouring herders to provide them with GUR allocations of sufficient area.

Whether the future holds a form of seasonal nomadic grazing or maintains GUR similar to the current system, there are possible changes that would be beneficial under either form of rangeland management. First among these will be an emphasis on increasing livestock productivity. This topic has been canvassed already in Section 15.5 above, but one aspect requires expansion. Short-duration rotational grazing is advocated strongly as a method to improve rangeland and livestock productivity, especially when the daily grazing area is restricted (McCosker, 1993). This leads to the question, 'How do you control where livestock graze each day?' Traditionally, grazing distribution was controlled by shepherding. This is still practised – on foot, on a motorbike, or on horseback or camel in some areas. However, the cost of a shepherd's labour, the cost of the fuel in the motorbike and the opportunity cost of the herder's own labour are all increasing. The rapidly increasing cost of iron ore and steel are starting to make fencing too expensive an option to contemplate. A future alternative that needs testing as an option for rotational grazing without fencing is the use of livestock attractants. These could be in the form of a portable watering facility (a tank and trough mounted on a farm trailer) or lick blocks containing an attractant and a recipe of minerals, vitamins, energy and non-protein nitrogen suitable for local conditions. The problem of mineral deficiencies is becoming more widely recognized (e.g. Nyima *et al.*, 2005). Thus, lick blocks on rangeland might solve the dual problems of grazing management and mineral deficiencies.

An effective, adequately resourced extension service is required to facilitate and undertake training and extension, as well as disease control and rangeland administration. It has been suggested that part of the extension service, particularly dealing with frequently asked questions, seasonal conditions or market information, could be performed efficiently as a 'phone-in' service (Peter Curry, Inner Mongolian Rangelands Management Project, personal communication). Herders and farmers would telephone a hotline to gain the information they require.

Whichever road the rangelands of China take into the future, there will be a requirement for objective rangeland monitoring. This is the focus of the following chapter (Chapter 16). Suffice it to say that there will always be a role for ground-based vegetation and soil monitoring to support analysis of satellite imagery. Future ground-based monitoring will need to be at a landscape scale in order to capture and understand the mosaic of surface conditions. In addition, monitoring will need to focus as much on environmental management as on productivity. Lastly, at least some of the ground-based monitoring should be aimed at informing herders and helping to promote sustainable rangeland management. For this purpose, a booklet containing photo standards for different amounts of feed on offer in different seasons and different rangeland types would have appeal. Comparing vegetation on the ground with the photographs in the booklet will help herders decide on the number of animal × grazing days a particular pasture offers.

15.7 Conclusion

Through the work of Sneath (1999, 2000) and others, livestock mobility has been identified as having a major influence on the condition of Inner Asian rangelands, and is more important than livestock numbers alone. Finding ways to increase livestock mobility on the rangelands of China should be one of the major objectives for those engaged in improving rangeland condition and making the use of this vast grazing resource sustainable.

References

ADB (2001) http://www.adb.org/Projects/PRC_GEF_Partnership/Desertification.pdf.

Belsky, J. and Blumenthal, D.M. (1997) Effects of livestock grazing on stand dynamics and soils in upland forests of the interior west. *Conservation Biology* 11(2), 315–327.

Brown, L.R. (1985) A false sense of security. In: Starke, L. (ed.) *State of the World 1985.* W.W. Norton and Co., New York, pp. 3–22.

Brown, L.R. (2001) *Dust Bowl Threatening China's Future.* Earth Policy Institute (http://www.earth-policy.org, accessed 10 March 2008).

Brown, L.R. (2004) *Troubling New Flows of Environmental Refugees.* Earth Policy Institute (http://www.earth-policy.org, accessed 10 March 2008).

Charney, J., Stone, P.H. and Quirk, W.J. (1975) Drought in the Sahara: a bio-geographical feedback mechanism. *Science* 187, 434–435.

Dalintai, A. (2005) Rethinking grassland desertification. *Journal of College of Finance and Economics of Guizhou* 3, 46–50 (in Chinese).

Dalintai, A. (2008) Rethinking overgrazing and strategies for its management in Inner Mongolia. In: Organizing Committee of IGC/IRC Congress, Hohhot, China (eds) *Proceedings of the International Rangeland Congress, Multifunctional Grassland in a Changing World,* Volume I. Guangzhoui Guangdong People's Publishing, PRC, Session A6, Paper 1676.

Gegenguva, O. (2002) Mongolian ecological culture in the context of ecological ethics. *Journal of Inner Mongolia University (Humanities and Social Sciences)* 34(4), 3–9 (in Chinese with English abstract).

Hawley, C. (2004) UN to combat growing deserts. *Pittsburgh Post-Gazette,*16 June. Associated Press, Pittsburgh, Pennsylvania.

Huang, X.M., Lyons, T.J. and Smith, R.C.G. (1995) Meteorological impacts of replacing native perennial vegetation with annual agricultural species. *Hydrological Proceedings* 9, 645–654.

Li, Q., Michalk, D., Chen, L., Zhang, P. and Tong, S. (2003) Analysis of constraints and management strategy for animal production in grasslands of Northern China. *Grassland Journal* 11(02), 47–53 (in Chinese).

McCosker, T. (1993) The principles of time control grazing. In: *Proceedings 3rd National Conference of the Beef Improvement Association of Australia, Armidale, September 1993*. Beef Improvement Association of Australia, Armidale, Australia, pp. 87–95.

Miller, D. (2002) The importance of China's nomads. *Rangelands* 24(1), 22–24.

Nyima, T., Luo, X., Yu, S. and Judson, G. (2005) *A Survey of the Mineral Status of Livestock in the Tibet Autonomous Region of China*. ACIAR Working Paper No. 59. Australian Centre for International Agricultural Research, Canberra.

Otterman, J. (1974) Baring high-albedo soils by overgrazing: a hypothesized desertification mechanism. *Science* 186, 531–533.

Pech, R., Arthur, A., Hinds, L. and Shi, D. (2005) *The Role of Small Mammals in Grassland Degradation in China: Defining the Problem*. Sino-Australian Workshop on Management of Grassland Livestock Systems and Combating Land Degradation in Northern China, 6–8 December 2005, Canberra (http://www.science.org.au/events/rangelands/pech.htm, accessed 5 March 2008).

Qingwu, B. (2001) Ecological insight of nomadic Mongolian. In: Zhongling, L. and Erdenebukh (eds) *Nomadic Civilisation and Ecological Civilisation*. Inner Mongolia University Publishing House, Hohhot, China, pp. 33–58 (in Chinese with English abstract).

Sagan, C., Toon, O.B. and Pollack, J.B. (1979) Anthropogenic albedo changes and the earth's climate. *Science* 206, 1363–1368.

SEPA (2004) Report on the State of the Environment in China, 2004 (http://english.sepa.gov.cn/SOE/soechina2004/rangeland.html, accessed 10 March 2008).

Sneath, D. (1999) Spatial mobility and Inner Asian pastoralism. In: Humphrey, C. and Sneath, D. (eds) *The End of Nomadism? Society, State and the Environment in Inner Asia*. Central Asia Book Series, Duke University Press, Durham, North Carolina, Chapter 6.

Sneath, D. (2000) A Study of Sustainable Development in Pastoral Inner Asia: The ECCIA Project. Environmental Research in Cambridge (http://www.iic.tuis.ac.jp/edoc/journal/ron/r4-1-1/r4-1-1d.html, accessed 10 March 2008).

Taylor, J. (2006) Rangeland policy, privatization and new ecology in Inner Mongolia. Presented at 'Survival of the Commons: Mounting Challenges and New Realities', the Eleventh Conference of the International Association for the Study of Common Property, Bali, Indonesia, 19–23 June 2006.

Wang, M.P., Zhao, C.Z., Long, R. and Yang, Y. (2006) Policy development process in China with special reference to rangeland policy. Proceedings in ICIMOD Regional Policy Workshop, Kathmandu, 18–20 September 2006.

Wilson, A.D. (1986) Principles of grazing management systems. In: Joss, P.J., Lynch, P.W. and Williams, O.B. (eds) *Rangelands: A Resource Under Siege*. Proceedings Second International Rangelands Congress, 1984. Australian Academy of Science, Canberra, pp. 221–225.

Yang, Y.Q., Hou, Q., Zhou, C.H., Liu, H.L., Wang, Y.Q. and Niu, T. (2008) Sand/dust storm processes in Northeast Asia and associated large-scale circulations. *Atmospheric Chemistry and Physics* 8, 25–33.

Zhang, M.A., Borjigin, E. and Zhang, H. (2007) Mongolian nomadic culture and ecological culture: on the ecological reconstruction in the agro-pastoral mosaic zone in Northern China. *Ecological Economics* 62, 19–26.

16 Monitoring and Evaluation as Tools for Rangeland Management

Aijun Liu
Inner Mongolia Academy of Animal and Agricultural Sciences, Huhhot, China

Synopsis
Prevention of land degradation relies on the assessment of the present condition of the land and on the monitoring of relevant and meaningful changes. Any efforts to induce change require a good knowledge of the current biophysical situation and some objective means of judging the degree of degradation so change can be measured. The challenge is to use these tools (remote sensing, GIS and decision-support systems) in a way that provides a feedback loop so that management can reflect the changing rangeland condition.

Keywords: remote sensing; GIS; '3S approach'; Grassland Monitoring Stations; land users; near real time; rangeland condition; regulatory framework; sustainable development; pastoral land; ecological health

16.1 A Brief History

The huge human and livestock population in the pastoral lands of China, and the relatively backward economic level of traditional animal husbandry impose a heavy load on the ecological system of rangeland. China has to apply technologies to monitor and evaluate rangeland resources to comply with the sustainable development strategy.

Chinese rangeland resources are characterized by diverse ecosystems that represent more than 40% of the national land area. The extent and diversity of rangelands provide numerous ecosystem services for human societies. Therefore, monitoring is required to document and anticipate ecosystem responses to various disturbances, direct management actions and promote wise stewardship.

As the Agenda 21 document (State Science and Technology Commission, 1994) points out, along with sustainable development, there are more and more demands for information concerning the environmental situation, especially the evaluation of rangeland resources and the overall ecological environment. Socio-economic development in China is now emerging from the era of target-driven decision making towards a sustainable development strategy (Uchida *et al.*, 2002).

Unfortunately, before the 1990s, pastoral land managers in both the public and private sectors knew little of the up-to-date information of relevance to their pastoral pursuits in the context of the economic and social changes that have accelerated over the past two decades. Many livestock producers are apt to be engaged in the industry based merely on their direct experience and to raise livestock with a view to maximizing return and exploiting the rangeland without the long-term future in mind. At that time, there was no effective and standard method for monitoring rangeland resources. Also, there was no monitoring 'toolbox' or methodology particularly suited for use in China's pastoral regions, particularly before 2000.

In response to growing awareness of environmental legislation and pastoral land management concerns, pastoral land managers showed increased interest in practical monitoring techniques for evaluating rangelands during the latter part of the 20th century (Li, 1994; Shi *et al.*, 1994; Liang *et al.*, 1996; Huang *et al.*, 1999; Chen and Liu, 2003; Liu and Xing, 2004; Su *et al.*, 2005; Liu and Han, 2007; Liu *et al.*, 2007). Despite this, there is no agreement yet on an effective and standard method for monitoring rangeland resources.

Considerable progress has been made in interpreting spatial and temporal patterns in remotely sensed data to identify 'grazing gradients' such as trends of increasing/decreasing cover with increasing distance from villages (Yang *et al.*, 2007). Apparent trends in foliage cover following above-average rainfall indicate the extent of 'landscape damage' attributable to grazing. Further developments have led to: (i) assessment of 'landscape resilience' (Ludwig *et al.*, 1997) by comparing the response of vegetation cover to rainfall with what would be expected with little grazing impact; (ii) application of these assessments on an experimental basis in some pastoral lands; and (iii) use of these techniques to monitor rangeland condition on selected sites. Thus, remote sensing provides the opportunity to monitor resource conditions over large areas in 'near real time' when degradation is occurring. Used in conjunction with ground-based monitoring, particularly by local grassland monitoring personnel at township or county level, remote sensing has the potential to provide advance warnings, allowing government and land-user action to minimize damage.

At present, a preliminary monitoring system has been developed for use at both the provincial and national levels. But establishing a valid, effective, timely and public-oriented rangeland monitoring–evaluating system still has a long way to go.

16.2 The Indicators and Criteria for Rangeland Monitoring and Evaluation

Monitoring change has been a central theme of rangeland management for many years. Improved understanding of changes in resource condition over time is needed by land managers and administrators, and long-term study of perennial vegetation dynamics and composition provides a useful indicator of broader changes in rangelands (Aguirre-Bravo, 1996; Aguirre-Bravo *et al.*, 2005).

Monitoring emphasizes change and, where possible, its causes. The important considerations are to ensure that significant changes are measured and that measured changes are real, not a consequence of sampling error or personal bias. Associated with detecting the magnitude and direction of change is taking the appropriate management action.

Rangeland degradation is often difficult to measure successfully over extensive land areas. Ground-based vegetation sampling techniques must be sensitive to changes in parameters capable of defining this degradation or lack thereof. These parameters include species cover, frequency, density and botanical composition, along with various surface soil characteristics (the relationship between them is shown in Fig. 16.1). Therefore, both the vegetation and the soil must be examined in a monitoring programme to determine if grazing practices are successful in maintaining the ecological health of the range landscape. Also, the socio-economic and policy issues must be considered.

The process of monitoring is defined as repeated assessments over time. Reliable rangeland monitoring demands that any short-term fluctuations disguising long-term trends be removed from the overall temporal sequence of vegetation maps. The task is made even more difficult by the fact that the fluctuations are spatial, as well as temporal (Squires, 2007). Some of the key factors in spatial–temporal variability are set out below.

1. *Temporal variation.* Rangelands are highly variable in space and time. Periodic drought and high year-to-year rainfall variability are superimposed on seasonal shifts in precipitation, temperature regimes and variation in day length and other factors of significance to plant and animal physiology.
2. *Altitudinal variation.* Rangelands may extend from highlands to lowlands and in many places altitudinal migration (transhumance) is practised to take advantage of the distribution and seasonality of the various forage resources.
3. *Spatial variation in grazing pressure.* Grazing is rarely even. The distribution of the various vegetation communities, proximity to villages or to watering

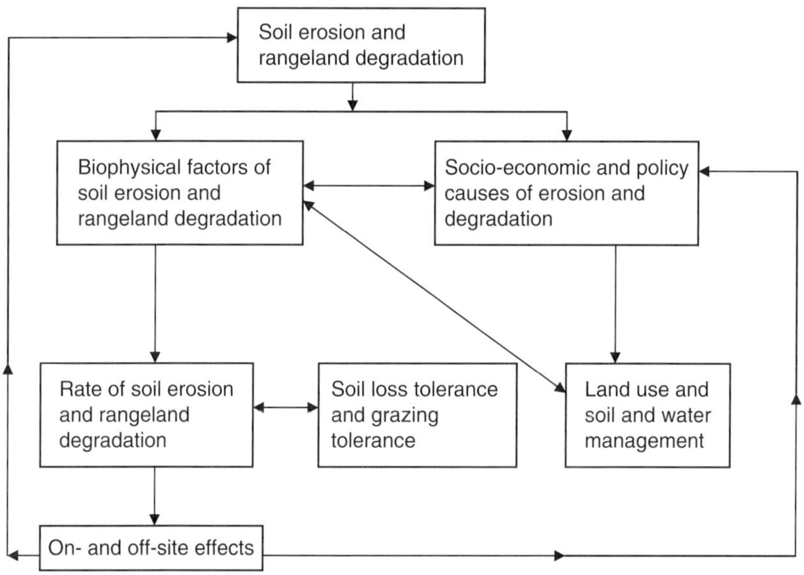

Fig. 16.1. The interplay of biophysical and socio-economic factors implicated in soil erosion and rangeland degradation.

points, etc., can influence this. Often, there are large 'sacrifice areas' near permanent water or villages.

4. *Temporal variation in grazing pressure.* As livestock go through their breeding cycle, the herd size rises and falls. In addition to changes in number of livestock, there is considerable variation in the demand for forage. These two influences lead to variation in grazing pressure throughout the year.

Relevant departments have understood fully that sustainable development of pastoral land is based on the dynamic monitoring and management of rangeland resources. Great progress has been achieved and successful use of remote sensing technology for application in pastoral management and environment fields include the prevention and alleviation of disasters such as fire and outbreaks of pests. Therefore, a valid, effective, timely and public-oriented rangeland monitoring–evaluating system has been or is being established in order to provide the scientific basis for the government's investment in rangelands and to promote the application of 3S (RS, GPS, GIS) technologies.

In 2000, the China Agricultural Ministry began publishing the National Rangeland and Ecological Monitoring Technology Rule, which describes an approach for evaluating the ecological health of rangeland ecosystems and identifies the need for a common means of evaluating rangeland health. Some indicators, flow charts and approaches were developed for monitoring and evaluating the rangeland condition and management in pasture areas. Many rules have been worked out. Some of them are concerned with rangeland monitoring and evaluation, such as:

1. Technical rule for rotational grazing of rangeland.
2. Parameters for degradation, sandification and salinization of rangelands (Agricultural Industry Criterion of the People's Republic of China (PRC)).
3. Technical Rules for Monitoring of Rangeland Resources and Ecology (Agricultural Industry Criterion of PRC).
4. Land desertification monitoring method.
5. Technical criteria for desertification monitoring for the sand source area around Beijing and Tianjing; based on remote sensing (being drafted) (Agricultural Industry Criterion of PRC).

In fact, no single monitoring technique can be prescribed to assess all rangelands or all aspects of change efficiently. Like those elsewhere, China's pastoral rangelands are characterized by a diverse assemblage of plant species, some abundant, others rare, distributed in a variety of spatial patterns expressed at different landscape scales. Plant communities also exhibit considerable temporal variation and individual species numbers may fluctuate substantially between years, or even seasons. Designing and maintaining rangeland monitoring programmes are therefore difficult, particularly given constraints such as staff turnover and budgetary issues. Sampling strategies (site location and number, measurement frequency, attributes measured) must be tailored to specific programme objectives, but be modified by logistical and budgetary constraints. It is important to define sampling regimes to accommodate inherent rangeland variability. The chosen methodology should be capable of addressing specific characteristics with sufficient precision to detect real change. This is still a challenge.

16.3 Rangeland Monitoring Technical System and Methodological Approach

Work by the China Agricultural Ministry since 2003 has led to the development of a powerful pasture assessment and monitoring system. More than 16 provinces have been involved with field surveying and monitoring; they include Inner Mongolia, Xinjiang, Qinghai, Gansu, Xizang (Tibet) and Sichuan, which collectively account for 88% of the area of national rangeland. Models were set up for calculating, processing and analysing field data. The rangeland monitoring technical system and its flow are shown in Fig. 16.2.

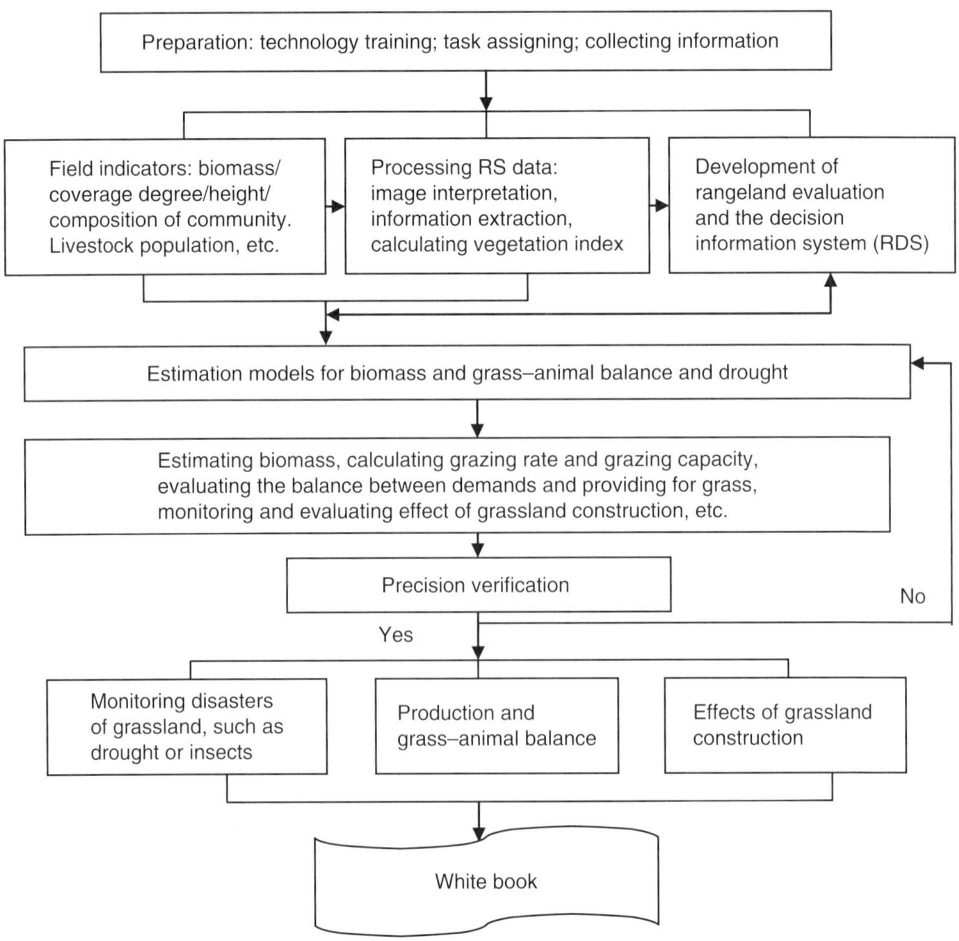

Fig. 16.2. Rangeland monitoring and evaluating programme and system in Chinese pastoral areas.

It also recommended that assessment be based on multiple indicators of basic ecosystem processes. The setting up of a national Rangeland Monitoring Information network, which is a public-oriented platform for evaluating the supervision of rangeland monitoring and management and making public the trends, in both quality and quantity of rangelands, and output from animal production in pastoral areas is also desirable. In addition, enhancing the decision-making ability in animal husbandry bureaux at national, provincial and county and community level was recognized as a priority. The task that remains is to develop and adopt a scientifically based and practical range-monitoring technique that is widely accepted by resource managers and readily employed to make sound management decisions. Monitoring of attributes other than biomass and productivity is also required (Chapter 15).

16.4 Experiences of Monitoring and Evaluation Based on RS and GIS

Since the 1990s, in a campaign against the degradation of the rangeland ecosystems that affects the pastoral industry so badly, the Chinese government has adopted a series of policies and measures and invested in relevant projects to improve rangeland conditions. Rangeland monitoring is intended to satisfy a wide range of potential users, from landowners to agency personnel and scientists. In order to meet the different needs of these groups, three levels of monitoring have been designed, large-scale (whole country), mid-scale (regional level) and small-scale (community level). Each level considers the same indicators, but with increasing detail, rigour and repeatability.

16.4.1 Field sampling and monitoring system

There have been more than 3000 field monitoring sites set up in more than 20 provinces, including Tibet, Xinjiang, Qinghai, Ningxia, Gansu and Inner Mongolia. Site selection took the following factors into account: (i) to have relevance in establishing baseline data; (ii) to be representative of a larger land area; (iii) to be located again easily; (iv) to reflect response to management changes; (v) to track change in areas of special concern; and (vi) to establish site potentials for other areas of land. Once a general site is identified, a transect starting point may be randomly located. A management system for field data collection and retrieval has been developed. All data for every year are uploaded to the system. Users can enquire about and obtain access to analysis of the data at short notice. Most importantly, estimation models of production have been established based on those data, and production and output estimates can be obtained. Field sampling is the basis for rangeland management and evaluation. The software, Management System of Field Data for Rangeland Monitoring V2.0, was developed in 2004 by the Rangeland Supervising and Management Department, China Agricultural Ministry (http://www.agri.gov.cn/).

16.4.2 Pastoral monitoring and evaluation

The development of the application of remote sensing in China still seems to be comparatively backward. Nevertheless, at the national level, its application has become more and more an integral part of monitoring, evaluation and planning.

In order to assess the status of rangeland resources at the national level, the integration of RS information derived from satellite data with geographical information (GIS) for mapping of land cover resulted in 2000 in the development of digital maps at a scale of 1:1 million. These maps replaced those made after the national survey in the 1980s (Su *et al.*, 2005).

Remote sensing and GIS are applied mainly in five fields, as described below.

Investigation and evaluation of rangeland resources

This is to verify rangeland resources as the basis of rational utilization and planning and to provide effective protection of pastoral land resources. China's pastoral land is vast and it is impossible for us to investigate and verify the resources situation through the traditional methods. In fact, we have made a great deal of effort to investigate rangeland resources at various degrees of detail in the past 60 years. These investigations included the rangeland resource types and land-use situation. In the future, the evaluation of rangeland resources will be included in the environmental assessment programme. The main products of this programme include: maps of rangeland types and a map of the location and extent of degradation at both the national and regional scales (http://www.grassland.net.cn/grass.asp).

Dynamic monitoring of rangeland resources

Timely and accurate remote sensing and geographical information are a major data source for monitoring local or regional change in the extent and condition of rangeland resources. MODIS has been used for this purpose in conjunction with other satellite data. Satellite data were found to be very useful and accurate for the assessment and monitoring of regional vegetation change and rangeland condition on a long-term and real-time basis. The Normalized Difference Vegetation Index (NDVI), which can be calculated directly from satellite data, is related to vegetation canopy characteristics such as biomass and percentage of vegetation cover. The NDVI is therefore representative of plants' photosynthetic efficiency. The approach and flow chart are shown in Fig. 16.3.

Since 2004, China's Agricultural Ministry has developed an operational system at the national level to monitor the seasonal growth of rangeland vegetation and dynamic changes that result from climate, land-use change or management. The location and extent of rangeland degradation are also mapped. The areas that are changing are rated on a five-point scale that ranges from 'perfect' to 'worst' (http://www.agri.gov.cn/).

Estimation of rangeland yields

One of the main aims of pasture assessment is to obtain information about biomass and, specifically, palatable biomass of a particular region or areas with similar geophysical characteristics. The biomass estimation is based on models (linear relationship between the NDVI value and biomass using documented parameters).

At a national level, the Agricultural Ministry has, since 2004, distributed estimates of biomass, as well as a situation report on forage supply and demand. Figure 16.4 is a map of national rangeland production according to RS and field data in 2007 that is distributed by the Rangeland Supervising and Management Centre of China's Agricultural Ministry (http://www.agri.gov.cn/).

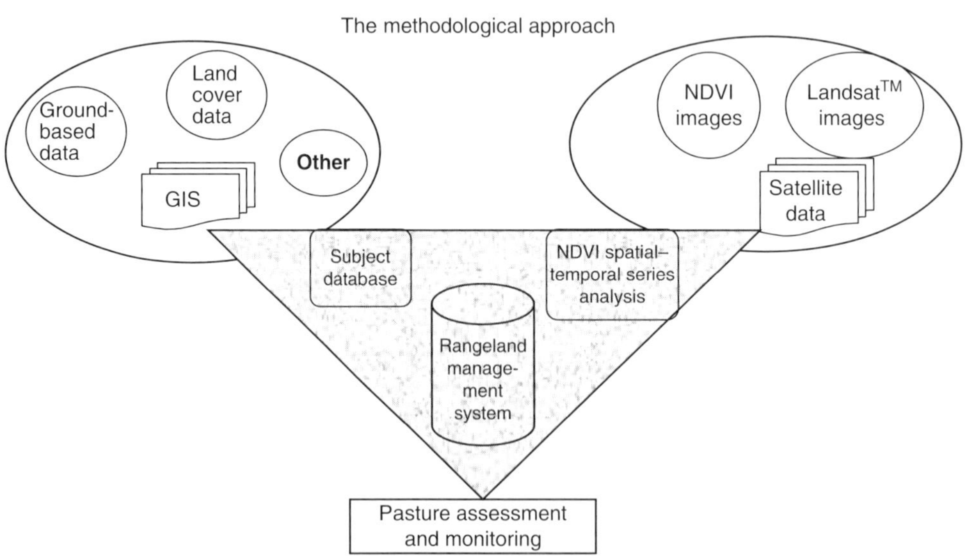

Fig. 16.3. Flow chart of dynamic monitoring of rangeland resources.

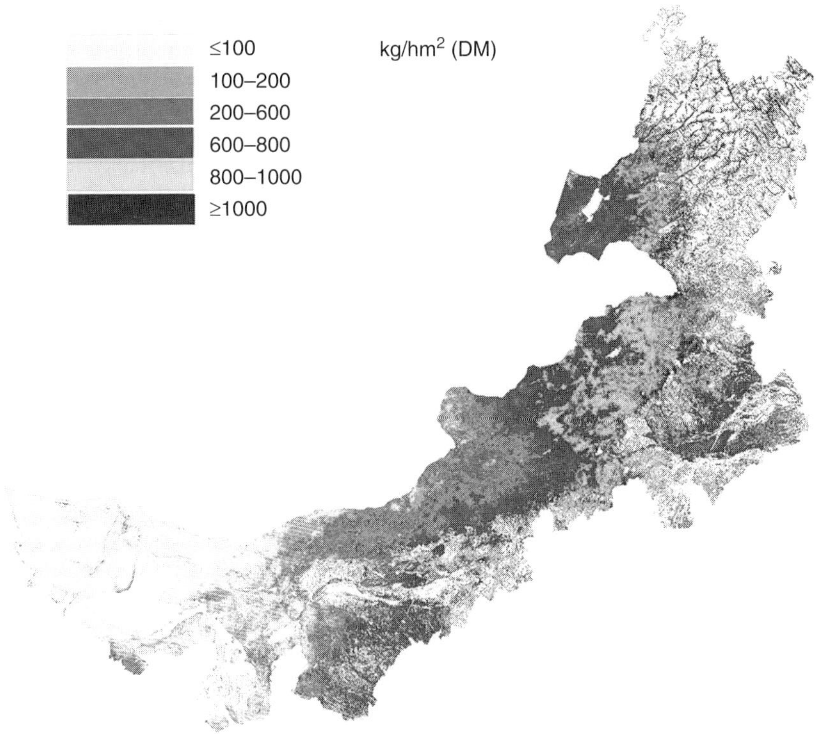

Fig. 16.4. The distribution map of standing biomass in Inner Mongolia in 2005 as derived from satellite imagery and field survey.

At a province level, Inner Mongolia has monitored and evaluated its rangeland every year using RS technology since 2003. Production models for Inner Mongolia grassland in 2005 according to RS and field data are shown in Fig. 16.4.

Defining the key phenological stages

According to RS definition of the phenological stage (Liu and Han, 2007), the green-up stage is coupled with a sequential increase of NDVI and the maturation stage is coupled with a sequential decrease of NDVI. Figure 16.5a shows that the change of NDVI (measured every 10 days) has a visible orderliness, which relates to the phenological stage of the rangeland, and Fig. 16.5b shows the change as detected by ground-based surveys every 10 days, which show the same phenological stage of the rangeland as that of NDVI.

The change of NDVI every 10 days can distinguish the change rules of the phenological stage, particularly two key stages: the green-up phase and the maturation phase. This can provide reference points of rest grazing for the decision-making department.

Grass–livestock balance evaluation

In 2001, the central and local government started implementing the Grass–Livestock Balance Policy to manage rangeland resources in a rational way. The local government (above county level), together with the rangeland administrative authority, checked and tried to normalize the livestock carrying capacity of pasture in the light of the carrying capacity standard established by the state's rangeland administrative department and according to the calculation of a rational grazing rate for natural grassland (Industry Criterion of PRC, NY/635-2002), which evaluates the balance between grass and livestock. Also, governments at all levels have taken measures to prevent overgrazing and supervise the rangeland contract operator to maintain the balance between grass

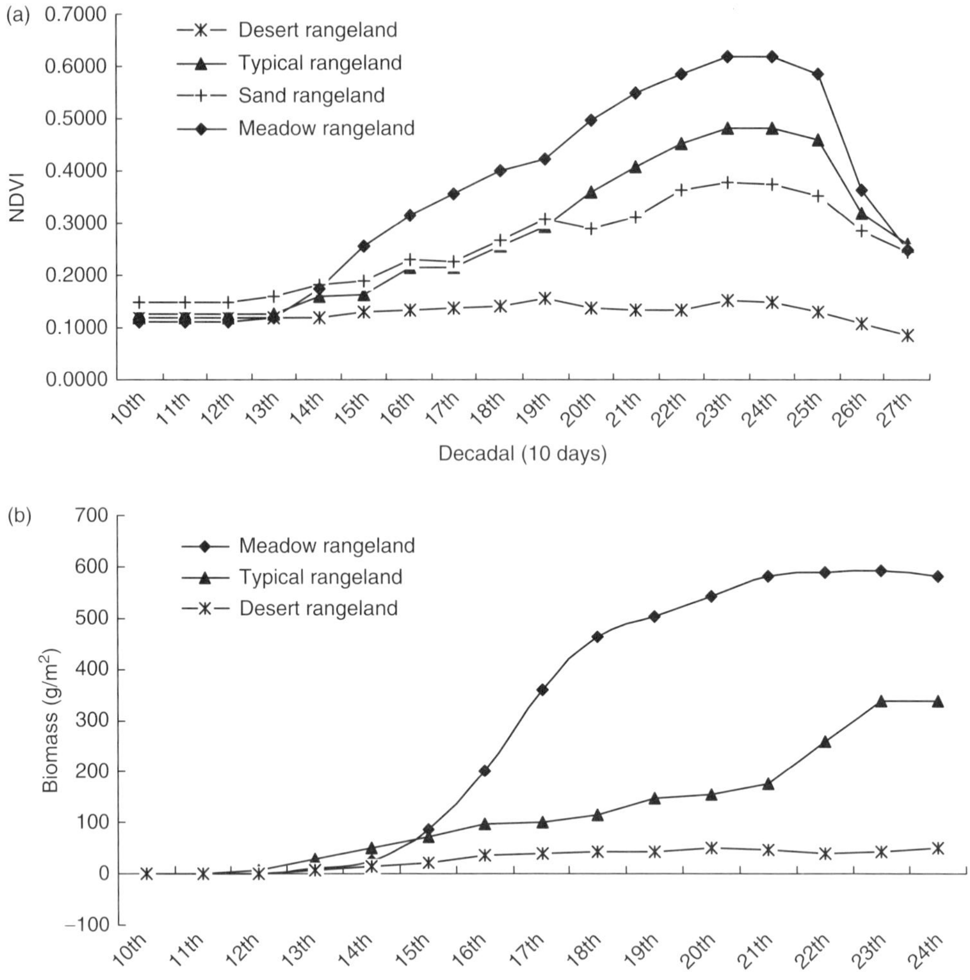

Fig. 16.5. Dynamic change for (a) decadal NDVI and (b) field data of types of rangeland under a total grazing ban.

and livestock. Each household has been assigned some land under the Grazing User Rights legislation as part of the Household Responsibility System (see Chapter 15). These contracts carry an obligation to provide care and stewardship of the land.

For example, Inner Mongolia has established a rangeland supervising and managing bureau at all levels from province to county. One important function of the bureau is to implement the Grass–Livestock Balance Policy. The agencies organize to examine, count and record the livestock population of each household every summer and autumn and distribute a map of the grassland–livestock balance (Fig. 16.6).

When they find that the number of livestock is over the standard, the householder has to either reduce the number of livestock or lease additional rangeland from other households, or face a fine. At present, there is no enforced reduction through confiscation and sale of seized livestock, so livestock numbers remain unaltered.

However, according to information reported in a survey, the implementation of this policy is not uniform or consistent. For example, the unified Grass–Livestock Balance Policy

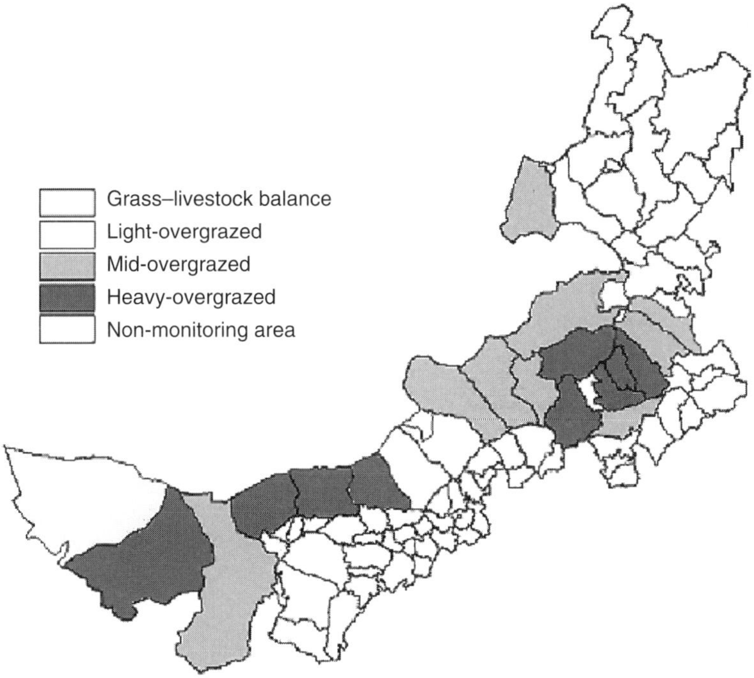

Fig. 16.6. A grassland–livestock balance map based on a survey of households in Inner Mongolia.

is not being implemented in Hulunbeier, Inner Mongolia, or in other provinces such as Gansu and Qinghai.

16.5 Rangeland Decision System (RDS)

The rangeland decision system (RDS), a tool-based GIS tailored especially for investigating the status of the rangeland, was developed by the General Station of Animal Husbandry of the Chinese Agriculture Ministry during 2003–2005. It has a wide range of applications because it allows a first-look overview of rangeland conditions in near real time. At present, the software has been spread throughout many provincial rangeland management departments in China.

The functions of RDS are as follows:

- gather information useful for management of rangeland resources;
- realize an operational dynamic monitoring capacity for large-scale rangeland resource inventory and environmental assessment; and
- build up a service system that provides information for government agencies.

The main objective of RDS is to support rangeland monitoring by integrating data from remote sensing, GIS and field observations. This provides significant multi-temporal and up-to-date information for the decision maker.

The first function of RDS is the storage and display of NDVI images for each 10-day period since 2001. For the whole region, or for any chosen subregion, the system provides possibilities for in-depth spatial analysis by visualizing relationships between NDVI classes and altitude, slope or aspect. Moreover, for each of the selected areas, or for the whole region, a time-series analysis can be carried out. A graphic utility shows the time series of NDVI for the selected region, as well as its minimum, maximum and average annual cycle for reference purposes.

16.6 Distribution of Monitoring and Evaluation Results

The Rangeland Supervising and Managing Department of China's Agricultural Ministry have gathered monitoring results of 16 provinces and have written a 'white book' (manual) of monitoring for rangeland resources and ecology, including mainly:

1. General report of monitoring for national rangeland resources and ecology.
2. Monitoring report and map of national rangeland production and grazing capability.
3. Monitoring report of grass–livestock balance of every province concerned.

http://www.agri.gov.cn/
http://www.gricaas.net/Gateway/News/Grass Industry
http://www.agri.gov.cn/
http://www.nmagri.gov.cn

16.7 Summary and Conclusions

Although the importance of monitoring is widely recognized and numerous protocols exist, the rangeland profession is currently in the process of designing and implementing a modified set of monitoring protocols to provide broader ecosystem assessments and address more directly the contemporary needs and expectations of society.

The major components of effective rangeland monitoring have been well defined and they include the following (Western, 2003):

1. Easily implemented, cost-effective and provide relevant information to various clients or stakeholder.
2. Provide anticipatory information about ecological processes that can inform management actions.
3. Predictable and responsive to single or multiple stresses and disturbances placed on ecosystems.
4. Applicable and adaptable to a broad range of climate, soil and vegetation variation and over a wide range of ecological scales (Dale and Beyeler, 2001).

The monitoring methodology must combine the characteristics of low cost, covering a vast area and capability of stratification, accounting for the mosaic of differing land types and yet retaining sufficient resolution to detect processes such as soil erosion. In addition, the timescales of climatic events must be considered.

A suitable monitoring system requires a spatial resolution of 1 m or less with aggregation or sampling spread over five orders of magnitude. Timescales range from a 1-day rainfall event to an upper limit of 10 years or so. Both large-scale aerial photography (say 1:1000) and (oblique) ground photography have a lower spatial resolution limit of, say, 10 cm, adequate to resolve all of the microscale features. Landsat MSS data are composed of pixels sized approximately 80 m × 60 m; however, the presence or absence of smaller higher-contrast features can be inferred. Landsat TM has a pixel size of 30 m × 30 m. SPOT (20 m × 20 m) has been used in some situations and AVHRR data of low resolution (1 km × 1 km) have been used for many studies because of their high temporal resolution (i.e. daily). Table 16.1 is a comparison of the utility of three sampling approaches for rangeland monitoring.

The best frequency of observations is provided by Landsat with coverage every 16 days (cloud cover permitting), declining to realistic estimates of once or twice per year using photography. An important feature here is not just the total number of possible assessments, but also the flexibility to choose the most appropriate times for measurement. Recently, the value of multi-spectral airborne videography with high resolution (0.3–1.0 m pixels) has become very important for the future monitoring of rangeland vegetation and soils. There is also potential to couple small-scale satellite imagery such as Landsat™ to these near-earth large-scale images for overall analysis and interpretation, or range condition and trend.

Multiple data sources provide a wealth of rangeland information. The success of rangeland monitoring can be enhanced with the development of a more encompassing conceptual framework. The major challenge for effective monitoring is to develop a comprehensive monitoring framework. However, it must also explicitly incorporate local and professional knowledge and comprehensive monitoring–management–policy networks (Briske, 2008).

Table 16.1. A comparison of the utility of three sampling approaches for rangeland inventory and monitoring.

Satellite imagery	Large-scale aerial photos	Ground-based methods
Landsat is limited by its relatively coarse resolution, but has the advantage of frequent acquisition, large area coverage and the unique capability of assigning a reasonably accurate geographical location for any detected features.	In many aspects, the best tool to assess sites in this situation because a skilled interpreter can extract and quantify variables that relate to erosion, its causal agents and the process of re-vegetation.	Can be dismissed as a quantitative and landscape assessment technique. It is restricted to an illustrative, qualitative role or to an approach for collecting detailed data on a few sites.
Many features could be detected by their spatial context, rather than brightness level alone.	Is excellent for assessing soil erosion. Resolution at scales up to 1:50,000 is sufficient to detect most erosion features and schemes for quantifying the assessments are easily constructed.	In spite of its popularity, is a very inadequate method of sampling and recording landscapes. The capability of resolving quite small features (e.g. seedling establishment, rodent damage) is counterbalanced by the inadequate sample size and distorted perspective.
Landsat is able not only to detect the amount of vegetation cover, but also to record the pulse of vegetation following a rainfall event.	The re-vegetation of areas can be assessed readily, accurately and quantitatively, but at considerable labour cost.	Within the limitations, records of re-vegetation can be made if permanent sites are established, although problems of exact relocation are appreciable.
The data generated are amenable to data management, especially through GIS.	Disadvantage lies in its restricted and costly coverage, subjectivity in interpretation and uncertainties associated with site relocation for monitoring.	Disadvantage lies in its restricted and costly coverage, subjectivity in interpretation.
In spite of the low spatial resolution, 'problem' areas can be detected by Landsat, although the identification of causal agents (human, grazing, etc.) may not be possible.	'Problem' areas can be seen easily and can be studied as a subset.	'Problem' areas can be studied in great detail, but at considerable cost.
The use of rectified image data means that the exact location of a feature in the image can be read by geographical coordinates.	The technique by itself cannot be used to give a precise geographical location to any point on the ground unless the aircraft is fitted with GPS or similar navigation systems.	Cannot by itself provide any geographical reference.
Because of their resolution limitations, Landsat data are not optimal to detect and assess invasion of pest species such as rodents or grasshoppers/locusts.	Most suited to assess the impact of pest species. The trade-off is between resolution and the area of the sample site.	Limited to selected facets only of any pest problem.
Problem areas (e.g. tracks precipitating the extensive blowout of dunes or scalding of swales, etc.) should be identifiable if they become extensive (more than 2–3 pixels wide).	Problem areas can be studied in detail.	Problem areas can be studied in detail.

Acknowledgement

I am especially grateful to Dr Victor Squires for his contribution to this chapter and for the guidance and help he provided.

References

Aguirre-Bravo, C. (ed.) (1996) *North American Workshop on Monitoring for Ecological Assessment of Terrestrial and Aquatic Ecosystems; 18–22 September 1995, Montecillo, Texcoco, Mexico.* General Technical Report RM-GTR-284, US Forest Service, Ft Collins, Colorado, 305 pp.

Aguirre-Bravo, C. *et al.* (eds) (2005) *Monitoring Science and Technology Symposium: Unifying Knowledge for Sustainability in the Western Hemisphere; 20–24 September 2004, Denver, Colorado.* Proceedings RMRS-P-37CD. US Department of Agriculture Forest Service, Fort Collins, Colorado, CD-ROM.

Briske, D.A. (2008) Information technologies for rangeland monitoring: what do they need to address? In: Organizing Committee of IGC/IRC Congress (eds) *Proceedings of the International Rangeland Congress, Multifunctional Grassland in a Changing World,* Volume I. Guangzhoui Guangdong People's Publishing House, PRC, pp. 633–637.

Chen, L. and Liu, G. (2003) Study on evaluation of the net primary production of vegetation in China by remote sensing. *Journal of Remote Sensing* 3, 129–135.

Dale, V.H. and Beyeler, S.C. (2001) Challenges in the development and use of ecological indicators. *Ecological Indicators* 1, 3–10.

Huang, J., Wang, R. and Hu, X. (1999) Study of model for yield estimation of grassland in the north of Xinjiang province. *Transactions of Zhejiang Agricultural University* 25(2), 125–129 (in Chinese with English abstract).

Li, J. (1994) Study on yield estimation of grassland on a large scale by NOAA/AVHRR data. *Natural Resources* 4, 365–368 (in Chinese with English abstract).

Liang, T., Chen, Q. and Wei, Y. (1996) Study on monitoring model of rangeland production dynamics in Fukang, Xinjiang. *Remote Sensing Technology and Application* 3, 27–32 (in Chinese with English abstract).

Liu, A. and Han, J. (2007) Remote sensing monitoring for the key phenological stages of rangeland. *Acta Agrectir Sinica* 3, 201–205 (in Chinese with English abstract).

Liu, A. and Xing, Q. (2004) Study on estimation model of production of rangeland based on MODIS-NDVI. *Prataculture Science* Supplement 2005(1), 15–19 (in Chinese with English abstract).

Liu, A., Wang, J. and Han, J. (2007) Study on method of estimating net primary production of rangeland by remote sensing. *Chinese Journal of Grassland* 1, 31–38 (in Chinese with English abstract).

Ludwig, J., Tongway, D., Freudenberger, D., Noble, J. and Hodgkinson, K. (1997) *Landscape Ecology, Function and Management: Principles from Australia's Rangelands.* CSIRO, Melbourne, Australia, 372 pp.

Shi, P., Li, B., Li, Z. and Hu, T. (1994) A study of technique on yield estimation of rangeland on a large scale by remote sensing. *Grassland Journal* 1994(1), 9–13 (in Chinese with English abstract).

Squires, V.R. (2007) Detecting and monitoring impacts of ecological importance in semi-arid rangelands. In: El-Beltagy, A., Mohan, C., Saxena, M. and Wang, T. (eds) *Human and Nature – Working Together for Sustainable Development of Drylands.* Proceedings of the Eighth International Conference on Development of Drylands, 25–28 February 2006, Beijing, China. ICARDA, Alleppo, Syria, pp. 718–723.

State Science and Technology Commission (1994) *A Government White Paper on China's Population, Environment and Development in the 21st Century.* China Environmental Sciences Publishing House, Beijing.

Su, D., Liu, J., Zhong, H., Xiang, B. and Shao, B. (2005) A study on remote sensing and other technologies applied in the quick survey of China's grassland resources. *Acta Agrectir Sinica* 2005(1), 4–9 (in Chinese with English abstract).

Uchida, S.A.T., Chen, Y. and Saito, G. (2002) *Application of Remote Sensing Technology for the Management of Agricultural Resources.* Agricultural Science and Technology Press, Beijing, pp. 233–237.

Western, D. (2003) Reflections on rangeland monitoring. In: Allsopp, N., Palmer, A.R. and Milton, S.J. (eds) *Rangelands in the New Millennium.* Proceedings of the VIIth International Rangeland Congress, Durban, South Africa, Vol. VII. NISC Pty, Durban, South Africa, pp. 681–682.

Yang, X., Zhang, K. and Ci, L. (2007) Vegetation-based analysis of *a priori* individual settlement-site-centered degraded gradient in semi-arid *Leymus chinensis* dominated steppe. *Journal of Vegetation Science* 19, 245–252.

17 How Can the Next Degradation Episode be Prevented?

Victor R. Squires[1] and Yang Youlin[2]
[1]*University of Adelaide, Australia;* [2]*UNCCD, Regional Coordination Unit, Bangkok, Thailand*

Synopsis

This chapter assesses whether the next degradation episode can be prevented. The historical case studies described in the book represent a failure to manage for the extreme climate variability that characterizes the vast rangeland/grassland/steppe/desert steppe (the pastoral rangelands) in northern and western China. Thus, they represent a historical 'test bed' for our current scientific understanding of pastoral rangelands and the government authorities' and land users' responses. The challenge is to overcome decades of neglect of rangelands as ecosystems, and their exploitation for economic gain, and of inappropriate policy interventions (with unintended consequences) against a background of negative attitudes towards herders.

Some views about the benefits of using drylands in a sustainable way are outlined and some likely consequences of allowing further dryland degradation to occur are discussed. The long-term future of traditional nomadic pastoralism is discussed in the light of environmental change, economic transition, social development and political reformation.

There are still a few options open to dryland inhabitants, but these are being reduced as each year passes. This chapter tries to set some parameters for future generations to ensure intergenerational equity on resource consumption, while at the same time allowing the present generation to prosper, or at least survive.

Keywords: environmental change; lessons learnt; underlying causes; policy interventions; drought; land degradation; recovery; intergenerational equity; future of nomadism; cultural homogenization

17.1 Key points

1. This book obtains and distils the most important concepts, findings and suggestions for future directions in China's pastoral lands. The volume provides a 'snapshot in time' of the pastoral lands in China's vast north and north-west at the start of the 21st century. It serves as a benchmark which not only helps us all to identity the underlying causes of the present-day situation, but also helps us see what needs to be done to improve the status of these important natural resources. A further step is to develop priority actions and to identify resources (human, financial and legislative) that can be applied to mitigate the current situation.

2. There exists a very broad spectrum of systems that are being applied now to exploit the pastoral rangelands (grassland/steppe/desert steppe) – many of which lower the efficiency of resource use and may even accelerate the degradation of the natural resource base.

3. Four important themes emerge from the analysis presented in this book:

(i) *The issue of sustainability*. There is a clear need to maintain the basic resources upon which all rangelands depend. The challenge is how to ensure that the level of production from pastoral rangelands matches the long-term maintenance of the resource base. There are so many areas in the pastoral rangelands that are not assessed routinely for changes in status and trend in condition of their resources (Chapter 16). The prevailing monitoring usually focuses only on biomass production and stops short of assessing changes in the health of the rangeland ecosystem.

(ii) *The importance of socio-economic processes* in resource management. In China, 'grassland science' has commonly dealt with only the biophysical constraints of livestock production on rangelands, and specifically on the plants, with little integration of the livestock into a 'system'. There have been numerous instances highlighted in the eight case studies that factors other than biophysical ones also limit management options and the sustainability of rangeland enterprises, industries and communities. It is clear from the detailed analysis of these eight case studies that socio-economic factors are crucial in determining whether particular technical solutions to rangeland problems can be employed in practice and on a large scale.

(iii) *The importance of dealing with interrelationships.* Rangelands do not exist in isolation from non-rangeland systems. This is true in biophysical terms because of the natural and anthropogenic interchange of materials and energy between rangelands and non-rangelands. It is also true in socio-economic terms because of the cultural and economic exchanges between rangeland and non-rangeland human communities and societies. Rangelands cannot be managed in isolation from the non-rangeland areas with which they interact. Increasingly, rangelands and the human communities of rangelands are influenced by decisions made principally by communities living outside the rangelands. Cultural homogenization is a factor, too. The pastoral rangelands of China still harbour relatively intact and different human populations and a diversity of cultures. However, homogenization of the cultures of the rangelands, akin to that already occurring in the cities, may occur as a result of a number of forces. These include uniformity of production methods to meet mainstream commodity standards at the expense of traditional methods and livestock types, constraining management and investment activities to stereotypes that meet the perceived needs. The exposure of rangeland to external media may also lead to greater homogeneity of expectations in terms of lifestyles and material goals.

(iv) *Institutional capacity for change* is a factor. The institutions that are charged with, or assume, management and development responsibilities with respect to rangeland ecosystems and societies have done badly in the past 70 years and are often poorly equipped to deal with the next few decades of dramatic change that must inevitably occur. Successive institutional change (from nomads to state control to privatization of livestock and private control over capital investment) has tended to intensify rangeland use, but without recognition of the environmental and socio-economic constraints under which rangeland users operate, resulting in marginalization of pastoral communities, increasing inequities and decreasing resource condition (Wu and Richard, 1999). The new, more sedentary, pastoral system that has been promoted in recent times lacks the flexibility that enables effective response to environmental changes. Furthermore, vulnerability to change may increase as a multi-resource economy is replaced with a livestock system that is characterized by rigid marketing, often with a single product and prices that do not reflect the true cost of production.

4. Furthermore, the ecological functions of rangelands are not consistent with those that lead to economic returns to the people who use the rangelands. In many rangelands, there are an inadequate understanding of how landscapes function and, often, poor commitment by various jurisdictions in appropriate monitoring that can provide management feedback for land users and decision makers.

17.2 Rangelands as Ecosystems Under Siege

The rangelands of China have been described by some commentators as a resource under siege. In fact, it seems that few of China's rangelands remain in the 'under threat' category; siege has already been laid and the battle for the resource base has begun. The socio-political-economic environment described in the eight case studies gives little cause for optimism, but there are improvements (recovery) that need to be replicated and scaled up to deal with the vast areas undergoing accelerated land degradation.

Documentaries shown on television reveal widespread destruction of plants and soils, and the ravages of dust and sandstorms and of floods are more common now and bring home to the largely urban population of China, and the world, that things have gone seriously wrong in China's pastoral rangelands. The eight case studies described here are a cause for alarm and reinforce the message that increasing numbers of people are seeing on television. But the analyses presented here of the rangeland resource, its functioning, its improvement and its management are encouraging. The siege can, and must, be lifted.

17.2.1 Understanding ecosystem function and process

Until recently, arid and semi-arid rangelands, more than any other system, have been associated with land degradation. The concept of 'desertification' originated from a perception of degrading fringes of arid rangelands and advancing deserts. In addition, the arid grazing systems and pastoral production modes have been described as inefficient and backward production systems. But these views need to be re-examined. First, there is now evidence that, in the arid, semi-arid and dry subhumid zones, the extent of land degradation is greatly exaggerated (Wang, 2006) and that we are dealing with a highly resilient ecosystem, as shown by the 'expanding and contracting' Sahara found by Tucker *et al.* (1991). Secondly, there is convincing evidence that traditional mobile production systems on pastoral rangelands can be highly efficient (Ellis and Swift, 1988). In addition, arid grazing systems often have multiple uses, with wildlife and other plants (e.g. medicinal) being important products.

The indisputable fact is that the total area of degraded rangeland in China increased by about 95% between 1988 and 1998, from about 65 million ha (Mha) to about 30 Mha, with a notable acceleration in the mid-to-late 1990s. From all accounts, it is still expanding. This land degradation is caused by a combination of natural and human factors such as inappropriate land-use policies, inadequate rangeland management and overharvesting of rangeland products. The human-induced factors are exacerbated by: (i) an overall poor understanding of the functioning and resilience of ecosystems; (ii) lack of awareness by various levels of administration and management of the medium- and long-term environmental impact of interventions, for example, technologies such as stall-feeding (see Box 17.1); and (iii) a failure to seek an objective analysis of the past decades of mismanagement.

Year-to-year changes in livestock number in many parts of China over the past decades are compared with rainfall and other indices in all of the eight case studies. Analysis reveals two main points: (i) there has been an overall increase in livestock numbers and a higher level of output of livestock products; and (ii) there is a confounding of the cause and effect relationships. The impact of climatic variability on livestock numbers and carrying capacity is often discussed with a view to assessing future stocking policies, but the indisputable fact is that many of the underlying assumptions are based on false reasoning. The first and most important assumption is that herders act irrationally by trying to maintain larger herds and by pursuing a mobile lifestyle (Heshmati and Squires, 2007). There is a lack of understanding of the fundamentals of livestock production in marginal areas, which is characterized by cycles of 'boom and bust'. Over the centuries, herders have devised coping strategies to allow them to survive. Many of these strategies are no longer relevant or suitable in the changing socio-economic climate. For example, degradation of both land and vegetation is occurring around settlements as a new form of 'village-based herding' emerges. Denuded areas are mostly within a radius of about 1–5 km of the villages (Yang *et al.*, 2007).

Several policy measures have been introduced and implemented over the past 50 years or so and these are summarized below:

1. 'Stabilization' of the system. Often, well-intentioned policies sought (and still seek) to stabilize the 'boom and bust' cycles that existed traditionally between humans, animals and

Box 17.1. Shifts in the number of livestock supported on the rangelands in Inner Mongolia.

Data on overall livestock population, as published in the statistical yearbooks of Inner Mongolia, show a continued increase but, as shown in Fig. 17.1, the livestock population *wholly dependent on natural grassland* in Inner Mongolia has decreased since 1990. This apparent anomaly is explained by the fact that many livestock now depend on energy inputs (fodder, grain) from outside the pastoral system. Moreover, this energy inflow from outside has been increasing rapidly under the Household Production Responsibility System (HPRS), for three reasons: (i) Herders have to buy forage from outside, such as in cropping areas, to maintain their herds during natural disasters, as the HPRS prevents them from moving livestock to other places to avoid disasters. (ii) In order to prevent serious livestock loss from natural disasters, herders, communities and local governments have invested significant effort in building systems to mitigate disasters. (iii) The area of natural grassland has continued to decrease and degrade due to reduction in grazing scale, repeated trampling by livestock and increase in conversion of rangeland to cropland to grow more fodder and grain. Therefore, the proportion of livestock living on energy from outside the pastoral system has increased sharply under the HPRS. Unfortunately, this cannot be shown through statistical data and this has led to the false presumption that overgrazing is universal on rangelands/grassland/steppe/desert steppe in Inner Mongolia. Measures taken to eliminate overgrazing and restore rangeland based on this false assumption are unable to achieve their objectives.

Dalintai (2005)

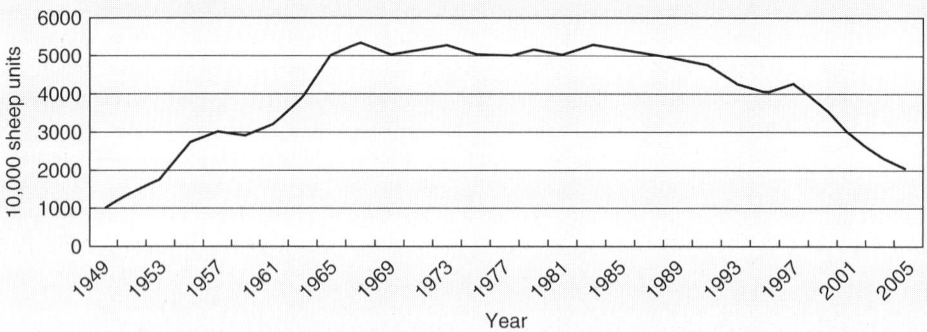

Fig. 17.1. The change of livestock population (sheep units) depending on natural grassland of pastoral area of Inner Mongolia (1949–2005).

vegetation in pastoral rangelands, through settlement of pastoralists and attempts to regulate and control the stocking rate.

2. Feed subsidies for disaster relief (leading to massive energy inputs from outside the pastoral system after widespread and severe snow disasters or during and after prolonged drought).

3. Changes in access to land; 'privatizing' communal areas through the HPRS system, carving them up into small plots, which do not allow the necessary mobility.

4. Applying inappropriate financial incentives that send the wrong signal.

5. Encroachment of crops (land conversion) into the 'key resource' sites of the herders.

17.2.2 Livestock and their impacts on the global environment

Livestock production, mainly as a result of pressures on this process, has become an important factor in environmental degradation. Large land areas have become degraded through overcropping, overgrazing and the concomitant loss and/or degradation of vegetation. All these pressures on the environment are the results of a metamorphosis, where the role of livestock has been altered due to rising and changing demands for livestock commodities by the burgeoning urban population and to a different perception of the environment that has arisen in recent decades.

Biodiversity may also be affected by extensive livestock production, although there are many cases showing increases in plant biodiversity in well-balanced grazing systems, especially those using multi-species (Milchunas and Lauenroth, 1993). The interaction between wildlife and domestic livestock in these ecosystems is complex. First, there is increasing evidence of complementarity and only limited competition of wildlife and livestock in grazing. The grazing 'overlap' between most wildlife species and domestic livestock is rather limited. The combination of livestock raising and wildlife management often results in an equal or better species wealth than any of these activities carried out individually. Pest species such as rangeland rodents are an exception and it is clear that considerable damage can be caused as local rodent population explosions occur. The scientific evidence is that overgrazing creates the very conditions that favour pika and some other rodents and that the populations soon rebound after 'control' measures are applied (Pech *et al.*, 2007; Arthur *et al.*, 2008). Long-term control depends on recreating the rangeland conditions that are less favourable to the rodents and result in the maintenance of low population densities.

However, some would question whether irreversible rangeland degradation is occurring at all, and just what the role of livestock is in that process. Livestock do not move, produce or reproduce without human intervention. Livestock do not degrade the environment – humans do. As a result of these misconceptions about livestock development, institutions and governments continue to miss opportunities which would permit the livestock sector to make its full contribution to human welfare and economic growth. It should be noted that livestock are an important source of gaseous emissions, contributing to global warming, which is projected to increase further by >2.0°C worldwide over the next 3–4 decades.

In essence, the conflict between livestock and environment is a conflict between different human needs and expectations. In many affected regions, livestock production is growing out of balance with the environment or is under so much pressure that it leads to environmental degradation. Finding the balance between increased food production and the preservation of the world's natural resources remains a major challenge. It is clear that food will have to be produced at less cost to the natural resource base than at present.

17.3 Setting the Parameters for the 21st Century

There are still a few options open to dryland inhabitants, but these are being reduced as each year passes. This section sets some parameters for future generations to ensure intergenerational equity on resource consumption, while at the same time allowing the present generation to prosper, or at least survive.

At a time when traditional lifestyles are under threat throughout the world, it seems appropriate to ask: 'What future is there for traditional (non-commercial) range/livestock production systems in western China?' The strongholds of nomadic and semi-nomadic pastoral systems (Xinjiang, Tibet, etc.) are undergoing rapid evolution as market-based systems overtake them.

At the same time, concerns about the sustainability of the resource base have emerged with the ratification, by many nations, of international conventions on *biodiversity* (CBD) and on *desertification and drought* (UNCCD). Many people are left to wonder if the environmental damage done to rangelands, especially on desert margins, will bring about the demise of traditional pastoralism. In general, ecosystems are grazed because they are not sufficiently productive or reliable to be cropped. This means that management must cope with low or unreliable production, complex semi-natural systems, large management units and greater economic risk.

The key constraints that have plagued pastoralism in the past include the technical (animal health and nutrition), the socio-political (land tenure, policy issues, religion) and the economic (marketing) issues. These constraints deserve some elaboration.

There are four recurring themes that deserve a lot more attention in China, which, to date, is hooked on technical 'fixes' without taking the socio-economics (apart from wealth creation) into account:

- the issue of sustainability (ecological, economic, cultural);
- the importance of socio-economic processes in resource management;

- the importance of dealing with interrelationships between the people who use the land and the means of utilization and exploitation of the resource base; and
- institutional capacity for change.

New socio-economic issues have arisen that impact on the way in which traditional societies view the future. Demands for education for their children, better health, higher expectations for their children and a desire for a more technologically based lifestyle (radio, television, satellite communications, motorized transport) have shifted priorities.

Changes in land tenure, security of access, the emergence of the market economy, rising domestic demands and growing exports as globalization of trade and commerce take hold have all played their part. These shifts call for a different set of institutions, markets and policies. They also call for the development and adaptation of new technologies to make livestock production environmentally more benign – the scope is enormous and so is the task.

Emerging problems, including the conservation of the resource base (including biodiversity issues), the globalization of the world economy, the breakdown of tradition and the potential impacts of climate change, prevail worldwide, particularly in affected developing countries. The challenge is to find ways of managing drylands that are more environmentally sustainable, economically viable and socially equitable than at present.

17.4 Problems and Prospects: Whither Traditional Pastoralism?

Over the longer term, the fundamental driving force on natural resources is population pressure, especially influences that impact on pastoral land users from outside the rangelands. While the population growth of the pastoral peoples has been rather low, the growth of non-pastoral groups in the arid and semi-arid regions has been among the highest in the world. This growth of other groups causes an increasing encroachment by crop farmers on the pastoral 'key resource' sites and constrains the critical mobility necessary to adjust to the disequilibrium conditions (Chapter 15).

The increased population pressure also leads to water development (reservoirs, artificial oases) and settlements in rangelands (Chapter 11). Also within the system, population pressure mounts and causes land degradation. Thus, in spite of the resilience of the system, many pastoralists face a downward spiral of increased crop encroachment, increased fuel-wood requirements and decreased grazing availability (Squires and Sidahmed, 1987). These forces contribute to impoverishment of the pastoral population and to rangeland degradation. This trend is being exacerbated by drought, and vulnerability to drought is one of the main indicators of long-term environmental and social sustainability of these arid grazing systems (see below).

The prevailing combination of poverty and high population growth, which characterizes many counties of the semi-arid regions of western China, cannot be broken easily. The *first priority* is an overriding need to stop the building up of further human pressure in arid and semi-arid zones through adequate control of population growth and creation of alternative employment generation policies.

As the *second priority*, external interventions in the system need to acknowledge the disequilibrium status of the pastoral systems in the arid and semi-arid zones (Chapter 3) and respond to their need for flexibility and mobility. This means that attempts to regulate stocking rates need to be stopped. Even apart from the technical flaws in the estimation of the carrying capacity, experience has shown that it is almost always impossible to enforce these stocking rates.

The *third priority* should be to empower traditional pastoral management and develop effective co-management regimes, forging partnerships between the state and a wide variety of users, with the state carrying the overall responsibility for arbitrating conflicting interests at the national level and facilitating negotiation between the multiple-level stakeholders. Access to land is to be based on customary resource user rights, however, avoiding rigid territorial boundaries.

The *fourth priority* is the identification of effective disaster relief policies (such as snow blizzards) and/or to revisit drought management, preparedness and mitigation policies.

Finally, the *fifth priority* should be to establish appropriate incentive policies by:

1. Increasing the costs of grazing on the rangeland/grassland/steppe/desert steppe, which can reduce animal pressure by promoting an earlier offtake.
2. Full-cost recovery, especially for water supply and animal health services. The water resource has been, in many cases, a free good supplied by the public sector (and frequently financed by the international donor community).
3. Removal of price distortions for other agricultural inputs, in order to reduce the conversion of pastoral key resources into marginal cropland.

In the light of the above observations and from an analysis of the eight case studies presented in this volume, it can be concluded that traditional nomadic pastoralism will become less and less important. There will always be those who wish to utilize (exploit) the otherwise unusable forage and water resources of the drylands, especially in the mountains, but social, economic and political pressures will see the demise of this way of life within the next generation.

The key areas where research is needed urgently emerge from the above recommendations. They include the identification of:

1. Appropriate indicators which provide reliable information on the resource trends in the arid and semi-arid areas.
2. Appropriate methodologies for economic appraisals of the investment in converting such 'key resources'.
3. Factors leading to strong pastoral institutions.
4. Sustainable drought preparedness plans, with particular emphasis on decentralized management and the design of appropriate rural credit and insurance schemes, and appropriate conflict resolution schemes.

17.5 What Can Be Learnt that Will Prevent Another Degradation Episode?

The analysis contained in this book confirms the importance of major issues related to pastoral rangelands in arid and semi-arid zones, such as the difficulty of clearly detecting degradation effects on production in highly variable environments (Chapter 16) and identifying the impact of global warming and climate change on degradation. This study emphasizes how, in a region of such spatial and temporal diversities, the decision makers (herders on a day-to-day, season-to-season basis) and land administrators and regulatory staff on a longer-term basis need to develop an understanding of the unique situation for each seasonal grazing land in each village when formulating a management strategy. These strategies cannot come from generalizations, even at the prefecture or county level. Because individual townships encompass much of this diversity, generalized production strategies are difficult to devise and rangeland (grassland/steppe/desert steppe) resources are difficult to describe and assess meaningfully.

A major issue apparent from these case studies was the lack of warning provided to governments at higher levels regarding the emerging resource degradation. With the development of remote sensing using satellite imagery and rangeland/grassland management techniques, the capability to assess rangeland/grassland condition comprehensively is much improved. Various national and local-level monitoring and the system of reporting and analysing outlined in Chapter 16 should ensure that degradation of the magnitude of past episodes (case studies Chapters 7–14) does not go undetected.

The use of inappropriate ecological models has perpetuated the concept of carrying capacity and the idea that land has some fixed capacity to carry and feed livestock. The introduction of maximum stock numbers in Inner Mongolia under the policy of 'Yi Cao Ding Mu', that is, 'to decide the number of household animals according to the availability of grass', as a mechanism to limit land degradation has proved ineffective. It has contributed to management problems and has been applied with little consideration of the biota and vegetation dynamics under grazing or of the land users (Williams, 1996; Zhang, 2006). Setting the rules on herd size, for example, has particular difficulties. Unlike other controls which may close rangelands to grazing, or limit grazing to animals of particular kinds or seasons of use, where breaches are obvious, with flock size limits on individual householders it is often not evident to the casual or even the official observer that breaches are occurring. Despite the most rigorous checks, misreporting of herd size is considered to be common in many quota-based systems.

Assessment of the status of the resource base in landscapes with great diversity of vegetation, landform and degradation and recovery process is extremely complex, requiring the cooperation of individual herder households to provide local knowledge to be melded with data obtained on the broader scale by remote sensing. Land degradation and desertification in China's pastoral regions are phenomena that cannot be represented by single variables or simple indexes. According to Reynolds and Stafford Smith (2002), there are two key propositions. The first is that the myriad of variables involved in land degradation (and the underlying processes they represent) can be classified into two, independent types, namely biophysical versus socio-economic, and speed of change, i.e. relatively fast versus slow processes. The second proposition is that the supply of desired ecosystem goods and services is governed by a *subset* of a few of those variables, which include both biophysical and socio-economic ones, and are always variables of relatively slow dynamics.

The areas affected most severely by degradation were those that were grossly overstocked compared with the calculated carrying capacity. Once these areas were severely degraded, they were less resilient in terms of soil stability and forage production and quality for many years, even decades. Recovery is usually slower than loss; that is, there is a hysteresis effect. As perennial grasses are lost, soil cover and litter generally decline, soil porosity is lost, there are fewer interruptions to surface water flow and, together, these changes result in reduced infiltration. When the perennial grass percentage declines below the threshold, the biophysical system is degraded and it will take considerable intervention or an unusual sequence of favourable seasons to return it to a productive state. This is so because the soil deterioration that accompanies loss of perennial grasses results in an environment that makes it difficult for perennial grass plants to germinate and establish (Ludwig and Tongway, 1997).

Understanding of the impact of climate change is still a topic of research, especially on the question of the extent to which biosphere feedbacks can amplify or dampen the effects of climate forcings, including the interdecadal variability of rainfall in arid environments (Chapter 3). Separation of the climatic and management signals in changes in resource condition is a major challenge, as is the detection of recovery and irreversible degradation. The sensitivity of elements of the animal husbandry industry to closer supervision by government agencies on the one hand and the lack of a willingness to take local knowledge into account on the other are also a factor.

There is strong anecdotal evidence from the case study areas, as well as research results from abroad, that the prime cause of rangeland degradation is the maintenance of excessive grazing pressure on sensitive grazing land types during drought periods. Management of sensitive grazing land during drought is therefore of overriding importance in *preventing* the onset of rangeland degradation on sensitive grazing land types. It follows that a drought assistance policy which creates an incentive for herders to reduce grazing pressure during the early stages of drought is a high priority. The combination of drought and high stock numbers results in the accelerated reduction and loss of perennial grasses. The pressure on any surviving grasses in the *post-drought* period leads to very little opportunity for the grasses to expand and provide adequate ground cover. This has resulted in large expanses of bare ground. These bare land surfaces offer little resistance to raindrop impact erosion and runoff. The increase in runoff results in serious loss of extremely limited resources such as seeds and nutrients and organic matter, placing the area in an artificial 'man-made drought' situation.

The eight degradation case studies described here involved severe and extended drought periods, which revealed and amplified the extent of resource damage. In single years prior to the initial drought year (taken as year zero), they experienced short droughts, as well as years of above-average conditions. The time series of livestock numbers indicate that patterns of livestock numbers alone cannot be used to identify the causes of degradation. The difficulty is that there is no obvious short-term ecological penalty for high utilization rates of rangeland (grassland/steppe/desert steppe), especially in the sequence of good years that often preceded the degradation case studies. However, it is clear that such stocking rates can 'prime' the system for degradation to occur during the subsequent drought.

If herders had increased herd/flock size in the sequence of good years in an attempt to maximize income, it is possible that the historical livestock numbers during the base period represent utilization rates well above the recommended 50% of forage growth. Single-year droughts during the pre-degradation episode would have imposed even higher grazing pressure, but any damage would have been masked by a return of favourable rainfall conditions. Retention of higher numbers of livestock into the drought led to even higher utilization of possibly >70%, sufficient to damage the perennial forage species severely and remove the protective layer of spoil. Retention (or replacement of stock lost in the drought) during the recovery phase would further exacerbate the grazing pressure on the remaining perennial forage plants and any regenerating seedlings.

The extended drought periods in each of the case study areas have provided a test of the capacity of grazing systems (i.e. land, plants, animals, people and the social structure) to handle stress. In most areas, evidence that degradation was already occurring was identified prior to commencement of the extended drought sequences. The sequence of dry years ranging from 2 to 4 years exposed and/or amplified the degradation processes. The unequivocal evidence was provided by: (i) the physical 'horror' of bare ground and dead livestock, erosion scalds and gullies, dust and sandstorms; (ii) the biological devastation and animal suffering/deaths or forced sales; and (iii) the financial and emotional plight of the herders and their families due to reduced production and loss of income.

A major question raised by the case studies is to what extent they could have been avoided or mitigated by better grazing management and fewer misguided policies. During and following the severe droughts experienced in the case study areas, it was debated whether the soil erosion and loss of 'desirable' perennial grasses and shrubs were the results of extremes of climatic variability or caused by too many animals. For example, some contemporary accounts place the blame on the climate rather than overstocking, but other observers say that overgrazing killed the desirable species over many thousands of hectares and, when the plant cover had disappeared, the soil was bare and unprotected and at the mercy of wind and water. Degradation in response to heavy utilization would appear to be unequivocal. While the vegetation changes appear reversible over time, the loss of soil through wind and water erosion is not. In reality, it is not one or the other. Climatic variability and grazing pressure interact (Chapters 5 and 6), particularly, when human intervention is misdirected during specific periods.

17.5.1 Commonalities to emerge: an opportunity to learn

The most important determinant of degradation was the rapid rise of human and animal populations in areas that previously supported a sparse population of herders from the various ethnic minorities (Tibetan, Mongolian, Kazak, Uygur, Yugur). Those areas were subject to massive inward migration and the conversion of all suitable (and some unsuitable) rangelands to cropland. This fact precluded herders from their customary 'refuge' area where water was abundant and forage could be kept in reserve for dry seasons.

Governmental policies were generally applied to agricultural regions and pastoral regions without regard to the differences in the reliability of the climate or the local knowledge that had allowed herders to utilize the harsh environments for millennia. Many policies had unintended consequences as China struggled with post-war reconstruction and food shortages. It is unlikely that many of the more extreme measures practised in the period from 1950 to 1980 will recur, but present-day policies that are counterproductive and even potentially devastating should be revisited and amended before it is too late.

Analysis of the conditions on the pastoral rangelands that led to the devastating degradation episodes, although by no means the same for each region, have many commonalties. It is this repetition of factors, common to events in different places and at different times, that suggests that future impacts may be reduced.

1. There was a general overexpectation of safe carrying capacity by herders, local administrative and technical staff and governmental departments at higher levels.
2. Livestock numbers increased in response to a period of mainly above-average rainfall that preceded the drought/degradation episode.

3. Intermittent dry seasons (or years) resulted in heavy utilization, damage to the 'desirable perennials' and, ultimately, the rangeland resource. This led to the collapse of the capability of the land to carry animals at the onset of severe drought.
4. Extreme levels of utilization in the first years of the drought by retaining herd/flock size caused the further loss of perennial species, exacerbating the effects of the drought in subsequent years.
5. Rapid decline in or generally low commodity prices (wool, meat, hides) resulted in herders retaining livestock in the hope of better prices or the fear of the high cost of restocking.
6. Continued retention of large herds/flocks through a prolonged drought compounded damage to the resource and delayed recovery or revegetation of degraded rangeland.
7. The sequence of drought years resulted in rapid decline in vegetative cover, which revealed the extent of previous resource damage and further accelerated the processes and intensity of degradation.
8. Partial recovery of rangeland/grassland occurred during sequences of above-average years, sometimes decades after the major degradation episode. This points to the merit of developing time-series data and looking at the sequence of good, average and dry years as part of the arsenal available to manage better for climate variability.

The present-day restrictions on herder mobility, the heavy reliance on the import of energy and feed supplements, the reduced area of rangeland, the ever-spreading area of degradation and the rising numbers of livestock suggest that China's pastoral rangelands are far from secure. However, there is enough accumulated experience now to prevent widespread catastrophe if the lessons of history are taken into account when charting the course of rangeland monitoring and management over the next few decades as China strives for sustainable use of the vast areas of rangeland, which includes true grassland, shrub/grass steppe and desert steppe.

The two most important 'take-home' messages are:

- Land degradation is a problem that encompasses both biophysical and socio-economic dimensions and cannot be comprehended adequately by focusing on just one of them.
- The need for prevention, an approach that is not only better but also more cost-effective than after-the-fact action (i.e. remediation).

References

Arthur, A.D., Pech, R.P., Davey, C., Jiebu, Zhang, Y. and Lin, H. (2008) Livestock grazing, plateau pikas and the conservation of avian biodiversity on the Tibetan plateau. *Biological Conservation* 141(8), 1972–1981.

Dalintai, A. (2005) Rethinking grassland desertification. *Journal of College of Finance and Economics of Guizhou* 3, 46–50 (in Chinese).

Ellis, J.E. and Swift, D.M. (1988) Stability of the African pastoral ecosystems: alternate paradigms and implications for development. *Journal of Range Management* 41, 450–459.

Heshmati, G.A. and Squires, V.R. (2007) A number of ecological propositions and their implications for rangeland management in drylands. In: El-Beltagy, A., Mohan, C., Saxena, M. and Tao, W. (eds) *Human and Nature – Working Together for Sustainable Development of Drylands.* Proceedings of the Eighth International Conference on Development of Drylands, 25–28 February 2006, Beijing, China. ICARDA, Aleppo, Syria, pp. 438–442.

Ludwig, J.A. and Tongway, D.J. (1997) A landscape approach to rangeland ecology. In: Ludwig, J., Tongway, D., Freudenberger, D., Noble, J.C. and Hodgkinson, K. (eds) *Landscape Ecology Function and Management: Principles from Australia's Rangelands.* CSIRO, Melbourne, Australia, pp. 1–12.

Milchunas, D.G. and Lauenroth, W.K. (1993) Quantitative effects of grazing on vegetation and soils over a global range of environments. *Ecological Monographs* 63, 327–366.

Pech, R.P., Jiebu, Arthur, A.D., Zhang, Y. and Lin, H. (2007) Population dynamics and responses to management of plateau pikas *Ochotona curzoniae. Journal of Applied Ecology* 44(3), 615–624.

Reynolds, J.F. and Stafford Smith, D.M. (eds) (2002) *Global Desertification: Do Humans Cause Deserts?* Dahlem University Press, Dahlem, Berlin.

Squires, V.R. and Sidahmed, A.E. (1987) Livestock management in dryland pastoral systems: prospects and problems. *Annals of the Arid Zone* 36(2), 79–96.

Tucker, C., Dregne, H.E. and Newcomb, W.W. (1991) Expansion and contraction of the Saharan Desert from 1980–1990. *Science* 253, 299–301.

Wang, T. (2006) *Desert and Desertification in China.* Science Press/Longman Books Co. Ltd, Beijing.

Williams, D.M. (1996) Grassland enclosures: catalyst of land degradation in Inner Mongolia. *Human Civilization* 55(3), 307–313.

Wu, N. and Richard, C. (1999) The privatization process of rangeland and its impacts on the pastoral dynamics in the Hindukush Himalayas: the case of Western Sichuan, China. In: Eldridge, D. and Freundenberger, D. (eds) *People and Rangelands.* Proceedings of VI International Rangelands Congress, Townsville, Australia. International Rangeland Congress, Inc., Aitkenvale, Australia, pp. 14–21.

Yang, X., Zhang, K. and Ci, L. (2007) Vegetation-based analysis of *a priori* individual settlement-site-centered degraded gradient in semi-arid *Leymus chinensis* dominated steppe. *Journal of Vegetation Science* 19, 245–252.

Zhang, Q. (2006) May they live with herds: transformation of Mongolian pastoralism in Inner Mongolia of China. MSc thesis, University of Tromso, Norway.

Index

Page numbers in *italics* refer to tables and figures.

Abandoned farmland 4, 6, 7, 12, 101, 103, 114–115, 134, 143, 204, 231
Aerial sowing 146, 173
Afforestation 4, 73, 146, 152, 166
impact of 51
windbreaks 70
Agenda 21 213, 235
Agroecological zones 46
Albedo 229
Animal Husbandry Bureau 18, 213, 225
Arable land area 26
Archival records 16, 17, 79, 119, 183
Artificial grassland (sown pastures) 100, 116, 117–118, 213
Artificial oasis 7, 152, 152, 155–156, 163, 170
Asian Development Bank (ADB) 222

Biodiversity 4, 8, 9, 50, 93, 95, *113*, 150, 159, 162, 166, 165, 176, 192, 199, 228, 251, 252
Biological crusts 70–71
Biomass 125

C_4 plants 125
Carbon dioxide
flux 8, 51
sequestration 208
soil organic matter 41, 84, 187, 208
Carrying capacity 4, 46, 98, 119, 121, 161, 176, 177, 220, 223, 226, 227
concept of 228
feed deficiency 176
herd size limitation 253–254
overloading 98, 110, 109, *110*, 121, 132, 226

Climate
change 50–51, 137–138, 140–142, 156, 201, 230
climatic zones 50
Hexi corridor 152–154, 156
Horqin 105
variability 9, 33–34, 50–52, 103, 106, 124, 140–142, 177, 200–203
warming trend 33–36, 39–40, 51, 106, 108, 109, 124, 141, 177, 186, *201*
Cold arid regions 62
alpine meadow 192, 185, 186
snow disasters 178, 190, 229
Collectivization 17, 221
communes 18, 165, 220, 221
Commodity prices 256
Conflict between economic development and environment 145, 187, 198
Cryptograms *see* Biological crusts
Cultural homogenization 248
Cultural Revolution 17, 18, 20, 155
Communal grazing 47, 182, 186

Decline of perennial plants in rangelands 48–49
Deferred grazing 100
Desert encroachment 4, 6–7, 11, 12, 51, 156, 165, 197, 222
Desertification control techniques 73–74
Desertification Law 228
Deserts
Badain Jaran (Badain Jilin) 152, 168, 173
Gobi 42, 139, 153, 173, 174
Kubuqi 138, 145
Kuerbantonggute (Guerbantongte) 90, 197
Mu Us 4, 6, 40

Deserts (*continued*)
  Taklamakan 42
  Tenggli 168
Drought 8, 9, 11, 13, 16, 50, 52, 76, 80, 103, 178, 198, 256
  crucible of 13
  drought index 120, 142
  frequency 10
  preparedness 253
  sequences 256
Dust- and sandstorm (DSS) 7, 12, 48, 64, 62, 64, 71, 76, 80, 171, 173, 198–199
Dynasties
  Han 41, 43
  Jin 119
  Qing 17, 25, 119

Ecological refugees 4, 11, 126, 179, 222
Ecological services 48, 185, 182–183, 184, 186–188, 191
Economic indicators 201, 224
Economic rent gradient 79
Ecosystem management 131–132, 131, 168, 169, 249
Environment Protection Bureau *see* SEPA
Environmental impact 9, 149
Eroding rainfall events 66
Exclosure – long-term 53, 99, 119
Extension service 225, 232
Extreme weather events 220, 229, 252
  snow disaster 190, 229, 230

Farming–nomadic interface zone 25, 26
Feed balance 47, 181–182, 204
Felsenmeer 8
Fencing 99, 132, 134, 220, 232
Fuel wood 7, 8, 9, 93, 110, 136, 138, 152, 161, 177, 204, 205

Glacier 42, 43, 51, 156, 202, 230
  meltwater 42, *43*, 156, 161, 172, *202*, 230
Gachas
  Gong Dalai 226
  Hatuhuduge 223–225
    household financial returns *224*
    land allocation and stocking rates *223*
Grain for Green 148, 169, 179, 182, 194
Grasslands of China 3
Grassland Monitoring Stations 19, 221, 244
Grassland Supervision Office 98, 205, 221, 244
  Rangeland police 20
Grazing Ban Programme 11, 12–13, 57, 87, 119, 180–181, 194, 199, 149, 169, 210, 213, 225, 242
  restructuring/land reform 17, 178, 181, 212, 213, 214
Grazing gradients 236
Grazing history 17, 127, 204
Grazing management 21, 45, 57, 178, 189, 230
  spring deferment 18, 230
Grazing systems 47, 131
  agro-pastoral 45, 129, 131
  open-access systems 21
  transhumance 8, 17, 45, 47, 122, 128, 131, 186, 204
Grazing user rights 130, 132, 178, 221, 223–225, 230, 231–232
Great Leap Forward 20
Great Wall 23, 219
Groundwater 36, 103, 105, 145, 150, 152, 156, 165, 168, 222

Hei River Water Allocation Scheme 166
Heilongjiang Province 170
Herders 4, 8, 9, 11, 24, 27, 57, 101, 127, 186, 194–195, 212, 213
  attitudes towards 23
  community participation in decisions 232, 254
  constraints on 213
  cooperatives 231
  experienced versus inexperienced 22
  household income 24, 224
  land allocation to 13, 19, 223–225, *227*, 230
  living standards 149, 180
  nomadic culture 8, 17, 122, 127, 130–131, 189, 195, 199–200, 203, 204, 219
  profile of 128
  resettlement 8, 12, 22–23, 27, 95, 194, 205–206, 212, 250
    implementation of 212
Historical records 5, 11, 15, 16, 122, 127–128, 154–155
Household responsibility system 18, 20, 21, 133, 186, 190–191, 205, 221, 223, 226, 231–232, 242
  contract system 96–98, 130, 132, 205
  property rights 205
  resource management 194
  setting stocking rates 225–227
  supervision systems *see* Grassland monitoring and supervision
Human impact on ecosystems 8, 11, 16, 18, 20, 22, 42, 63, 91, 98, 107, 119, 154, 161, 171, 183
Hydrology 163, 164, 167, 222

Indigenous knowledge 127, 130
Infiltration rates 65
Infrastructure loss 103
Inland river basins 6, 156–157
Inner Asia 226, 230
Insects 9, 222, 229
  see also Locusts

Institutional capacity for change 248
Intergenerational equity 248
Investment, lack of 206

Japan 149

Lakes 6, 41
  drying of 4, 6–7, 33, 162, 164, 165, 170
Land allocation 18, 223–225, *227*, 230
Land reclamation *see* Rangeland conversion
Land tenure 21, 24, 25, 96, 96–97, 101, 178, 219–220, 222, 226, 252, 256
Land-use change 41, 96–97, 156, 160
  fragmentation of rangelands 24–25, 26
  intensification 24–25, 26, 47–48
Law of Diminishing Returns 226
Livestock
  breed improvement 179
  changes in balance between species 228
  livestock–rangeland balance 98, 121–122, 123, 151, 214, 223
  losses 47, 123, 178, 190
  mobility 101, 226, 231, 233, 256
  ownership 18, 21, 23
  population 11, 12, *41*, 46, 103, 121, 142–143, 157, 189
  production quotas as a contributor to overstocking 220
  products, statistics for 13, 201
  replacement after disaster 255
  watering facilities 221
Locally adapted species
  livestock 184
  plants 81, 82–83
Locusts 55, 57, 194

Market economy 24, 101, 226, 228, 231
  price distortion 253
Medicinal plants 139, 161, 170, 177, 204, 205
Migration 8, 23, 24, 26, 57, 101, 119, 155, 170, 174, 179, 222
  ecological refugees 222
  outmigration 57, 132
  seasonal *see* Transhumance
Mineral deficiencies in livestock 232–233
  mineral block licks 232
Miniaturization of plants 49
Mining 40–41, 73, 136, 145, 149, 156, 160, 181
  contamination of water 168
Ministry of Agriculture 237, 240
Minority nationalities 45, 57, 153, 154, 155, 174, 181, 199, *200*, 202, 255
Mongolia 23, 120, 172
Monitoring 16, 21 94, 168, 205, 221, 223, 233, 235–246
  exclosures 57, 99, 210, *211*
  field sampling 233, 239, 245
  inventory 239
  modelling 16, 238, 253
  monitoring and measurement 214, 233
  NVDI 240, 243
  phenological stages 241
  protocols 244
  Rangeland Decision System 243–244, 239, 243
  remote sensing 4, 94, 177, 214, 236, 237, 244, 253
  spatial variation 236, 238
  standards 233, 236, 237, 239
  temporal variation 236, 253
  toolbox 235
  utility of methods 245
Mountain ranges
  Altai Shan 173
  Helan 172
  Himalaya 7, 185
  Kunlun 199
  Lagangguiri 7
  Minshan 185
  Pamir 185, 199
  Qilian Shan 4, 43, 47, 156
  Qinghai–Tibet plateau 4, 45, 152, 184–195
  Tianshan 4, 42, 211, 212, 197, 202, 211
  Xing'an 104, 105
  Zhuozi 188
Mudslides 7

New China 16, 17, 200, 220
  post-war reconstruction 255
Nomads 17, 23, 24, 25, 45, 127–129, 186, 199, 204
  belief systems 219
  history 219
  lifestyle 149

Over-use of water 40, 103, 151–152
  artificial oases, expansion of 7, 152, 152, 155–156, 163, 170
  groundwater depletion 145, 150, 165, 168
  over-irrigation, poor drainage 165
  reservoir construction 4, 6, 163, 168, 170

Pastoral lands
  areal extent and distribution 3, 4, 5, 6, 45–46, *46*, 78
  defined 3, 45
Pesticide 168
Plant responses to grazing 48–49
Policy 4, 8, 11, 15, 20, 18, 96, 101, 119, 133, 155, 195, 220
  livestock development – national policy 204, 220
  makers 13, 231

Policy challenges
  autonomous regions 195, 220
  design and implementation gap 205, 220
  fragmented organizational structures 24, 225, 249
  regulatory frameworks 17, 18, 24, 98, 205, 220
Population density 91, 103, 107, 136, 142, 157, 202–203, 250
  changes in numbers of herders and livestock *97*, 103, 119, 136, 142
Population movement *see* Ecological refugees
Population pressure 22, 25, 26, 142, 152, 202–203
Populus euphratica (*diversifolia*) 4, 6, 7, 153, 173, 197, 204
Potential evapotranspiration 36, 141
Poverty 4, 23, 26, 157, 167, 224, 252
Privatization of rangelands 20, 21, 24, 25, 195, 252
Property rights/tenure 21, 24, 26, 101, 178, 205, 219–234
  institutional arrangements 222
Provinces and autonomous regions
  Gansu Pronvince 4, 9, 17, 36, 43, 46, 47, 52, 62, 151, 172, 238, 242
    Dunghuang county 72
    Gangzhou county 165, 166
    Gaotai county 165, 169
    Hexi Corridor 4, 11, 39, 42, 47, 53, 151–170
    Jingchang 168
    Lianzhou 168
    Linze county 165, 169
    Louyu watershed 67
    Minle 165
    Minqin county 167, 168, 170
    Subei county 53
    Su'nan county 220
    Tianshui county 67
    Xifeng county 67
    Yongchang 155
  Hebei Province 36, 120
    Bashang Plateau 36
  Heilongjiang Province 70
  Inner Mongolia Autonomous Region 3, 26, 33, 34, 36, 41, 46, 49, 51, 52, 62, 87, 103, 136, 152, 221, 222, 230, 238, 253
    Alashan Plateau 11, 170, 171–183, 223
    Alashan Plateau, Alxa League 223–225
    Erjina county 153, 165, 170, 171, 174, 176
    Horqin Sandy Land 4, 36, 38, 49, 103–119
    Hulunbeier Grassland 36, 91–101
    Hulunbeier Grassland, areal distribution of grassland types *93*
    Hulunbeier Grassland, Chinbaerhuqi *92*, *97*
    Hunshandake Sandy Land 38
    Jilin 104
    Ordos Plateau 11, 136–150, 172
    Ulanqub 36
    Xilingol 120–135, 181
  Liaoning Province 104, 111
  Ningxia Autonomous Region 18, 20, 150, 172, 222
    Shapatou 71–72, 79
  Qinghai Province 46, 152, 188, 190, 230, 238
    Xinhai county *189*
  Shaanxi Province 4, 11
    Anshai county 66
    Suide county 66
  Shanxi Province 4
  Sichuan 46, 62, 186, 238
  Tibet Autonomous Region 4, 8, 40, 46, 172, 219, 238, 251
    Cuana county 11
    Dingjie county 7
    Dingri county 7
    Great Basin 12
    Lazi county 7
    Qadam Basin 36
    Sajia county 7
    Shiquanhe Basin 12
    Turpan Basin 199
  Xinjiang Agricultural University 213
  Xinjiang Autonomous Region 4, 17, 26, 46, 197–215, 238, 251
    Altai Prefecture 200
    Bole Prefecture 200
    Changji Prefecture 200
    Junggar Basin 11, 37, 54, 198
    Tacheng Prefecture 200
    Turpan Basin 199
    Xinyuan county 26
    Yili Prefecture 200
  Yunnan Province 46
  Zhejiang Province 67
Public awareness 149

Qinghai–Tibetan Plateau 4, 36, 39, 54, 184

Rangelands
  defined 3
  ecosystems under siege 248
  extent and distribution 3, 4, 5, 41, 78, 124–125
  linkages 248
  resilience 249
Rangeland conversion 3, 8, 17, 20, 23, 26, 65–66, *92*, 95, *97*, 108–109, *110*, 134, 136, 151, 155, 171, 177, 204, 205
  as agent in accelerated land degradation *see* Abandoned farmland
  interactions with wind erosion 64
Rangeland degradation 3, 4, 5, 9, 11, 13, 48, 98, 158–159
  distribution and causes 4, 25, 52, 91, 93, 106, 123, 143–145, *158*, 159–161, 175, 178, 188, 190, 200, 203, *204*, 231, 254
  impacts of 157–159, 161–162
  processes of 76–80, 206
  terminology 76, 77–78

Rangeland Ecological Displacement Theory 213
Rangeland Law (Grassland Law) 18, 20, 24, 25, 27, 94, 101, 221
  concerns about enforcement of 13, 18, 20–21, 221–222, 225
Rangeland productivity 5, 124–125, 175, 176
  comparison of 101
  loss of 50, 139, 162, 171, 190, 192
  nutritive value of 207
Rangeland protection 220
  rational use of 57, 222
Rangeland rehabilitation 77–80, 115–116, 210, 256
  constraints 80–83
  principles 81–86
  processes 76–78, 93, 210
  remediation 79, 146, 231
    artificial re-vegetation 80–82, 134, 179, 180
    chemical methods 71
    engineering methods 72
  site improvement 83, 90
    aerial sowing 73, 146, 169
    natural regeneration 11, 81, 86–87, 114, 117–118, 212
    shallow tillage 100–101
Rangeland soils
  Alpine 186–187
  Horqin 106–107
  Ordos 138, 145
  Qinghai–Tibet Plateau 186
  Xilingol 123–124
  Xinjiang 198–199
Recovery processes 13, 53, 76–80, 87–88, 208–211
  definitions 77–78
Replication and scaling up 13
Research priorities 253–254
Restoration ecology 77–78
Return of cropland to forest or grassland *see* Grain for Green
Return of grazing land to grassland 21, 213
Riparian and riverine habitats 7, 152, 161
Rivers (He)
  Baiyinhe 220
  Brahmaputra 185
  Dong 173
  Duoxiongzanbu 8
  Hai 36
  Hailaer 93, 94
  Hei 4, 162–166, 170
    Hei River Water Allocation Scheme 166
  Hotan 6
  Huang *see* Yellow
  Hutuo 36
  Kaxgar 6
  Kuitan 204
  Liaohe 104, 108
  Liyuan 152
  Luancang *see* Mekong
  Luanhe 120
  Manasi 204
  Mekong 43, 184, 185, 191
  Nan 173
  Pengqu 8
  Ruoshui 36
  Salween 185–186
  Shiyang 4, 36, 166–170
  Shule 152
  South 67
  Tarim 6, 7, 197
  Urumqi 43, 202, 204
  Wulagaihe 120
  Wulungu 197
  Xi 173
  Xiliaon 103
  Xilin 50, 120
  Yangtze 43
  Yarkant 6
  Yeerqian 204
  Yellow 36, 137, 173, 175, 184, 187, 191, 193
    zero flow 36
  Yeluzangbu 7
  Yimin 94
  Yongding 36
  Zhifang 67
Rodents 56, 189, 190, 222, 229, 231, 251
  impact on rangelands 112, 191
  role in ecology of rangelands 191
Roof of the world 185, 199
Rotational grazing 3, 99, 132, 134, 146, 220, 232
  benefits of 98–100
Runoff 66, 67, 68
Rural credit 101, 119
Russia 92

Sandy land 4, 103, 184, 192
  sandification 136, 149, 156, 160, 192, *193*
Seasonal pastures 18, 21–22, 47, 128, 160, 186, 211, 219, 231, 232
  four-season pastures 128–129, 211–212
  two-season pastures 47
Sedentary grazing systems 129–130
SEPA 9, 222
Shepherding 232
Shrub encroachment 120
Siberia 138, 171, 173
Snowline retreat 42, 156
Socio-economics 8
Socio-economics of herding 15, 57
Soil erosion 26
  area affected 61–62, *62*, 95
  chequerboards *72*, 74, 146
  climatic interactions 63
  control of 70–75
  effects of slope 66
  mechanisms 63–66, 76–77
    basal sapping 64, *65*

Soil erosion (*continued*)
chemical 68
freezing and thawing 63, 68
gravitational 63–66
rills 67
sandy blowouts 63
soil loss 69
susceptibility to 63, 67
water erosion 9, 62, 76, 91, 194, 197
wind erosion 8, 9, 62–63, 40, 69, 70, 76, 91, 94, 192, 197
Soil mechanical composition *115*, 208
bulk density 116, 208, *209*
Soil nutrients 69, 72, 111, 112, 113–114, 117, *118*, 149, 177, 190, 193, 208, *210*
magnesium in soil 68–69
Soil salinity 7, 61, 68, 112, 113–114, 155, 157, 166
Soil seed banks 207, 209, *211*, 231
Soil water 116
State Council 220
National People's Congress 221
Steppe types 52, 124–126
Stocking capacities and levels of overstocking *see* Rangeland productivity
Stocking rates 13, 223–224, 226–228
effect on productivity 225–227
net income versus 227
Successional pathways 103, 112–113, 130–133, 161–162, 169–170, 206–207
Supplementary feeding 18, 47, 100–101, 123, 230, 231, 232
hay fields 122, 123, 200, 213
stall feeding 100–101, 169, 180, 205–206, 232
Sustainability 8, 20, 22, 76, 184, 223, 247, 237, 247

Three-circle pattern 147–148
Three Norths Region 3, 22, 99
Three Responsibilities – One Reward 220
Three Rivers Headwater Area 36
Thresholds of irreversible change 16
Tibetan sheep 187
Time series
Alashan 38
area of rangeland converted to cropland 94, *205*
climate 9, *10*, 11, *33*, 34
North China 35, 50
glacier meltwater 203
human population *41*, 142, *174*, *176*
livestock numbers *97*, 122, *142*, *157*, *189*, *203*, *250*
rainfall
Inner Mongolia 34
Hexi Corridor 39
Horqin *107*, *109*
Hulunbeier 37
Junggar Basin 37
Xinjiang *202*
temperature
Ordos *141*, *201*
Xilingol 39, 124
Yellow River annual runoff 40
Toxic plants 40, 55, 62, 192, 194
Trampling 171, 178, 193, 228
Tube wells 18, 168
Two Rights and One System 96

UNCCD 251
UNDP 148
Urbanization 169
Usufruct 96

Vegetation types 52
alpine meadow 126, 187–188, 202
halophytes 113–114
hydrophyte 166
hydrosere 113
psammosere 113, 124
shrubs 9, 53–55, 77, 83, 106, 120, 148, 204
role of 54–55
shrub encroachment 120
steppe types 124–126, 200
desert steppe 126
meadow steppe 52, 126
needle grass steppe 52
temperate steppe 52
typical steppe 126
temperate meadow 52
Vegetative cover 40, 49, 53, 79, 99–100, 171, 176, 197, 228, 229, 256
grass basal cover 77, 164, 171
secondary bare land 49, 184, 193, 194

Water Abstraction Permit System 167
Water allocation 165, 166, 170
Water-efficient drip irrigation 148
Water pricing 26
Watershed 3, 6, 9, 48, 66, 67, 81, 83, 193–194
Western Region Development Programme 57
Wetland 6, 192
Wildlife 3, 9, 16, 22, 166, 185, 189
habitat 3, 185
Wind 104, 186
as erosive agent 61, 68
windbreaks 70, 229
wind speed 7, 8, 24, 33, 71, 104, *106*, *109*, 111, 117–118, 123
influence of artificially planted vegetation *70*, 118

Yaks 46, 184, 185, 187, 190
role in herder's livelihood 189